Tyrosine Hydroxylase
from discovery to cloning

A Tribute to Toshiharu Nagatsu

Professor Toshiharu Nagatsu

TYROSINE

HYDROXYLASE

FROM DISCOVERY TO CLONING

A TRIBUTE TO TOSHIHARU NAGATSU

Editors: M. Naoi and S.H. Parvez

Utrecht, The Netherlands, 1993

VSP BV
P.O. Box 346
3700 AH Zeist
The Netherlands

First published in 1993

ISBN 90-6764-154-5

CIP-DATA KONINKLIJKE BIBLIOTHEEK, DEN HAAG

Tyrosine

Tyrosine hydroxylase : from discovery to cloning / ed. M. Naoi and S.H. Parvez. - Utrecht [etc.] : VSP. - Ill.
ISBN 90-6764-154-5 bound
NUGI 821
Subject headings: neurochemistry / biochemistry / tyrosine hydroxylase

Typeset in Lithuania by TEV, Vilnius.

CONTENTS

EDITORIAL

Toshiharu Nagatsu - A tribute on his 60th birthday

Toshiharu Nagatsu, MD, and Ph.D., is well known to tyrosine hydroxylase, the key enzyme in the biosynthesis of catecholamines. This enzyme belongs to a group of mono-oxygenase, so its recommended name is tyrosine 3-mono-oxygenase. To celebrate his 60th birthday, this book was meant as a tribute but we soon discovered that it has become not only a tribute to a person, but also a tribute to the history of the research of this important enzyme in neurochemistry. Research on this enzyme extends to widely in basic and clinical fields, such as neurology, psychiatry and genetics, and its functions have been examined under physiological and pathological conditions. This tribute is also indicative of the contribution to the field of neuroscience made by Toshiharu Nagatsu, not only as a scientist but as a person.

Toshiharu Nagatsu was born on 20 October, 1930 and graduated from Nagoya University School of Medicine where he received his Doctorate of Medicine in 1955. He started his research in the Department of Psychiatry (postgraduate course at Nagoya University School of Medicine) in 1956 and specialized in nerochemistry. In 1960 he received his Doctorate of Philosophy from Nagoya University. He was the appointed as Research Associate for Biochemistry at the same University. In 1962 he began his research work on catecholamines in NIH, Bethesda, USA until 1964. During this period he found and characterized tyrosine hydroxylase with Professor Sidney Udenfriend. The history of this discovery and the following studies on tyrosine hydroxylase is summarized by Dr Nagatsu himself in this book. In 19966 he was appointed as Professor in Aichi-Gakuin University School for dentistry, Nagoya, later moving in 1976 to a new position at Life Chemistry, Tokyo Institute of Technology, Yokohama. He worked in TIT and began his work on Parkinson's disease and other neurodegenerative diseases in collaboration with Professor Hiroshi Narabayashi in Tokyo. During thee years he spent many periods in other universities and institutes in the USA and other countries - many of his previous coworkers have kindly contributed their recent advances to this book. After he was appointed Professor to the Department of Biochemistry, Nagoya School of Medicine, Nagoya,he began the molecular studies on this hydroxylase and other related enzymes - his recent results in this new field of study are also summarized here. In 1991 having served as Dean of Nagoya University School of Medicine for two years, he moved to a new position in Molecular Genetics II, Neurochemistry Institute for Comprehensive Medical Science, School of Medicine, Fujita Health University to extend even further his studies on tyrosine hydroxylase and other enzymes from molecular, biochemical and pharmacological aspects.

As you see here, he is beloved by all friends who generously contributed their recent results and helped us publish an up-to-date review of tyrosine hydroxylase. Together with all his friends, especially the contributors to this book, we wish him a pleasant 60th birthday. In Japan we say that a human being can renew life every 60 years, and we are that Dr Nagatsu will continue to succeed in enriching this important field of neuroscience with his novel and fascinating discoveries for a new cycle of life.

Tyrosine Hydroxylase:
Its properties and the regulation of the activity

Tyrosine Hydroxylase, pp. 3–6
M. Naoi *et al.* (Eds)
© VSP 1993

TYROSINE HYDROXYLASE: HISTORICAL REFLECTIONS

TOSHIHARU NAGATSU

Institute for Comprehensive Medical Science, School of Medicine,
Fujita Health University, Toyoake, Aichi, 470-11, Japan

Tyrosine hydroxylase (TH, tyrosine 3-monooxygenase) was discovered in 1964 in the laboratory of Sidney Udenfriend at National Institutes of Health (NIH) (Nagatsu *et al.*, 1964a, b). TH catalyzes the first step in the biosynthesis of catecholamine (dopamine, norepinephrine, and epinephrine). Catecholamines are synthesized from tyrosine: tyrosine → L-3,4-dihydroxyphonylalanine(dopa) → dopamine → norepinephrine → epinephrine, and TH catalyzes the hydroxylation of tyrosine to dopa. The work on discovering a new enzyme, TH, was started in 1963, when I was a NIH international postdoctoral fellow and Morton Levitt was a graduate student. At that time, among the four enzymes involved in the catecholamine biosynthesis, only the enzyme responsible for converting tyrosine to dopa was elusive. Tyrosinase was assumed to catalyze this reaction, but was not found in catecholamine-containing tissues including the brain. This reaction was also assumed to be non-enzymatic *in vivo*, since it can easily observed *in vitro* under various conditions. However, it was difficult to believe that such an important step in catecholamine biosynthesis occurs non-enzymatically.

Therefore, assuming that such an enzyme may exist in catecholamine-containing tissues, we first developed a highly sensitive isotopic assay for the enzyme activity.

L-(^{14}C)tyrosine was used as a substrate, and L-(^{14}C)dopa, enzymatically formed, was isolated on an alumina column and assayed. We initially suffered from high blank values, but we solved this problem by purifying the substrate, L-(^{14}C)dopa, on an alumina column and then on an cation-exchange column, to remove the contaminating dopa like substance. We started our initial work to discover the enzyme in tissue slices and minces of the brain stem, and of peripheral sympathetically innervated tissues such as heart and spleen, and adrenals of the guinea pig. These tissues were known to contain catecholamines. The tissue slices and minces were thought to contain necessary cofactors with the enzyme, and we found a substantial formation of dopa from L-(^{14}C)tyrosine. We observed a nice linearity between the amounts of (^{14}C)dopa formed and those of tissues used. We identified the formation of dopa, dopamine, and norepinephrine by electrophoresis and chromatographies. Dopa was not formed in the absence of slices or minces. However, we also found a significant nonenzymatic hydroxylation of tyrosine to dopa in heated tissue slices and minces. Heated enzyme preparations were generally used for control incubation. This may be the reason why the presence of this enzyme went unnoticed for so long. Sidney Udenfriend suggested to use D-(^{14}C)tyrosine for the blank. As he predicted, we found the absolute stereospecificity of this enzyme which permitted the use of D-(^{14}C)tyrosine as a control, and we became convinced that we were really detecting a new enzyme.

We then found that this enzyme was both membrane-bound and soluble, but the activity in the soluble fraction was too small to be purified. We tried to solubilize the enzyme from the particulate fraction, but solubilization was not successful.

We found bovine adrenal medulla contained a large amount of the enzyme activity in the soluble fraction, and we could isolate the enzyme by ammonium sulfate fractionation.

We tried to identify the cofactor of this enzyme. After testing many probable cofactor substances, the preparations were shown to require for activity a tetrahydropterin as a cofactor and molecular oxygen. Gordon Guroff who were also working at the laboratory of Sidney Udenfriend kindly supplied me with 6,7-dimethyl-5,6,7,8-tetrahydropterin, and we could see a marked activation of the preparations. Thus we concluded that this enzyme is a pterine-requiring monooxygenase. The requirement of a tetrahydropterin as a cofactor of the hydroxylating process was first discovered with rat liver phenylalanine hydroxylase by Seymour Kaufman in the NIH in 1959 (Kaufman, 1959).

We then found that the activity of the partially purified preparations was inhibited by iron-chelating agents such as α, α-dipyridyl and stimulated by adding Fe^{2+} to the incubation mixture. We therefore assumed this enzyme to be an iron enzyme.

Another significant finding in this work was the inhibition of the enzyme by the products catecholamines (Nagatsu et al., 1964), and we proposed the possibility of the feed-back inhibition in vivo, which is now of great interest for the short term regulation of catecholamine biosynthesis. Later we found that catecholamines inhibit the enzyme in competition with a tetrahydropterin cofactor (Udenfriend et al., 1965; Nagatsu et al., 1972).

The product, dopa, was determined to be L-form based on the finding that the dopa formed could be decarboxylated to dopamine by aromatic L-amino acid decarboxylase.

One of the keys to discover the enzyme was the first development of a highly sensitive and relatively simple assay method, i.e. radioassay of the (^{14}C)dopa formed after removing the substrate (^{14}C)tyrosine. We (Nagatsu et al., 1964c) then established another simple method. The availability of L-[3,5-^{3}H]tyrosine suggested a simple procedure for assay based on replacement of tritium in the following reaction sequence:

$$L\text{-}[3,5\text{-}^3H]tyrosine + 1/2\,O_2 \text{ hydroxylase } L\text{-}[5\text{-}^3H]dopa + {}^3H$$

$$^3H + H_2O \rightarrow H^+ + {}^3HHO$$

The ^3HHO can be separated from substrate and product very simply and subjected to radioassay. The enzyme activity is now mostly measured by HPLC-electrochemistry (Nagatsu *et al.*, 1979).

Our original paper (Nagatsu *et al.*, 1964b) has been frequently cited and became a "citation classic" in 1980 (Nagatsu, 1980). The reason why our publication is so frequently cited may be the great physiological significance of this enzyme, its widely-applicable assay procedure, and its interesting properties as a pterin-requiring monooxygenase.

I recall distinctly how much we were pleased to see the enzyme activity of TH in front of the liquired scintillation counter at midnight. The best reward for scientists may be the pleasure and excitement of new findings.

TH has been homogeneously purified and characterized from various animal tissues including human brains and adrenals. TH was found to be regulated not only acutely by feed-back inhibition and phosphorylation/dephosphorylation, but also chronically by enzyme induction (Mueller *et al.*, 1969).

Since 1985, molecular biological approach to TH has made it possible to determine the amino acid sequence from the nucleotide sequence of complementary DNA (cDNA). In 1987, Grima *et al.*, (1987) and we (Kaneda *et al.*, 1987) found the presence of 4 types of human tyrosine hydroxylase messenger RNAs (mRNAs). We purified tyrosine hydroxylase from human adrenals (Mogi *et al.*, 1984) and human brains (striatum) (Mogi *et al.*, 1986), and found both active and less active forms. The 4 types of mRNAs of human tyrosine hydroxylase were expressed in COS cells, and showed enzyme activity. The type 1 human TH had the highest homospecific activity (activity per enzyme protein), the values for the other enzymes ranging from -0.3 to -0.4 (Kobayashi *et al.*, 1988). Transgenic mice carrying multiple copies of human tyrosine hydroxylase (TH) gene expressed tissue-specifically mainly type 2 and type 1 human tyrosine hydroxylase at high-level (Kaneda *et al.*, 1991). We (Ichikawa, Ichinose, Nagatsu, 1990) found multiple mRNAs of monkey tyrosine hydroxylase. Two types of TH mRNA corresponding to type 1 and type 2 were detected in adrenal gland and brain of two species of monkeys, but only a single form of TH mRNA was detected in *Sunkus murinus* and rat. We therefore suggest that the multiplicity of TH mRNA is primate specific. The functional significance of multiple forms of human TH should be elucidated under physiological and pathological conditions.

REFERENCES

Grima, B., Lamouroux, A., Boni, C., Julien, J.-F., Javoy-Agrid, F. and Mallet, J. (1987). A single human gene encoding multiple tyrosine hydroxylases with different predicted functional characteristics. *Nature* **326**, 701–707.

Kaneda, N., Kobayashi, K., Ichinose, H., Kishi, F., Nakazawa, A., Kurosawa, Y., Fujita, K. and Nagatsu, T. (1987). Isolation of a novel cDNA clone for human tyrosine hydroxylase: alternative RNA splicing produces four kinds of mRNA from a single gene. *Biochem. Biophys. Res. Commun.* **146**, 971–975.

Kaneda, N., Sasaoka, T., Kobayashi, K., Kiuchi, K., Nagatsu, I., Kurosawa, Y., Fujita, K., Yokoyama, M., Nomura, T., Katsuki, M. and Nagatsu, T. (1991). Tissue-specific and high-level expression of the human tyrosine hydroxylase gene in transgenic mice. *Neuron* **6**, 583–594.

Kaufman, S. (1959). Studies on the mechanism of the enzymatic conversion of phenylalanine to tyrosine. *J. Biol. Chem.* **234**, 277–288.

Kobayashi, K., Kaneda, N., Ichinose, H., Kishi, F., Nakazawa, A., Kurosawa, Y., Fujita, K. and Nagatsu, T. (1987). Isolation of a full length cDNA clone encoding human tyrosine hydroxylase type 3. *Nucl. Acids Res.* **75**, 6733.

Kobayashi, K., Kaneda, N., Ichinose, H., Kishi, F., Nakazawa, A., Kurosawa, Y., Fujita, K. and Nagatsu, T. (1988a). Structure of the human tyrosine hydroxylase gene: alternative splicing from a single gene accounts for generation of four mRNA types. *J. Biochem.* **103**, 907–912.

Kobayashi, K., Kiuchi, K., Ishii, A., Kaneda, N., Kurosawa, Y., Fujita, K. and Nagatsu, T. (1988b). Expression of four types of human tyrosine hydroxylase in COS cells. *FEBS Lett.* **238**, 431–434.

Mogi, M., Kojima, K. and Nagatsu, T. (1984). Detection of inactive or less active forms of tyrosine hydroxylase in human adrenals by a sandwith in human adrenals by a sandwich enzyme immunoassay. *Anal. Biochem.* **138**, 125–132.

Mogi, M., Kojima, K., Harada, M. and Nagatsu, T. (1986). Purification and immunochemical properties of tyrosine hydroxylase in human brain. *Neurochem. Int.* **8**, 423–428.

Mueller, R. A., Thoenen, H. and Axelrod, J. (1969). Adrenal tyrosine hydroxylase comprensatory increase in activity after chemical synpathectomy. *Science* **163**, 468–469.

Nagatsu, T. (1980). This week's citation classic. In: *Contemporary Classics in the Life Science*, **Vol. 2**: *The Molecules of Life*. J.T. Barrett (Ed.). ISI Press, Philadelphia, pp. 146–146.

Nagatsu, T., Levitt, M. and Udenfriend S. (1964). A rapid and simple radioassay for tyrosine hydroxylase activity. *Anal. Biochem.* **9**, 122–126.

Nagatsu, T., Levitt, M. and Udenfriend, S. (1964a). Conversion of L-tyrosine to 3,4-dihydroxy-phenylalanine by cell-free preparations of brain and sympathetically innervated tissues. *Biochem. Biophys. Res. Commun.* **14**, 543–549.

Nagatsu, T., Levitt, M. and Udenfriend, S. (1964b). Tyrosine hydroxylase. The initial step in norepinephrine biosynthesis. *J. Biol. Chem.* **239**, 2910–2917.

Nagatsu, T., Mizutani, K., Nagatsu, I., Matsuura, S. and Sujimoto, T. (1972). Pteridines as cofactor or inhibition of tyrosine hydroxylase. *Biochem. Pharmacol.* **21**, 1945–1953.

Nagatsu, T., Oka, K. and Kato, T. (1979). Highly sensitive assay for tyrosine hydroxylase activity by high-performance liquid chromatography. *J. Chromatogr.* **163**, 247–252.

Udenfriend, S., Zaltzman-Nirenberg, P. and Nagatsu, T. (1965). Inhibitors of purified beef adrenal tyrosine hydroxylase. *Biochem. Pharmacol.* **14**, 837–845.

Tyrosine Hydroxylase, pp. 7–14
M. Naoi *et al.* (Eds)
© VSP 1993

TYROSINE HYDROXYLASE AND FACTORS INFLUENCING ITS ACTIVITY

SIDNEY UDENFRIEND

Department of Neurosciences, Roche Institute of Molecular Biology,
Roche Research Center, Nutley, NJ, USA

It is both a pleasure and a bit sad for me to write this introductory section to the festschrift honoring Toshi Nagatsu on his 60th birthday. It is a pleasure because writing about Toshi's stays in my laboratory (one from 1962–1964 and the second in 1972) bring to mind the vitality he brought with him during his visits. The sadness is that we are both getting older and, except for occasional short visits, we will never again experience the excitement of working together.

When Toshi came to my laboratory at the NIH (in 1962) he was a young post-doc. Today he is a chairman of a prestigious department of biochemistry and a world renowned scientist who has made many important scientific contributions. However, he is best known for those in the area of catecholamines, particularly with respect to tyrosine hydroxylase. I would like, therefore, to recount the story of Toshi's role in the detection and initial characterization of tyrosine hydroxylase and discuss some subsequent work relating to this enzyme in which I was involved. Before I begin I would like to point out to the reader that I left the amine field about 15 or more years ago and have been involved in other areas of research since then. My writings on tyrosine hydroxylase are therefore of a historical nature. However, through Toshi I feel that I am still contributing to this important area of neuroscience. As I will point out later, however, one of my current interests may lead me back to catecholamine biosynthesis.

I will first lay the groundwork preceding Toshi's arrival in my laboratory at the NIH. My interest in catecholamines was stimulated by Dr James Shannon prior to his becoming the legendary Director of the National Institutes of Health. It was he who made me aware of the sympathetic nervous system and of the physiologic significance of norepinephrine, which had been discovered only a few years earlier by von Euler (1948). Because of my prior experience with amino acids I was quickly able to develop procedures that enabled me to demonstrate the ability of whole animals to produce norepinephrine from radiolabeled phenylalanine and tyrosine (Udenfriend *et al.*, 1953). This was in accord with the pathway of biosynthesis predicted by Blaschko (1957): phenylalanine → tyrosine → DOPA → dopamine → norepinephrine.

The first step in this scheme involves the hydroxylation of phenylalanine to yield tyrosine. We were able to isolate phenylalanine hydroxylase from liver and to partially characterize it (Udenfriend and Cooper, 1952). Shortly thereafter Kaufman (1962) demonstrated the requirement of tetrahydropteridine as a cofactor. However, it became apparent that phenylalanine hydroxylase was not really the starting point for catecholamine synthesis. The enzyme is mainly present in liver with little, if any, in sympathetic tissue. In addition tyrosine is a normal dietary constituent and adequate amounts of the latter are present in blood and tissues. Toshi Nagatsu came to my laboratory at the NIH as a postdoc in 1962, when I had turned my attention to the conversion of tyrosine to DOPA. After he had developed methods for assaying DOPA formation (Nagatsu *et al.*, 1964a), we were faced with the unusual task of proving that hydroxylation of tyrosine to DOPA required an enzyme. This may sound surprising today. However, DOPA decarboxylase had been partially purified (Lovenberg *et al.*, 1962) as well as dopamine β-hydroxylase (Levin *et al.*, 1960), the problem with tyrosine hydroxylation was that it could be catalyzed by tissues even after they had been boiled. Many in the field argued that if autoxidation of tyrosine took place so rapidly why did one need an enzyme? However, Toshi and I firmly believed that nature would not utilize autoxidation as the initial step in the biosynthesis of an important nonregulator. We therefore continued our attempts to demonstrate an enzymatic requirement.

After many unsuccessful experiments and weeks of deliberation the key experiment that Toshi and I devised was to use both D- and L-tyrosine as substrates. We reasoned that enzymatic hydroxylation to DOPA should be stereospecific as compared to autoxidation, which should lack this specificity. The first problem we encountered was to obtain radiolabeled D-tyrosine. Commercial sources, at that time, supplied only L-tyrosine and the racemic mixture DL-tyrosine in radioactive form. Although it is not shown in any of our publications, Toshi prepared the D-amino acid by destroying the L-isomer in the racemate with L-amino acid oxidase. He then used D-amino acid oxidase to characterize the resulting D-isomer. I should point out that it would be difficult to repeat this experiment today because DL-tyrosine is no longer available from commercial sources, at least in the United States. I'll never forget the excitement in our laboratory when Toshi ran the key experiment and obtained the results shown in Table 1. We realized, at last, that we were on the right track. He showed further that tyrosinase, which had been characterized earlier and is involved in melanin formation, oxidized both D- and L-tyrosine to DOPA. Working night and day, in the characteristic Nagatsu manner (perpetual motion) he purified the enzyme from bovine adrenal medulla and characterized it in just a few months (Nagatsu *et al.*, 1964b). He also demonstrated the presence of tyrosine hydroxylase in subcellular particles of brain and peripheral sympathetically innervated tissues.

Table 1.
Conversion of tyrosine to DOPA by normal and heated slices of guinea pig tissues

Tissue	Substrate	DOPA formed	
		Normal	Heated
		pmol/g	
Adrenal	L-Tyrosine	3.52	0.74
	D-Tyrosine	0.12	0.84
Spleen	L-Tyrosine	0.82	0.34
	D-Tyrosine	0.05	0.24

(Taken from Nagatsu *et al.*, 1964b).

The properties of the pure enzyme determined by Toshi Nagatsu (tissue levels, K_m, V_{max}) already made it apparent that tyrosine hydroxylation could be the rate limiting step in norepinephrine biosynthesis. In fact, this issue was raised in our initial publication on the purification of tyrosine hydroxylase (Nagatsu *et al.*, 1064b). Investigations on the rate-limiting step in norepinephrine synthesis were begun in my laboratory while Toshi Nagatsu was still characterizing tyrosine hydroxylase. In 1963, with Sydney Spector and other colleagues, we had shown that the isolated guinea pig heart can synthesize norepinephrine when perfused with radioactive tyrosine (Spector *et al.*, 1963). This indicated that the sympathetic innervation of the heart contains all the enzymes necessary for the formation of the sympathetic transmitter from dietary sources. It was possible, therefore, to perfuse guinea pig hearts with each of the three precursors and monitor norepinephrine formation. We reasoned that from the known K_m and V_{max} values that had been determined for each of the purified enzymes (Table 2), it should be possible to determine which of the three was rate-limiting in perfusion experiments (Levitt *et al.*, 1965). Accordingly, radiolabeled tyrosine, dopa, and dopamine were perfused, individually, over a range of concentrations, and rates of norepinephrine formation were measured for each. As shown in Fig. 1 norepinephrine formation increased on increasing the concentration of each substrate in the perfusion fluid. However, only with tyrosine was saturation achieved. A maximal rate of norepinephrine formation was attained at a tyrosine concentration of *c.* 5×10^{-5}M. Normal tyrosine levels in guinea pig tissues are also in the range of 10^{-5}M. The maximal rate of norepinephrine formation attained with tyrosine, *c.* 0.2 μg/g/hr, approximated the value that had been determined previously for the rate of formation of norepinephrine in the guinea pig heart *in situ* (Spector *et al.*, 1963). From the data in Fig. 1 the K_m for perfused tyrosine was estimated to be about 2×10^{-5} M. This compares to a K_m of 5×10^{-5} for purified bovine tyrosine hydroxylase. Thus, the enzyme with the lowest K_m (Table 2) is rate-limiting in the intact heart. This probably signifies that there is less tyrosine hydroxylase in the heart or less of the active form of the enzyme and explains why the K_m for the overall conversion of tyrosine to norepinephrine in the perfused organ is approximately the same as the constant for the pure hydroxylase.

Table 2.
Kinetic constants of enzymes involved in norepinephrine biosynthesis

Enzyme	$K_m \times 10^{-5}$	V_{max}
Tyrosine hydroxylase	5	150
Aromatic L-amino acid decarboxylase	40	33 000
Dopamine-β-hydroxylase	580	50 000

V_{max} are in picomoles of substrate converted per milligram of enzyme protein per hour. (Taken from Udenfriend, 1966).

Being the rate-limiting step in the formation of norepinephrine tyrosine hydroxylase is obviously the most susceptible to regulatory mechanisms. Not long after this was proposed we were able to demonstrate that the increased turnover and excretion of catecholamines that occurs as a result of exercise and exposure to cold is due to increased activity of tyrosine hydroxylase (Gordon *et al.*, 1966). It became evident later that phosphorylation of tyrosine hydroxylase was the rapid means of regulating activity of the enzyme, as occurs in acute stress or acute sympathetic stimulation (Ames *et al.*, 1978; Joh *et al.*, 1978). Prolonged sympathetic stimulation is now known to increase tyrosine hydroxylase activity through increased transcription to the corresponding mRNA (Mueller *et al.*, 1969; Dairman and Udenfriend, 1970; Joh *et al.*, 1973). Additional evidence that tyrosine hydroxylase is rate-limiting *in vivo* is that inhibition of the enzyme, such as by the inhibitor α-methyl tyrosine, specifically depletes tissue stores of norepinephrine in animals (Fig. 2) (Spector *et al.*, 1965) and lowers the excretion of catecholamine metabolites in man (Sjoerdsma *et al.*, 1965). This is not readily accomplished with inhibitors of the decarboxylase or dopamine-β-hydroxylase (Udenfriend *et al.*, 1966). Because tyrosine hydroxylase is rate-limiting, "norepinephrine formation, *in vivo*, is most susceptible to physiologic and pharmacologic regulation at this step" (Udenfriend, 1966).

As I stated earlier, one of my current areas of interest may prove to be important in catecholamine biosynthesis. This relates to the transport of tyrosine into neurons. In the paper in which we presented the first evidence that tyrosine hydroxylase was rate-limiting with respect to norepinephrine biosynthesis (Nagatsu *et al.*, 1964b) we added a caveat: "The apparent K_m for conversion of tyrosine to norepinephrine by the perfused heart would, of course, include the phenomenon of transport. Should the latter prove to be a catalyzed process with a K_m comparable to that of tyrosine hydroxylase (about 10^{-5}M) it will be necessary to

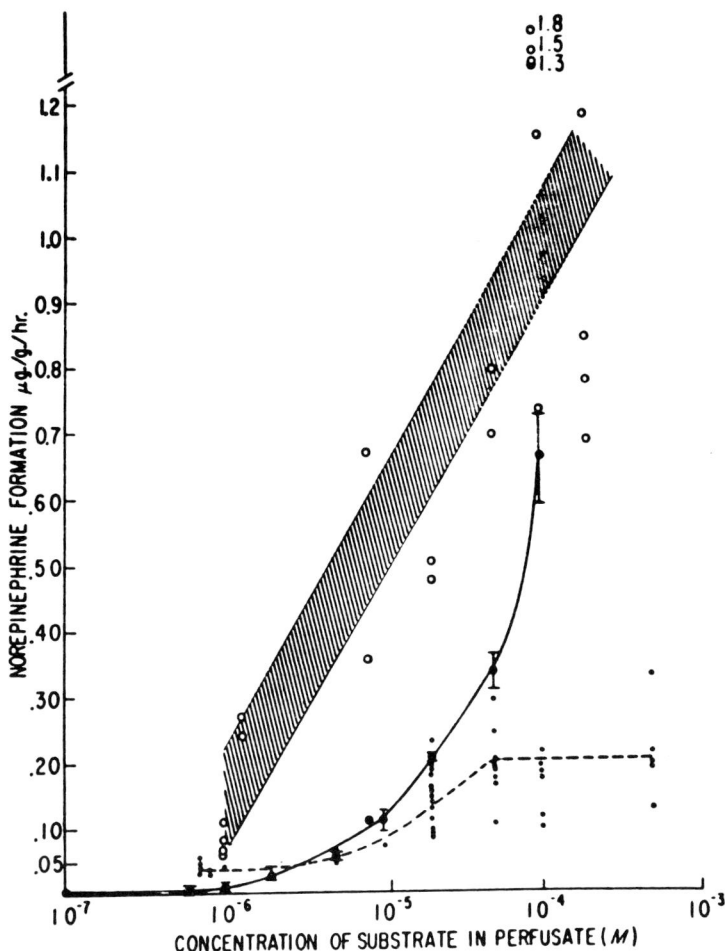

Figure 1. Formation of norepinephrine from its precursors in the isolated perfused guinea pig heart. o – – – – – – – o, tyrosine; •————————•, dopa; o///////o, dopamine (from Udenfriend, 1966).

reconsider the rate-limiting nature of tyrosine hydroxylase". It is now apparent that transport into cells of most nutrients, including amino acids, is catalyzed by specific transport systems. Uptake of tyrosine into most cells is catalyzed by the "L transporter", a Na^{2+} independent system that catalyzes the transport of the aromatic and long chain aliphatic amino acids (Shotwell *et al.*, 1983; Christenson, 1985). Many years ago we presented evidence that the uptake of tyrosine into brain "involves a catalytic mechanism" (Udenfriend, 1964–1965). This is undoubtedly true of all neuronal tissues. The apparent K_m for uptake of aromatic amino acids into cells is about 10^{-5}M which is comparable to the K_m for tyrosine hydroxylase. Since fasting plasma levels in rats are about 1.6×10^{-5}M,

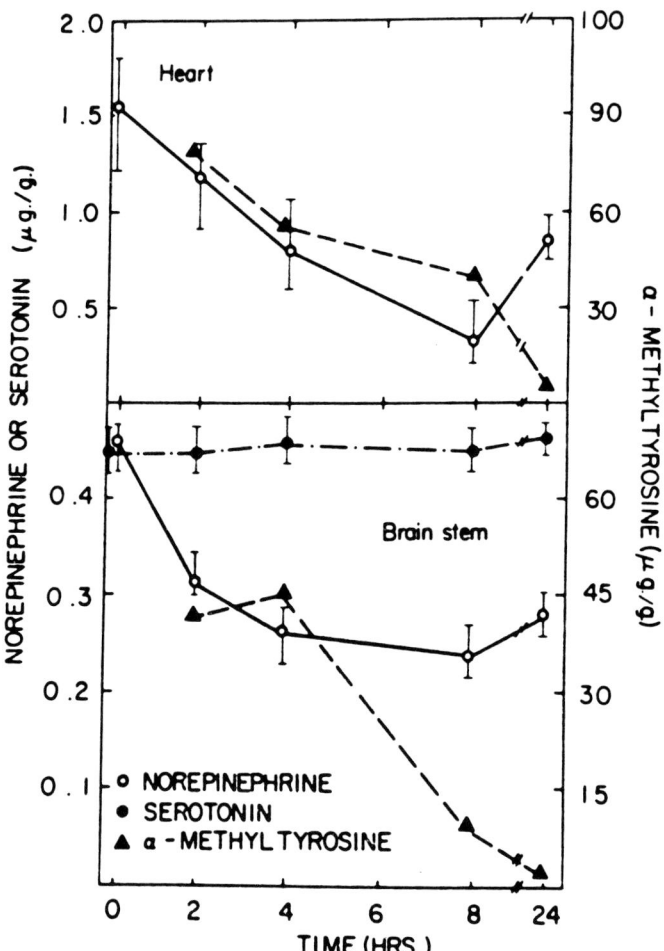

Figure 2. Effect of a single dose of α-methyltyrosine (80 mg/kg) on tissue levels of norepinephrine and serotonin.

it is quite possible that uptake of tyrosine by sympathetic nervous tissue could be rate-limiting. This might be the case in starvation or when other amino acids compete for uptake of tyrosine by the "L transporter" (Udenfriend *et al.*, 1966). Competitive inhibition of tyrosine transport apparently occurs in patients with phenylpyruvic oligophrenia, where plasma levels of phenylalanine are very high and in maple syrup urine disorder where the levels of branched chain amino acids are elevated. Amino acid drugs such as α-methyldopa can also compete for uptake of tyrosine. It remains to be seen whether tyrosine transport into sympathetic neurons is rate-limiting under physiologic conditions.

To investigate the role of transport in neurotransmitter biosynthesis requires the characterization of the transporter protein. In animals glucose is the only nutrient for which transporter proteins have been characterized thus far. However, amino acid transporters will soon be cloned and characterized. With Dr Suresh Tate we have been able to demonstrate the presence of an mRNA in animal cells that, when injected into *Xenopus laevis* oocytes, induces expression of a protein responsible for tyrosine uptake (Tate *et al.*, 1989). We have succeeded in preparing a library from which mRNA transcripts can be obtained that express the transporter in oocytes. Cloning of the transmitter has already been accomplished (Tate *et al.*, 1992). We have recently demonstrated the transporter mRNA in brain using Northern analysis (Yan *et al.*, 1992).

REFERENCES

Ames, M. M., Lerner, P. and Lovenberg, W. (1978). Tyrosine hydroxylase activation by protein phosphorylation and end product inhibition. *J. Biol. Chem.* **253**, 27–31.

Blaschko, H. (1957). Formation of catecholamines in the animal body. *Brit. Med. Bull.* **13**, 162–165.

Christenson, H. N. (1985). On the strategy of kinetic discrimination of amino acid transport systems. *J. Membr. Biol.* **84**, 97–103.

Dairman, W. and Udenfriend, S. (1970). Increased conversion of tyrosine to catecholamines in the intact rat following elevation of tissue tyrosine hydroxylase levels by administered phenoxybenzamine. *Mol. Pharmacol.* **6**, 350–356.

von Euler, U. S. (1948). Identification of the sympathomimetic substance in nerves of cattle (Sympathin A) with laevonoradrenaline. *Acta. Physiol. Scand.* **16**, 63–74.

Gordon, R., Spector, S., Sjoerdsma, A. and Udenfriend, S. (1966). Increased synthesis of norepinephrine and epinephrine in the intact rat during exercise and exposure to cold. *J. Pharmacol. Exper. Therap.* **153**, 440–447.

Joh, T. H., Geghnan, C. and Reis, D. (1973). Immunochemical demonstration of increased accumulation of tyrosine hydroxylase protein in sympathetic ganglia and adrenal medulla ellicited by reserpine. *Proc. Natl. Acad. Sci. USA* **70**, 2767–2771.

Joh, T. H., Park, D. H. and Reis, D. J. (1978). Direct phosphorylation of brain tyrosine hydroxylase by cyclic AMP-dependent protein kinase — mechanism of enzyme activation. *Proc. Natl. Acad. Sci. USA* **75**, 4744–4748.

Kaufman, S. (1962). In: *Oxygenases*. O. Hayashi (Ed.). Academic Press, New York, p. 129.

Levin, E. Y., Levenberg, B. and Kaufman, S. (1960). The enzymatic conversion of 3, 4-dihydroxyphenylethaline to norepinephrine. *J. Biol. Chem.* **235**, 2080–2086.

Levitt, M., Spector, S., Sjoerdsma, A. and Udenfriend, S. (1965). Elucidation of the rate-limiting step in norepinephrine biosynthesis using the perfused guinea pig heart. *J. Pharmacol. Exper. Therap.* **148**, 1–8.

Lovenberg, W., Weissbach, H. and Udenfriend, S. (1962). Aromatic L-amino acid decarboxylase. *J. Biol. Chem.* **237**, 89–93.

Mueller, R. A., Thoenen, H. and Axelrod, J. (1969). Adrenal tyrosine hydroxylase compensatory increase in activity after chemical sympathectomy. *Science* **163**, 468–469.

Nagatsu, T., Levitt, M. and Udenfriend, S. (1964a). Conversion of L-tyrosine to 3, 4-dihydroxyphenylalanine by cell-free preparations of brain and sympathetically innervated tissues. *Biochem. Biophys. Res. Comm.* **14**, 543–549.

Nagatsu, T., Levitt, M. and Udenfriend, S. (1964b). Tyrosine hydroxylase. The initial step in norepinephrine biosynthesis. *J. Biol. Chem.* **239**, 2910–2917.

Shotwell, M. A., Kilberg, M. S. and Oxender, D. L. (1983). The regulation of neutral amino acid transport in mammalian cells. *Biochim. Biophys. Acta* **737**, 267–284.

Sjoerdsma, A., Engelman, K., Spector, S. and Udenfriend, S. (1965). Inhibition of catecholamine synthesis in man with α-methyltyrosine, an inhibitor of tyrosine hydroxylase. *Lancet*, 1092–1094.

Spector, S., Sjoerdsma, A., Zaltzman-Nirenberg, P. and Udenfriend, S. (1963). Norepinephrine synthesis from tyrosine C^{14} in isolated perfused guinea pig heart. *Science* **139**, 1299–1301.

Spector, S., Sjoerdsma, A. and Udenfriend, S. (1965). Blockade of endogenous norepinephrine synthesis by α-methyltyrosine, an inhibitor of tyrosine hydroxylase. *J. Pharmacol. Exper. Therap.* **147**, 86–95.

Tate, S. S., Urade, R., Getchell, T. V. and Udenfriend, S. (1989). Expression of the mammalian Na$^+$-independent L-system amino acid transporter in *Xenopus laevis* oocytes. *Arch. Biochem. Biophys.* **275**, 591–596.

Tate, S. S., Yan, N. and Udenfriend, S. (1992). Expression cloning of a Na$^+$-independent neutral amino acid transporter from rat kidney. *Proc. Natl. Acad. Sci. USA* **89**, 1–5.

Udenfriend, S. (1966). Biosynthesis of the sympathetic neurotransmitter, norepinephrine. *The Harvey Lectures* **60**, 57–83.

Udenfriend, S. and Cooper, J. (1952). The enzymatic conversion of phenylalanine to tyrosine. *J. Biol. Chem.* **194**, 503–511.

Udenfriend, S., Cooper, J. R., Clark, C. T. and Baer, J. E. (1953). Rate of turnover of epinephrine in the adrenal medulla. *Science* **142**, 394–396.

Udenfriend, S., Zaltzman-Nirenberg, P., Gordon, R. and Spector, S. (1966). Evaluation of the biochemical effects produced *in vivo* by inhibitors of the three enzymes involved in norepinephrine biosynthesis. *Mol. Pharmacol.* **2**, 95–105.

Yan, N., Mosckovitz, R., Udenfriend, S. and Tate, S. S. (1992). Distribution of mRNA of a Na$^+$-independent neutral amino acid transporter cloned from rat kidney and its expression in mammalian tissues and *Xenopus laevis* oocytes. *Proc. Natl. Acad. Sci. USA*, (in press).

Tyrosine Hydroxylase, pp. 15–22
M. Naoi *et al.* (Eds)
© VSP 1993

THE INFLUENCE OF NERVE GROWTH FACTOR ON THE LEVEL AND ACTIVITY OF TYROSINE HYDROXYLASE

GORDON GUROFF

Section on Growth Factors, National Institute of Child Health and Human Development, National Institutes of Health, Bethesda, MD 20892, USA

Abstract—Nerve growth factor is known to increase the amount of tyrosine hydroxylation in its target cells. Its ability to do this is due to its action on tyrosine hydroxylase at two different levels. Treatment of tyrosine hydroxylase-containing cells with nerve growth factor increases the phosphorylation of the enzyme within minutes. This acute action is followed, after several hours or days, by a longer-term, selective increase in the amount of enzyme protein in the cells. The molecular mechanisms by which this action of nerve growth factor is carried out are the subject of substantial study and some measure of controversy.

Key words: nerve growth factor; tyrosine hydroxylase; PC12 cells; phosphorylation; sympathetic neurons.

Nerve growth factor (NGF) is a peptide required for the survival and development of certain sensory and sympathetic neurons (Levi-Montalcini and Angeletti, 1968; Greene and Shooter, 1980). It also has effects on the chromaffin cells of the adrenal medulla (Unsicker *et al.*, 1978; Aloe and Levi-Montalcini, 1979), on specific populations of neurons in the central nervous system (Gnahn *et al.*, 1983; Hefti *et al.*, 1985), and on certain tumor lines as well (Waris *et al.*, 1973; Fabricant *et al.*, 1977).

The biological activity of nerve growth factor resides in a molecule of 26 kD, a non-covalently-bound dimer, the monomer of which exhibits some sequence homology with insulin (Frazier *et al.*, 1972). The gene for nerve growth factor has been identified in both mouse and human (Scott *et al.*, 1983; Ullrich *et al.*, 1983), and it has been shown that there is 86% homology between these species. Kinetic studies have shown that there are two classes of receptors for nerve growth factor on the plasma membranes of most nerve growth factor-responsive cells, the high-affinity and the low-affinity. These two classes have kDs of 10^{-11}M and 10^{-9}M, respectively, and the low-affinity receptor is usually ten times more abundant than the high-affinity receptor. The gene for the low-affinity receptor (now called pp75) has been cloned (Johnson *et al.*, 1986; Radeke *et al.*, 1987) and it proved to be a rather simple molecule with a single membrane-spanning region and no kinase domain. The search for the high-affinity receptor led to the identification of the protein product of the trk proto-oncogene as a binding site for nerve growth factor. Some data indicates that the trk protein is itself the high-affinity site (Klein *et al.*, 1991); other data suggests that the trk protein and the pp75 each bind with low-affinity, but that together they form the physiologically-active high-affinity site (Kaplan *et al.*, 1991).

Prominent among the cells on which nerve growth factor acts is the remarkably informative model, PC12 (Greene and Tischler, 1976; Dichter *et al.*, 1977). This clone, isolated originally from a pheochromocytoma, a tumor of the rat adrenal medulla, grows readily in standard culture as a small (6–14 μm in diameter), round, fluorescent cell with a doubling time of 48–96 h. The native cellular fluorescence is due to the presence of catecholamines, largely dopamine. When treated with nanomolar quantities of nerve growth factor, these chromaffin-like cells are converted into a sympathetic neuron-like phenotype. Within a few hours or days, they stop dividing and become electrically excitable. They extend neurites and will form cholinergic synapses with appropriate muscle cells in culture (Schubert *et al.*, 1977). These cells, when so treated, are very much like mature sympathetic neurons, the most striking difference being that the differentiation is freely reversible. If nerve growth factor is removed from the medium, the neurites disintegrate, the cells round up and begin to divide, and, within a few days, the chromaffin-like phenotype is fully restored.

The action of nerve growth factor on many of its targets, including PC12 cells and sympathetic neurons, involves an enhancement of the differentiative properties of the cells. Since several of the targets of nerve growth factor synthesize and secrete catecholamine neurotransmitters, and since the rate-limiting step in the biosynthesis of catecholamines is catalyzed by the enzyme tyrosine hydroxylase (Nagatsu *et al.*, 1964), it is not surprising that the effect of nerve growth factor on this pivotal enzyme has been the subject of a number of investigations. It has been observed that nerve growth factor influences both the activity and the amount of this enzyme in cells upon which it acts. The activity of the enzyme changes rapidly in the cells by means of a nerve growth factor-induced alteration in the phosphorylation of the protein. The level of the enzyme changes more slowly and is due to a nerve growth factor-induced alteration in the transcription or translation of the tyrosine hydroxylase gene.

The initial observations on the effect of nerve growth factor on the levels of tyrosine hydroxylase were presented by Thoenen *et al.* (1971). In this study, newborn rats were treated with nerve growth factor for several days and the specific activities of the several catecholamine-synthesizing enzymes in superior cervical ganglia were measured. The data showed that the specific activity of tyrosine hydroxylase increased some 5-fold and that the increase was quite selective in that the specific activities of the other enzymes in the pathway, dopamine beta-hydroxylase and dopa decarboxylase, increased to a much lesser extent, 3-fold and 50%, respectively. The authors ruled out direct effects of nerve growth factor on the solubilized enzyme as a cause of the increased activity, but were not able to shed any further light on the mechanism by which nerve growth factor produced this increase.

In pursuit of such information, MacDonnell *et al.* (1977) studied the biosynthesis of tyrosine hydroxylase in organ cultures of superior cervical ganglia from animals that had been treated with nerve growth factor. When animals were given a single subcutaneous administration of nerve growth factor and the ganglia removed 24 h later, the synthesis of tyrosine hydroxylase increased from 0.11 to 0.42% of the total soluble protein of the ganglia. If actinomycin was given to the animal just before the nerve growth factor, this rise in tyrosine hydroxylase synthesis was inhibited by almost 90%. The data were interpreted to indicate that nerve growth factor increases the amount of tyrosine hydroxylase in the ganglia by a mechanism involving RNA synthesis, more than likely by increasing the transcription of the tyrosine hydroxylase gene.

A different conclusion was reached by Rohrer *et al.* (1978) working with sympathetic ganglia from adult rats treated with nerve growth factor *in vitro*. In these studies, conditions were developed that permitted the maintenance of tyrosine hydroxylase levels in cultured ganglia for at least 48 h in the absence of nerve growth factor. When nerve growth factor was added to the cultures, the activity of tyrosine hydroxylase increased some 50% and that increase was not inhibited by either alpha-amanitin or actinomycin D. The data were taken to indicate that the selective induction of tyrosine hydroxylase by nerve growth factor does not require RNA synthesis. Since it had previously been shown (Otten and Thoenen, 1977) by the same group using similar methodology that the selective increase was blocked by cycloheximide, the conclusion would seem to be that nerve growth factor increases the stability or the translation efficiency of the tyrosine hydroxylase mRNA, but does not alter the amount of tyrosine hydroxylase mRNA being transcribed.

While the nerve growth factor-evoked increase in incorporation of radiolabel into tyrosine hydroxylase *in vitro* (Max *et al.*, 1978) was somewhat smaller than that seen when nerve growth factor was given *in vivo* (MacDonnell *et al.*, 1977), the data in each study is clear. The *in vivo* experiments indicate that nerve growth factor alters the transcription of the tyrosine hydroxylase gene in sympathetic neurons; the *in vitro* experiments indicate that it does not. The resolution of these contradictory findings is not presently apparent.

The findings with the PC12 cell model have been somewhat more consistent. Although nerve growth factor-induced increases in the levels of tyrosine hydroxylase have been reported in certain clones of pheochromocytoma (Goodman and Herschman, 1978; Hatanaka, 1981), little or no change, or even a decrease, in level was seen in the fully-differentiated parent PC12 (Greene and Tischler, 1976). However, a detailed analysis of the changes in mRNAs occurring when PC12 cells were treated with nerve growth factor (Leonard *et al.*, 1987) revealed a rather complex course of regulation for the tyrosine hydroxylase mRNA. Within 4 h after treatment of the cells, the tyrosine hydroxylase mRNA increased about 2-fold and then dropped rather precipitously to about half the control value and stayed there for the remaining two weeks of the experiment.

The control of the tyrosine hydroxylase gene by nerve growth factor at the molecular level is apparently quite complex (Gizang-Ginsberg and Ziff, 1990). Nerve growth factor induces transcription of the gene with delayed early kinetics. The induction depends upon new protein synthesis and is preceded by the expression of the early response genes such as c-fos, which are also induced by nerve growth factor (Greenberg *et al.*, 1985). Indeed, the data show that an element of the tyrosine hydroxylase promoter, designated TH-FSE because of its extensive homology with the fat-specific element seen on genes induced in adipose tissue differentiation, is a binding site for c-fos. The model that has been proposed is that the increase in tyrosine hydroxylase transcription seen after a few hours of nerve growth factor treatment is due to the nerve growth factor-dependent induction of c-fos, and the subsequent binding of c-fos and c-jun to the TH-FSE. The later decrease in transcription, which occurs after the first 24 h of treatment, is accompanied by a change in the proportion of fos family members in the cells, and the fra proteins become more abundant that the fos proteins. The repression, then, of tyrosine hydroxylase transcription is thought to be caused by a change in the composition of the complex binding to the TH-FSE element. It is suggested that complexes containing fra proteins rather than fos proteins decrease rather than increase the transcription of the tyrosine hydroxylase gene.

The first observation on the nerve growth factor-stimulated phosphorylation of tyrosine hydroxylase was made by Halegoua and Patrick (1980). In these studies it was shown that labeling PC12 cells with [^{32}P]orthophosphate for 1 h or more led to a 2.8-fold increase in the incorporation of radioactivity into a band with a molecular weight of 60 000 D. The molecular weight, and the additional finding that the band increased when the cells were treated with dexamethasone, suggested that the band contained tyrosine hydroxylase. The suggestion was validated through the use of a specific tyrosine hydroxylase antibody.

Somewhat later, Greene *et al.* (1984) demonstrated that nerve growth factor treatment has an acute effect on tyrosine hydroxylase activity. The data presented show that the activity of tyrosine hydroxylase increases some 60% within 2–5 min of exposure to nerve growth factor. Kinetic measurements indicated that there is no alteration in the binding of either the pteridine cofactor of the enzyme or the catecholamine end-product, simply a change in the maximal velocity attending the catalysis. The increase in activity is inhibited by the presence of tyrosine hydroxylase antibodies, but is not prevented by the simultaneous presence of inhibitors of protein synthesis. Thus, this increase in activity is distinct from the increase in enzyme level produced by nerve growth factor over a longer time course and due to an increase in the synthesis of the enzyme. The authors conclude that the activation they have observed is most likely due to an increase in the phosphorylation of the enzyme.

Although nerve growth factor treatment causes both increased phosphorylation and increased enzyme activity, it has been a bit difficult to prove cause and effect. In a study of the activations caused by various agents (Lee *et al.*, 1985), the time course of activation was compared with previous data on the time course of phosphorylation. The data show that the two events are not strictly superimposable. The hydroxylase is activated 60% within 10 min and this activation is half complete in 2–3 min; the phosphorylation exhibits a minute or two lag, but then increases rapidly and is complete within 8 min. The authors, while acknowledging this inconsistency, suggest that either full enzyme activation does not require phosphorylation of all the available sites, or that some of the phosphorylations are involved in things other than enzyme activity, or that, finally, some aspect of the different methodology used to measure phosphorylation and activity results in inherently different data.

McTigue *et al.* (1985) reached the firm conclusion that phosphorylation and activation were linked through study, they compared the phosphopeptides from tyrosine hydroxylase phosphorylated in whole nerve growth factor-treated cells with the peptides phosphorylated when tyrosine hydroxylase was activated *in vitro* by various kinases. Since a specific peptide, called T1, was phosphorylated *in vitro* whenever the enzyme activity increased, and T1 was phosphorylated in cells treated with nerve growth factor, the conclusion that phosphorylation caused increased activity seemed justifiable. The comparison between the phosphopeptide patterns from cells treated with nerve growth factor and those treated with agents that raise cAMP levels led the authors to the further suggestion that cAMP-dependent kinase was activated by nerve growth factor treatment and phosphorylated tyrosine hydroxylase.

This latter point, however, is far from settled. More recently, Vulliet *et al.* (1989) described a kinase in PC12 cells that they find physically associated with the tyrosine hydroxylase molecule. The activity of this enzyme is increased about 2-fold within 3 min of nerve growth factor treatment. It has a molecular weight of some 45 kD and phosphorylated tyrosine hydroxylase on serine 8. The specificity of the enzyme requires that a proline be present in the sequence on the carboxyl-terminal side of the serine phosphorylated, thus leading to the designation of the enzyme as proline-directed protein kinase. Since, in previous studies, the authors had found that tyrosine hydroxylase is also phosphorylated on serine 8 in intact nerve growth factor-treated cells, they suggest in this study that the enzyme they have identified is responsible for the nerve growth factor-induced increase in tyrosine hydroxylase phosphorylation.

However, in most recent work (Haycock, 1990; Mitchell *et al.*, 1990) it has been found that tyrosine hydroxylase in nerve growth factor-treated PC12 cells is most strongly phosphorylated on serine 31. The stimulation of the phosphorylation of this site is reported to be 6-fold in one study and more than 3-fold in the other. This site is also involved in the stimulation of tyrosine hydroxylase phosphorylation used by K^+, by phorbol esters, and by calcium ionophores. This site is not phosphorylated by any of the well-characterized cellularkinases, nor is it phosphorylated, in the intact tyrosine hydroxylase molecule, by the proline-directed kinase described above. The nature of the kinase phosphorylating serine 31, and presumably stimulated by treatment of the cells with nerve growth factor, is presently unknown, but the sequence surrounding serine 31 has a proline on the carboxyl-terminal side, so the possibility exists that a second proline-directed protein kinase, stimulated by nerve growth factor treatment, is responsible. There is presently no information on the nature of the molecular changes that are initiated by phosphorylation and that lead to the increase in hydroxylase activity.

Thus, nerve growth factor treatment of responsive cells increases tyrosine hydroxylase action by two different mechanisms. The first is a short-term elevation of tyrosine hydroxylase activity due to an increase in tyrosine hydroxylase phosphorylation. The second is a longer-term elevation of tyrosine hydroxylase level due to an increase in the synthesis of the enzyme. The physiological meaning of the increased tyrosine hydroxylation seems straight-forward. It is rather generally accepted that nerve growth factor serves as a trophic signal from target tissues to catecholamine-producing neurons, acting to attract and maintain sympathetic innervation of these targets. The presence of nerve growth factor, then, in the synaptic region could produce both a rapid and a persistent increase in the synthesis of the catecholamine neurotransmitters and a consequent enhancement of catecholamine transmission. This increased synaptic function might then contribute to the maintenance of the requisite synaptic connection.

Acknowledgement

It is a great pleasure to congratulate Dr Toshiharu Nagatsu in this fashion. His scientific accomplishments are remarkable, his energy is extraordinary, and his students do him honor.

REFERENCES

Aloe, L. and Levi-Montalcini, R. (1979). Nerve growth factor-induced transformation of immature chromaffin cells *in vivo* into sympathetic neurons: Effect of antiserum to nerve growth factor. *Proc. Natl. Acad. Sci. USA* **76**, 1246–1250.

Dichter, M. A., Tischler, A. S. and Greene, L. A. (1977). Nerve growth factor-induced increase in electrical excitability and acetylcholine sensitivity of a rat pheochromocytoma cell line. *Nature* **268**, 501–504.

Fabricant, R. N., De Larco, J. E. and Todaro, G. J. (1977). Nerve growth factor receptors on human melanoma cells in culture. *Proc. Natl. Acad. Sci. USA* **74**, 565–569.

Frazier, W. A., Angeletti, R. H. and Bradshaw, R. A. (1972). Nerve growth factor and insulin. Structural similarities indicate an evolutionary relationship reflected by physiological action. *Science* **176**, 482–488.

Gizang-Ginsberg, E. and Ziff, E. B. (1990). Nerve growth factor regulates tyrosine hydroxylase gene transcription through a nucleoprotein complex that contains c-fos. *Genes Dev.* **4**, 477–491.

Gnahn, H., Hefti, F., Heumann, R., Schwab, M. E. and Thoenen, H. (1983). NGF-mediated increase in choline acetyltransferase (ChAT) in neonatal rat forebrain: evidence for a physiological role of NGF in the brain. *Brain Res.* **285**, 45–52.

Goodman, R. and Herschman, H. R. (1978). Nerve growth factor-mediated induction of tyrosine hydroxylase in a clonal pheochromocytoma cell line. *Proc. Natl. Acad. Sci. USA* **75**, 4587–4590.

Greenberg, M. E., Greene, L. A. and Ziff, E. B. (1985). Nerve growth factor and epidermal growth factor induce transient changes in proto-oncogene transcription in PC12 cells. *J. Biol. Chem.* **260**, 14101–14110.

Greene, L. A. and Tischler, A. S. (1976). Establishment of a noradrenergic clonal line of rat adrenal pheochromocytoma cells which respond to nerve growth factor. *Proc. Natl. Acad. Sci. USA* **73**, 2424–2428.

Greene, L. A., Seeley, P. J., Rukenstein, A., DiPiazza, M. and Howard, A. (1984). Rapid activation of tyrosine hydroxylase in response to nerve growth factor. *J. Neurochem.* **42**, 1728–1734.

Greene, L. A. and Shooter, E. M. (1980). Nerve growth factor: Biochemistry, synthesis, and mechanism of action. *Ann. Rev. Neurosci.* **3**, 353–402.

Halegoua, S. and Patrick (1980). Nerve growth factor mediates phosphorylation of specific proteins. *Cell* **22**, 571–581.

Hatanaka, H. (1981). Nerve growth factor-mediated stimulation of tyrosine hydroxylase activity in a clonal rat pheochromocytoma cell line. *Brain Res.* **222**, 225–233.

Haycock, J. W. (1990). Phosphorylation of tyrosine hydroxylase *in situ* at serine 8, 19, 31, and 40. *J. Biol. Chem.* **265**, 11682–11691.

Hefti, F., Hartikka, J., Eckenstein, F., Gnahn, H., Heumann, R. and Schwab, M. E. (1985). Nerve growth factor increases choline acetyltransferase but not survival or fiber outgrowth of cultured fetal septal cholinergic neurons. *Neuroscience* **14**, 55–68.

Johnson, D., Lanahan, A., Buck, C. R., Sehgal, A., Morgan, C., Mercer, E., Bothwell, M. and Chao, M. (1986). Expression and structure of the human NGF receptor. *Cell* **47**, 545–554.

Kaplan, D. R., Martin-Zanca, D. and Parada, L. F. (1991). Tyrosine phosphorylation and tyrosine kinase activity of the trk proto-oncogene product induced by NGF. *Nature* **350**, 158–160.

Klein, R., Jing, S. Q., Nanduri, V., O'Rourke, E. and Barbacid, M. (1991). The trk proto-oncogene encodes a receptor for nerve growth factor. *Cell* **65**, 189–197.

Lee, K. Y., Seeley, P. J., Muller, T. H., Helmer-Matyjek, E., Sabban, E., Goldstein, M. and Greene, L. A. (1985). Regulation of tyrosine hydroxylase phosphorylation in PC12 pheochromocytoma cells by elevated K^+ and nerve growth factor. Evidence for different mechanisms of action. *Mol. Pharmacol.* **28**, 220–228.

Leonard, D. G., Ziff, E. B. and Greene, L. A. (1987). Identification and characterization of mRNAs regulated by nerve growth factor in PC12 cells. *Mol. Cell. Biol.* **7**, 3156–3167.

Levi-Montalcini, R. and Angeletti, P. U. (1968). Nerve growth factor. *Physiol. Rev.* **48**, 534-569.

MacDonnell, P. C., Tolson, N. and Guroff, G. (1977). Selective de novo synthesis of tyrosine hydroxylase in organ cultures of rat superior cervical ganglia after *in vivo* administration of nerve growth factor. *J. Biol. Chem.* **252**, 5859–5963.

Max, S. R., Rohrer, H., Otten, U. and Thoenen, H. (1978). Nerve growth factor-mediated induction of tyrosine hydroxylase in rat superior cervical ganglia. *J. Biol. Chem.* **253**, 8013–8015.

McTigue, M., Cremins, J. and Halegoua, S. (1985). Nerve growth factor and other agents mediate phosphorylation and activation of tyrosine hydroxylase. A convergence of multiple kinase activities. *J. Biol. Chem.* **260**, 9047–9056.

Mitchell, J. P., Hardie, D. G. and Vulliet, P. R. (1990). Site-specific phosphorylation of tyrosine hydroxylase after KCl depolarization and nerve growth factor treatment of PC12 cells. *J. Biol. Chem.* **265**, 22358–22364.

Nagatsu, T., Levitt, M. and Udenfriend, S. (1964). Tyrosine hydroxylase. The initial step in norepinephrine biosynthesis. *J. Biol. Chem.* **239**, 2910–2917.

Otten, U. and Thoenen, H. (1977). Effect of glucocorticoids on nerve growth factor-mediated enzyme induction in organ cultures of rat sympathetic ganglia: enhanced response and reduced time requirement to initiate enzyme induction. *J. Neurochem.* **29**, 69–75.

Radeke, M. J., Misko, T. P., Hsu, C., Herzenberg, L. A. and Shooter, E. M. (1987). Gene transfer and molecular cloning of the rat nerve growth factor receptor. *Nature* **325**, 593–597.

Rohrer, H., Otten, U. and Thoenen, H. (1978). On the role of RNA synthesis in the selective induction of tyrosine hydroxylase by nerve growth factor. *Brain Res.* **159**, 436–439.

Schubert, D., Heinemann, S. and Kidokoro, Y. (1977). Cholinergic metabolism and synapse formation by a rat nerve cell line. *Proc. Natl. Acad. Sci. USA* **74**, 2579–2583.

Scott, J., Selby, M., Urdea, M., Quiiroga, M., Bell, G. I. and Rutter, W. J. (1983). Isolation and nucleotide sequence of a cDNA encoding the precursor of mouse nerve growth factor. *Nature* **302**, 538–540.

Thoenen, H., Angeletti, P. U., Levi-Montalcini, R. and Kettler, R. (1971). Selective induction by nerve growth factor of tyrosine hydroxylase and dopamine-beta-hydroxylase in the rat superior cervical ganglia. *Proc. Natl. Acad. Sci. USA* **68**, 1598–1602.

Ullrich, A., Gray, A., Berman, C. and Dull, T. J. (1983). Human beta-nerve growth factor gene sequence is highly homologous to that of mouse. *Nature* **303**, 821–825.

Unsicker, K., Krisch, B., Otten, U. and Thoenen, H. (1978). Nerve growth factor-induced outgrowth from isolated rat adrenal chromaffin cell: Impairment by glucocorticoids. *Proc. Natl. Acad. Sci. USA* **75**, 3498–3502.

Vulliet, P. R., Hall, F. L., Mitchell, J. P. and Hardie, D. G. (1989). Identification of a novel proline-directed serine/threonine protein kinase in rat pheochromocytoma. *J. Biol. Chem.* **264**, 16292–16298.

Waris, T., Rechardt, L. and Waris, P. (1973). Differentiation of neuroblastoma cells induced by nerve growth factor *in vitro*. *Experientia* **29**, 1128–1129.

Tyrosine Hydroxylase, pp. 23–36
M. Naoi *et al.* (Eds)
© VSP 1993

ALLOSTERIC REGULATION OF TYROSINE HYDROXYLASE ACTIVITY BY ITS COFACTOR TETRAHYDROBIOPTERIN

MAKOTO NAOI,[1] MIDORI MINAMI,[2] WAKAKO MARUYAMA[2] and HASAN PARVEZ[3]

[1] *Department of Biosciences, Nagoya Institute of Technology, Nagoya, Japan*
[2] *Department of Neurology, Nagoya University School of Medicine, Nagoya, Japan*
[3] *Unite de Neuropharmacologie, Universite de Paris XI, Orsay Cedex, France*

Abstract—Tyrosine hydroxylase is the rate-limiting enzyme in the catecholamine biosynthesis and the activity is regulated by a rapid and a slow mechanism. A new type of regulation of tyrosine hydroxylation has been found. The activity of tyrosine hydroxylase was regulated by allosteric effect of its cofactor, 5,6,7,8-tetrahydrobiopterin. Positive cooperativity toward the cofactors, $(6R)$-L-*erythro*-5,6,7,8-tetrahydrobiopterin [$(6R)BH_4$] and $(6S)$-L-*erythro*-5,6,7,8-tetrahydrobiopterin [$(6S)BH_4$] was observed. In addition, $(6R)BH_4$, the naturally-occurring cofactor, affected the affinity of the enzyme to its substrate, L-tyrosine. Some of neurotoxin candidates reduced the enzyme activity via the allosteric regulation of the enzyme activity by biopterin cofactor. A dopamine-derived tetrahydroisoquinoline, 6,7-dihydroxy-1,2,3,4-tetrahydroisoquinoline (salsolinol), inhibited the enzyme activity and the allostery induced by biopterin cofactors disappeared. Such a conformational change induced by the allosterism of biopterin regulates catecholamine biosynthesis in the brain and its regulation is further affected by other neurotransmitters or neurotoxins.

Key words: tyrosine hydroxylase; salsolinol; tetrahydrobiopterin; allosteric effect.

Since the discovery of tyrosine hydroxylase (tyrosine, tetrahydropteridine: oxygen oxidoreductase (3-hydroxylating), EC 1.14.16.2, TH) by Nagatsu *et al.* (1964), its enzymatic and kinetical properties have been intensively studied (Kato *et al.*, 1980; Oka *et al.*, 1981, 1982). Tyrosine hydroxylation in catecholamine neurons is regulated by two mechanisms. One mechanism is the one that involves an acute increase and decrease in the activity of TH, in response to the release of the catecholamines from the nerve terminals. This regulation is mediated by the covalent modification of the TH protein by phosphorylation (Goldstein *et al.*, 1984), or by inhibition with the end product, dopamine (Nagatsu *et al.*, 1972;

Markey *et al.*, 1980), or other neurotransmitters (for reviews, see, Goldstein and Greene, 1987; Zigmond *et al.*, 1989). Another mechanism involves an increase in tyrosine hydroxylation, observed after a delay and continued for a few days. This mechanism is caused by an increase in the amount of the TH protein, via transcriptional control (Mallet *et al.*, 1983; Leonard *et al.*, 1987).

TH is a mono-oxygenase, which catalyzes the hydroxylation of tyrosine (TYR) using molecular oxygen and a cofactor, 5,6,7,8-tetrahydrobiopterin. The therapeutic effects of the natural cofactor, (6*R*)-L-*erythro*-5,6,7,8-tetrahydrobiopterin [(6*R*)BH$_4$] to the patients with depression (Ito *et al.*, 1988) or Parkinsonism (Narabayashi *et al.*, 1982) suggest that the concentration of (6*R*)BH$_4$ regulates the rate of DOPA biosynthesis in the brain. TH exists mainly in a form of tetramer *in vivo* (Markey *et al.*, 1980; Kojima *et al.*, 1984), but the kinetical studies on TH polymers have not shown whether any factor, such as biopterin, can regulate TH activity by an allosteric effect. In concern to biopterin cofactors, the naturally-occurring cofactor (6*R*)BH$_4$ and its un-natural enantiomer, (6*S*)-L-*erythro*-5,6,7,8-tetrahydrobiopterin [(6*S*)BH$_4$] differently affected on TH activity. From *in vitro* studies with purified TH (Oka *et al.*, 1984), the synthetic cofactor (6*S*)BH$_4$ also had a cofactor activity. On the other hand, *in vivo* stimulating activity toward tyrosine hydroxylation was observed only with (6*R*)BH$_4$ (Matsuura *et al.*, 1986). The difference between *in vivo* and *in vitro* results may be due to different intracellular concentrations of (6*R*)BH$_4$ and (6*S*)BH$_4$, or due to different effects of biopterins on TH as cofactors or as other regulatory factors. This paper represents the allosteric regulation of tyrosine hydroxylase polymers by biopterins. Figure 1 shows the chemical structures of these two enantiomers of tetrahydrobiopterins.

ALLOSTERY OF BIOPTERIN COFACTOR

The effect of biopterin cofactor on TH activity was examined by use of the partially purified enzyme. PC12 cells cultured as reported previously (Maruyama *et al.*, 1991) were used as the enzyme source and TH was purified by use of high-performance liquid chromatography (HPLC) on a DEAE column, as described

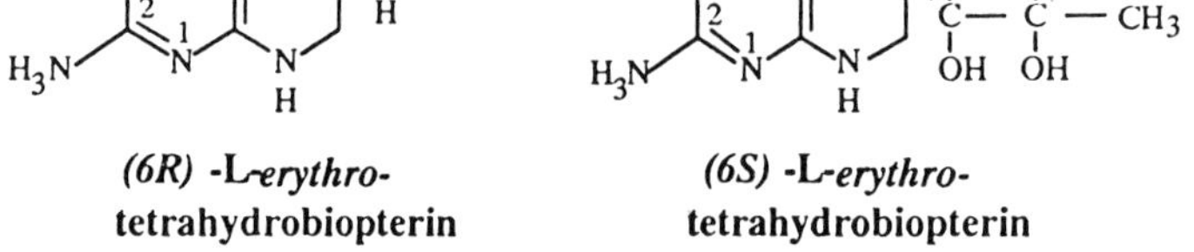

(6R) -L-*erythro*-
tetrahydrobiopterin

(6S) -L-*erythro*-
tetrahydrobiopterin

Figure 1. Chemical structures of two enantiomers of tetrahydrobiopterins.

in our papers (Minami *et al.*, 1992a, b). (6R)BH$_4$ and (6S)BH$_4$ were kindly donated by Dr S. Matsuura, Fujita Health University (Toyoake, Japan). TH activity was assayed by measurement of DOPA formation in the presence of an inhibitor of DOPA decarboxylase with HPLC (Naoi *et al.*, 1988).

By kinetical analysis the activity of TH was found to be regulated by the concentration of the biopterin cofactor. The enzyme activity measured with various concentrations of (6R)BH$_4$ was plotted against the cofactor concentration. Figure 2 presents the positive cooperativity for (6R)BH$_4$, the naturally occurring cofactor. Kinetical properties were analyzed by non-linear least squares error (Damped–Newton) method. Kinetics of TH were described by Hill equation

$$v = \frac{V_{\max} \times s^H}{K_m^H + s^H}$$

where s, v, and H denote the substrate concentration, velocity and Hill's coefficient, respectively.

$V_{\max}$, K_m values, and Hill coefficient of (6R)BH$_4$ and (6S)BH$_4$ with 100 μM of TYR are given in Table 1. The high values of Hill coefficient demonstrate the allosteric effect in concern with biopterin cofactors, especially in case of (6R)BH$_4$. Hill coefficient obtained with (6R)BH$_4$ was higher than that with (6S)BH$_4$; 2.52 $\pm$ 0.13 and 1.76 $\pm$ 0.19, respectively.

With respect to the kinetics for TYR by use of both (6R)BH$_4$ and (6S)BH$_4$, allosteric effect was not observed. Table 2 shows the changes in K_m and $V_{\max}$ values for TYR with various concentrations of the naturally-occurring cofactor

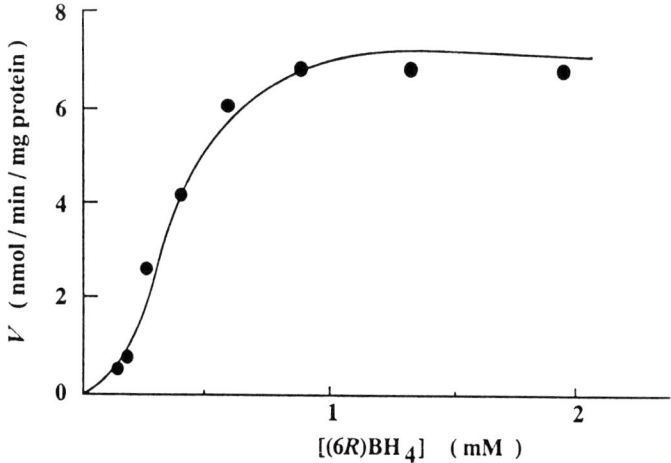

Figure 2. Effect of concentration of (6R)BH$_4$ on TH activity with 100 μM TYR. The reaction velocity plotted against the (6R)BH$_4$ concentration. Each value represents the velocity observed in the duplicate measurements.

$(6R)BH_4$. V_{max} value for TYR depended on the concentration of $(6R)BH_4$ and increased in a dose-dependent way, but K_m value was almost the same with $(6R)MH_4$ from 0.1 mM to 0.5 mM.

Table 1.
V_{max} and K_m values, and Hill coefficient for $(6R)BH_4$ and $(6S)BH_4$ of tyrosine hydroxylase prepared from PC12 cells using 100 μM TYR

	V_{max} (nmol/min/mg protein)	K_m (μM)	Hill coefficient
In terms of $(6R)BH_4$	7.3 ± 0.5	351 ± 64	2.52 ± 0.13
$(6S)BH_4$	8.0 ± 1.5	470 ± 190	1.76 ± 0.19

Each value represents mean ± SD.

Table 2.
Effects of $(6R)BH_4$ concentration on K_m and V_{max} value of tyrosine hydroxylase in terms of L-tyrosine

Concentration of the cofactor		K_m (μM)	V_{max} (nmol/min/mg protein)
$(6R)BH_4$	1.5 mM*	3.8 ± 0.5	11.13 ± 0.34
	1.0 mM*	34.3 ± 4.4	7.52 ± 0.75
	0.50 mM*	11.5 ± 2.1	3.23 ± 0.30
	0.25 mM*	11.5 ± 2.5	2.38 ± 0.28
	0.10 mM*	13.3 ± 4.5	1.29 ± 0.25

Each value represents mean ± SD.
*Substrate inhibition was observed.

Figure 3 presents these kinetical properties by Lineweaver–Burk plot. Not only the affinity to the cofactor but also the binding property to TYR was affected, which may be due to the conformational change induced by $(6R)BH_4$. Furthermore, substrate inhibition by TYR was observed. With a relatively low concentration of TYR (50 μM) where the substrate inhibition was not observed, the positive cooperativity for $(6R)BH_4$ was also observed, so that the allosteric effect in concern with $(6R)BH_4$ is not due to this substrate inhibition.

On the other hand, using $(6S)BH_4$ as the cofactor, substrate inhibition by TYR was not observed. Figure 4 presents these kinetical properties for TYR with $(6S)BH_4$ by Lineweaver–Burk plot. Moreover, V_{max} values were lower than those with $(6R)BH_4$ at each concentration of the cofactor as shown in Table 3. K_m values for TYR were 1.02 ± 0.28 μM and 0.96 ± 0.43 μM with 0.25 mM and 0.10 mM of $(6S)BH_4$, respectively, while K_m values for TYR at the same concentrations of $(6R)BH_4$ were 10 times higher than those obtained

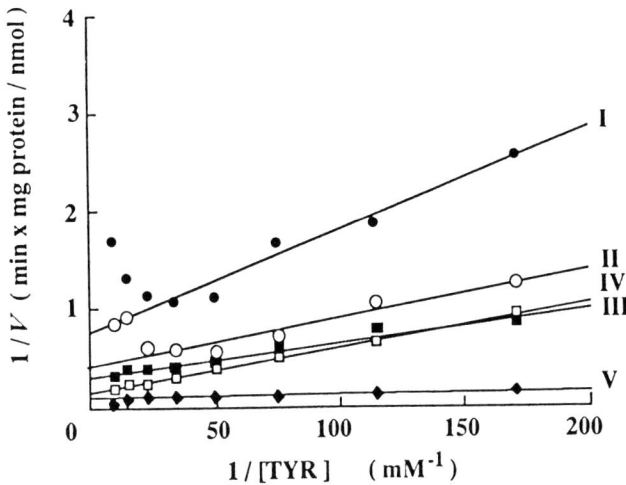

Figure 3. Effect of concentration of TYR on TH activity with different concentrations of $(6R)BH_4$. The reciprocal of the reaction velocity was plotted against that of the TYR concentration, according to Lineweaver and Burk. Curve I: with 0.1 mM of $(6R)BH_4$; Curve II: with 0.25 mM; Curve III: with 1.0 mM; Curve IV: with 0.5 mM; Curve V: with 1.5 mM.

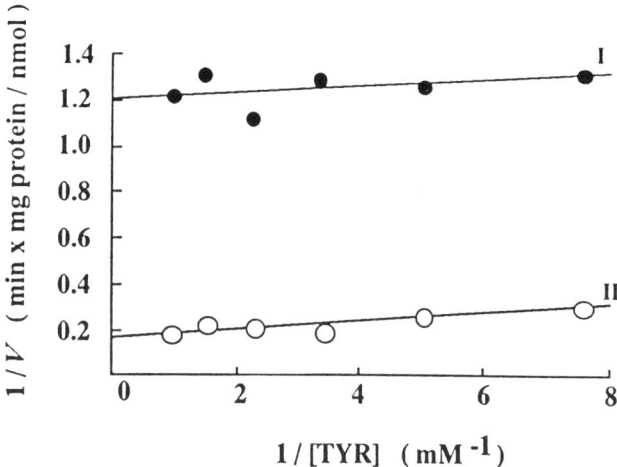

Figure 4. Effect of TYR concentration on TH activity with different concentrations of $(6S)BH_4$. The reciprocal of the velocity was plotted against that of TYP concentration according to Lineweaver and Burk. Curve I: with 0.25 mM, and II: with 1 mM of $(6S)$ BH_4.

with $(6S)BH_4$. In Table 4, the K_m values for TYR with 2 mM of either $(6R)BH_4$ or $(6S)BH_4$ are presented, and the K_m value for the former is about 5 times larger that that for the latter.

It is indicated that biopterin is a regulatory factor which induces a conformational change of TH polymers, and the affinity of TH for the cofactor depends on the concentration of the cofactor itself. The concentration of $(6R)BH_4$ used in these experiments is within the range of the $(6R)BH_4$ concentration in the nerve terminals of dopaminergic neurons (Levine et al., 1981). The kinetical properties for TYR were also affected by this conformational change. The larger K_m value of TYR with $(6R)BH_4$ suggests that the TH activity may be effected by the intracellular concentration of TYR. Furthermore, binding of TYR depends on the stereochemical structure at C-6 of tetrahydrobiopterin. Biopterin is also a cofactor of tryptophan hydroxylase (Lovenberg et al., 1967; Ichiyama et al., 1970; Friedman et al., 1972) and phenylalanine hydroxylase (Kaufman, 1963). In concern with phenylalanine hydroxylase, $(6R)BH_4$ was suggested to be a regulatory factor of its phosphorylation (Phillips and Kaufman, 1984). On the other hand, the concentration of $(6R)BH_4$ regulates the activity of tryptophan hydroxylase because $(6R)BH_4$ seems to be subsaturated in vivo (Sawada et al., 1986). It is indicated here that $(6R)BH_4$ is also subsaturated for TH and that not only the concentration but also the allosteric effect of $(6R)BH_4$ may contribute to the regulation of DOPA biosynthesis by inducing the conformational change of TH polymers. Such a direct effect of $(6R)BH_4$ on the affinity to the cofactor itself has never been observed. At present, the effect of $(6R)BH_4$ on the quaternary structure of TH polymers has not been clarified from the point of protein chemistry. The effects of the conformational change on TH activity in vivo are left for further studies.

NEUROTOXINS AFFECT ALLOSTERY BY BIOPTERIN

Since the discovery of 1-methyl-4-phenyl-1,2,3,6-tetrahydropyridine (MPTP) as a neurotoxin which causes Parkinsonism (Langston et al., 1983), various kinds of neurotoxic compounds have been found to cause cell death of dopaminergic neurons or to inhibit the enzymes related to catecholamine metabolism. Among these substances, such as heterocyclic amines (Ichinose et al., 1988) and 1,2,3,4-tetrahydroisoquinoline (TIQ) (Nagatsu and Yoshida, 1988), N-methyl-1,2,3,4-tetrahydroisoquinoline was found to be a neurotoxin synthesized in vivo from TIQ (Naoi et al., 1989a, b, c). Since TH is the rate-limiting enzyme of catecholamine biosynthesis (Nagatsu et al., 1964), studies on the effects of such naturally-occurring inhibitors on this enzyme under physiological conditions have a great importance.

1-Methyl-6,7-dihydroxy-1,2,3,4-tetrahydroisoquinoline (salsolinol, SAL) is one of dopamine-derived tetrahydroisoquinolines, synthesized from acetaldehyde or pyruvic acid and dopamine in vivo (Dostert et al., 1988). SAL was found to exist in the brain, but it cannot cross the blood-brain barrier, so SAL in brain should be synthesized *in situ* (Origitano *et al.*, 1981). SAL possesses an asymmetric center at C-1 and exists as *R* and *S* enantiomers. Although the *S* enantiomer [(*S*)SAL] is present in foods such as port wine or dried banana, it has been indicated that the endogenous salsolinol in the brain is the *R* enantiomer [(*R*)SAL] (Strolin-Benedetti *et al.*, 1989). To give rise to the stereospecificity at the C-1 position, synthesis of (*R*)SAL by the non-enzymatic condensation of dopamine with pyruvic acid should be followed by enzymic decarboxylation and reduction. It is known that the substrate TYR is nearly saturated but the cofactor (6*R*)BH$_4$ is not saturated in the brain tissue. It is necessary to study the effects of (*R*)SAL on TH activity under physiological conditions to get information of the effects of SAL *in vivo*. The structures of two isomers of salsolinols are presented in Fig. 5.

Both isomers of salsolinols inhibited TH activity in a dose-dependent way. With 1 mM (*R*)SAL, TH activity decreased to 74% and 61% of the control value with 1 mM (6*R*)BH$_4$ and (6*S*)BH$_4$, respectively. The effects of salsolinol isomers on the allostery of (6*R*)BH$_4$ are shown in Fig. 6. Both isomers of salsolinols changed the kinetic properties of TH significantly, measured with various concentrations of (6*R*)BH$_4$ from 2 mM to 0.12 mM and with a sufficient concentration of a substrate TYR. The enzyme activity was reduced significantly. In addition the allostery of (6*R*)BH$_4$ disappeared.

The effects of salsolinols on the Hill coefficient, V_{max} and K_m, values of TH toward biopterin cofactors are summarized in Table 3. The V_{max} value was reduced by either isomers of salsolinol. In term of (6*R*)BH$_4$, reduction of TH activity was more manifest with (*R*)SAL than with (*S*)SAL. On the other hand, with another cofactor (6*S*)BH$_4$, the reduction was almost the same with both SAL isomers. The K_m values were also reduced by both isomers of SALs, and the reduction was almost the same with these isomers. In the case of control sample, Hill coefficient was 2.52, which represents a positive cooperativity towards (6*R*)BH$_4$. With 500 μM (*R*)SAL, Hill coefficient was estimated to be about 1 and an allostery as observed with control sample was not observed. Similar results were obtained in concern with (*S*)SAL.

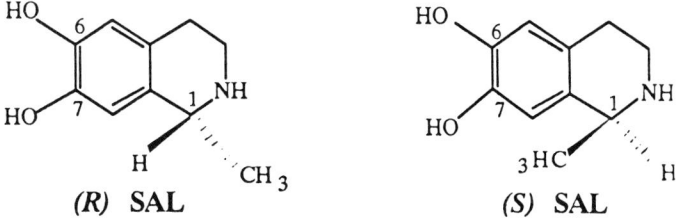

Figure 5. Chemical structure of (*R*)salsolinol and (*S*)salsolinol.

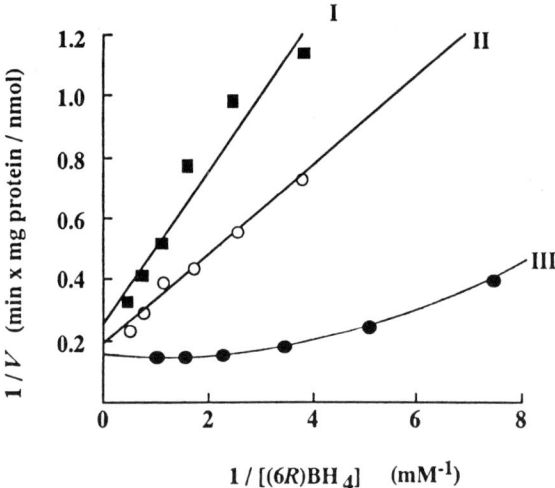

Figure 6.   Lineweaver–Burk plot of TH activity in terms of $(6R)BH_4$ with 100 μM TYR in the absence and presence of salsolinols. Curve I: Control, Curve II: with (R)SAL 500 μM, Curve III: (S)SAL 500 μM.

The effects of salsolinols on kinetic properties of TH in term of TYR were also examined. The TH activity measured with 2 mM $(6R)BH_4$ are represented by Lineweaver–Burk plot in Fig. 7. The activity of TH was reduced by both isomers of salsolinol; the K_m and V_{max} values reduced.

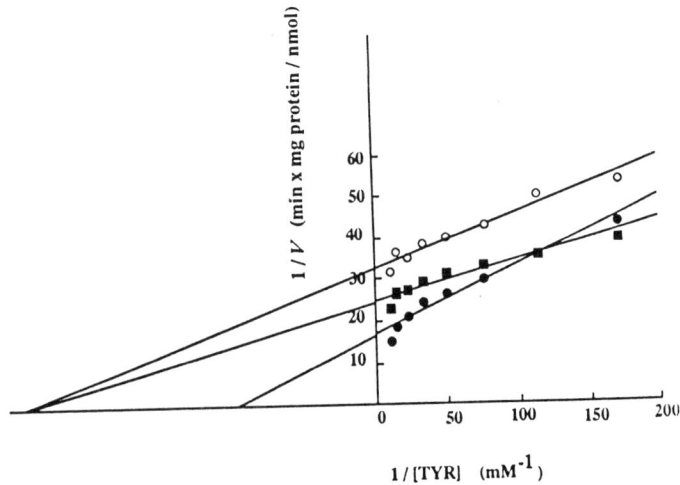

Figure 7.   Lineweaver–Burk plot of TH activity in terms of $(6R)BH_4$ with 100 μM TYR in the absence and presence of salsolinols. Curve I: Control, Curve II: with (S)SAL 500 μM, Curve III: with (R)SAL 500 μM.

Table 3.
Effects of salsolinols on Hill coefficient, V_{max} and K_m values of tyrosine hydroxylase for biopterin cofactors with 100 μM of tyrosine.

Concentration of salsolinol	Hill coefficient	V_{max} (pmol/min/mg protein)	K_m (μM)
In terms of (6R)BH$_4$			
Control	2.52 ± 0.13	308 ± 19	351 ± 64
(R)SAL 500 μM		161 ± 29	96 ± 25
(S)SAL 500 μM		220 ± 17	76 ± 9
In terms of (6S)BH$_4$			
Control	1.76 ± 0.19	304 ± 57	470 ± 190
(R)SAL 100 μM		146 ± 15	230 ± 70
(S)SAL 100 μM		151 ± 11	161 ± 45

Each value represents mean ± SD.

The results on the Hill coefficient, V_{max} and K_m values are also summarized in Table 4. Using (6R)BH$_4$ as a cofactor, R and S isomer of salsolinols caused similar results; V_{max} and K_m values decreased. The V_{max} value was decreased more markedly with (R)SAL than with (S)SAL, but reduction in K_m values was not observed with (R)SAL and (S)SAL, in contrast to the results with (6R)BH$_4$. Using (6S)BH$_4$, the K_m and V_{max} values were reduced, and the reduction is almost the same with either isomers of SALs. Figure 8 shows the effects of R and S salsolinols on the activity of TH, using (6S)BH$_4$ as a cofactor.

Table 4.
Effects of salsolinols on V_{max} and K_m values of tyrosine hydroxylase for L-tyrosine

Concentration of salsolinol	V_{max} (pmol/min/mg protein)	K_m (μM)
with (6R)BH$_4$: 2 mM		
Control	308 ± 24	8.0 ± 1.5
(R)SAL 500 μM	185 ± 2	6.7 ± 0.2
(S)SAL 500 μM	131 ± 3	2.9 ± 0.3
with (6R)BH$_4$: 2 mM		
Control	58.1 ± 3.4	9.3 ± 1.2
(R)SAL 100 μM	30.5 ± 0.9	4.0 ± 0.5
(S)SAL 100 μM	40.3 ± 1.3	4.0 ± 0.5

Each value represents mean ± SD.

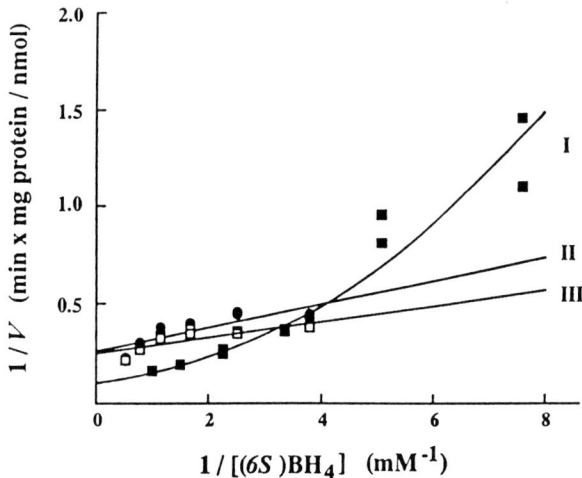

Figure 8. Lineweaver–Burk plot of TH activity in terms of $(6S)BH_4$ with 100 μM TYR in the absence and presence of salsolinols. Curve I: Control, Curve II: with (R)SAL 100 μM, Curve III: (S)SAL 100 μM.

It has been reported that racemic (RS)-salsolinol inhibits TH *in vitro* with K_i value of about 14 μM, and that the inhibition was competitive with an artificial pteridine cofactor $(6RS)$-methyltetrahydropterin (MePH$_4$) (Weiner and Collins, 1978). But endogenously occurring salsolinol is believed to be the R enantiomer, while alcohol or some specific foods such as dried banana, contain the S enantiomer (Strolin-Benedetti *et al.*, 1989). For the study of inhibition under the physiological conditions, it is therefore worthwhile to examine kinetics of each enantiomers separately.

As described in the first part of this paper, it is indicated that the allostery of TH polymers may contribute to the regulation of tyrosine hydroxylation by changing the affinity to the cofactor. Such allostery disappeared with (R) and (S) salsolinols. Moreover, both enantiomers reduced V_{max} and K_m values for biopterin cofactors especially in terms of the naturally occurring cofactor $(6R)BH_4$. Salsolinols caused the conformational change of TH polymers, which makes it impossible to vary the affinity of TH to $(6R)BH_4$ in a dose dependent way in concern to the cofactor itself.

With respect to the kinetic of TYR, salsolinols reduced V_{max} and K_m values and the inhibition was uncompetitive. When $(6S)BH_4$ was a cofactor, there were no differences in the kinetics of inhibition caused by (R)SAL and (S)SAL. On the other hand, using $(6R)BH_4$ as a cofactor, V_{max} and K_m values in the presence of (R)SAL were higher than those values in the presence of (S)SAL. Thus, conformation of TH polymers with $(6R)BH_4$ could selectively respond to the stereochemical structures of R and S isomers of salsolinol. The asymmetric center of salsolinol at C-1 is related to the binding site of TYR.

It is known that the concentration of $(6R)BH_4$ is 1–10 μM in human caudate nucleus (Nagatsu, 1983). With respect to the substrate TYR, it has been found that the concentration is about 100 μM in rat brain (Naoi *et al.*, in preparation). From the synthetic pathway *in vivo*, SAL may increase in intoxicated alcoholics and decrease by degeneration of dopaminergic neurons, as in the case of Parkinson's disease (Dostert *et al.*, 1988). There are several contradictory reports on the occurrence of endogenous SAL in brain. Sjoequist *et al.* (1982) reported that caudate, putamen, hippocampus, and cortex of intoxicated alcoholics contain higher levels of SAL than of sober alcoholics. However, the concentration of SAL in these brain regions of non-alcoholics were almost the same to the value of intoxicated alcoholics. The SAL concentration was high in dopamine-rich regions such as caudate or putamen, up to 0.2 μM. TH activity of the control is lower than that with salsolinols at low concentrations of biopterin cofactors. This is due to the disappearance of the allostery of TH polymers toward biopterins. Biosynthesis of DOPA is considered to be regulated mainly by the concentration of $(6R)BH_4$ and the affinity of TH toward it. It is indicated that salsolinols may alter such regulation by their disallosteric effect on TH polymers. Intracellular level of $(R)SAL$ may be different from the concentration in brain homogenates, so that further studies on uptake and accumulation of $(R)SAL$ in dopaminergic neurons are necessary to judge the importance of inhibition which is due to the conformation change caused by $(R)SAL$ *in vivo*.

MPTP was found to reduce TH activity (expressed as specific activity; TH activity/cell protein) selectively at concentrations much lower than those required to be cytotoxic. It is proved by culture of a dopamine cell model, rat clonal pheochromocytoma PC12 cells in the presence of MPTP or its oxidized product 1-methyl-4-phenylpyridinium ion (MPP^+) (Naoi *et al.*, 1988b, c). These results indicate that TH is specifically sensitive to dopaminergic neurotoxins, such as MPTP, MPP^+, TIQs, and heterocyclic amines (Naoi *et al.*, 1988d). Reduction of TH activity by culture with heterocytic amines was found to be due to reduced affinity of TH to the biopterin cofactor, which may be further ascribed to the conformational changes of TH polymers (Maruyama *et al.*, 1991). The allosteric effect of biopterin cofactor described here suggest that the allosterism and its alteration by a toxic compound are one of the most important factors to regulate the TH activity under physiological and pathological conditions in the brain.

Acknowledgments

This work was supported by a Grant-in-Aid for Scientific Research on Priority Area from the Ministry of Education, Science and Culture, Japan. The authors wish to thank Miss Aya Yoshida for her capable technical assistance.

REFERENCES

Bradford, M. M. (1976). A rapid and sensitive method for the quantitation of microgram quantities of protein using the principle of protein dye binding. *Anal. Biochem.* **72**, 248–254.

Dostert, P., Strolin-Benedetti, M. and Dordain, G. (1988). Dopamine-derived alkaloids in alcoholism and in Parkinson's and Huntington's diseases. *J. Neural Transm.* **74**, 61–74.

Goldstein, M. and Greene, L. A. (1987). Activation of tyrosine hydroxylase by phosphorylation. In: *Psychopharmacology: The Third Generation of Progress.* H. Y. Meltzer (Ed.). Raven, New York, pp. 75–80.

Goldstein, M., Lee, K. Y., Sabban, E. L., Seeley, J. and Greene, L. (1984). Studies on phosphorylation of tyrosine hydroxylase. In: *Catecholamines; Basic and Peripheral Mechanism.* E. Usdin, Carlsson, A. A. Dahlstrom and J. Engel (Eds). Alan R. Liss, New York, pp. 189–193.

Ichiyama, A., Nakamura, S., Nishizuka, Y. and Hayaishi, O. (1970). Enzymatic studies on the biosynthesis of serotonin in mammalian brain. *J. Biol. Chem.* **245**, 1699–1709.

Ichinose, H., Ozaki, N., Nakahara, D., Naoi, M., Wakabayashi, K., Sugimura, T. and Nagatsu, T. (1988). Effects of heterocyclic amines in food on dopamine metabolism in nigro-striatal dopaminergic neurons. *Biochem. Pharmac.* **37**, 3289–3295.

Ito, T., Fujita, K., Matsuura, S. and Nagatsu, T. (1988). Treatment of depression with (6*R*)-tetrahydrobiopterin, the natural cofactor of tyrosine hydroxylase and tryptophan hydroxylase. *Biogenic Amines* **5**, 489–493.

Kato, T., Oka, K., Nagatsu, T., Sugimoto, T. and Matsuura, S. (1980). Effects of structures of tetrahydrobiopterin cofactors on tyrosine hydroxylase. *Biochim. Biophys. Acta* **612**, 226–232.

Kaufman, S. (1963). The structure of the phenylalanine-hydroxylation cofactor. *Proc. Natl. Acad. Sci. USA* **50**, 1085–1093.

Kojima, K., Mogi, M., Oka, K. and Nagatsu, T. (1984). Purification and immunochemical characterization of human adrenal tyrosine hydroxylase. *Neurochem. Int.* **6**, 475–480.

Langston, J., W., Ballard, P., Tetrud, J. W. and Irwin, I. (1983). Chronic Parkinsonism in human due to a product of meperidine-analog synthesis. *Science* **219**, 979–980.

Levine, R. A., Millor, L. P. and Lovenberg, W. (1981). Tetrahydrobiopterin in striatum: localization in dopamine nerve terminals and role in catecholamine synthesis. *Science* **214**, 919–921.

Leonard, D. G. B., Ziff, E. B. and Greene, L. A. (1987). Identification and characterization of mRNAs reglurated by nerve growth factor in PC12 cells. *Mol. Cell Biol.* **7**, 3156–3167.

Lovenberg, W., Jequir, F. and Sjoerdsma, A. (1967). Tryptophan hydroxylase; measurement in pineal gland, brain stem, and carcinoid tumor. *Science* **155**, 217–219.

Mallet, J., Faucon Biguet, N., Buda, M., Lamouroux, A. and Samolyk, D. (1983). Detection and regulation of the tyrosine hydroxylase mRNA levels in rat adrenal medulla and brain tissues. *Cold Spring Harbor Symp. Quant. Biol.* **48**, 305–308.

Markey, K. A., Kondo, S., Shenkman, L. and Goldstein, M. (1980). Purification and characterization of tyrosine hydroxylase from a clonal pheochromocytoma cell line. *Mol. Pharmacol.* **17**, 79–85.

Maruyama, W., Minami, M., Ota, A., Takahashi, T., Takahashi, A., Nagatsu, T. and Naoi, M. (1991). Reduction of enzymatic activity of tyrosine hydroxylase by a heterocyclic amine, 3-amino-1,4-dimethyl-5H-pyrido(4,3-b)indole (Trp-p-1), was due to reduced affinity to a cofactor biopterin. *Neurosci. Lett.* **125**, 85–88.

Matsuura, S., Murata, S., Sugimoto, T., Sawada, M. and Nagatsu, T. (1986). Preparation and cofactor activity of (6*S*)-tetrahydrobiopterin. *Chemistry Express* **1**, 403–406.

Minami, M., Takahashi, T., Maruyama, W., Takahashi, A., Nagatsu, T. and Naoi, M. (1992a). Alloster effect of tetrahydrobiopterin cofactors on tyrosine hydroxylase activity. *Life Sci.* **50**, 5–20.

Minami, M., Takahashi, T., Maruyama, W., Takahashi, A., Dostert, P., Nagatsu, T. and Naoi, M. (1992b). Inhibition of tyrosine hydroxylase by *R* and *S* enantiomers of salsolinol, 1-methyl-6,7-dihydroxy-1,2,3,4-tetrahydroisoquinoline. *J. Neurochem.* **58**, 2097–2101.

Nagatsu, T. (1983). Biopterin cofactor and monoamine-synthesizing mono-oxygenases. *Neurochem. Int.* **5**, 27–38.

Nagatsu, T., Levitt, M. and Udenfriend, S. (1964). Tyrosine hydroxylase. The initial step in norepinephrine biosynthesis. *J. Biol. Chem.* **239**, 2910–2917.

Nagatsu, T., Mizutani, K., Nagatsu, I., Matsuura, S. and Sugimura, T. (1972). Pteridines as cofactor or inhibitor of tyrosine hydroxylase. *Biochem. Pharmacol.* **21**, 1945-1953.

Nagatsu, T. and Yoshida, M. (1988). An endogenous substance of the brain, tetrahydroisoquinoline, produces parkinsonism in primates with decreased dopamine, tyrosine hydroxylase and biopterin in the nigrostriatal regions. *Neurosci. Lett.* **87**, 178–182.

Naoi, M., Takahashi, T. and Nagatsu, T. (1988a). Simple assay procedure for tyrosine hydroxylase activity by high-performance liquid chromatography employing coulometric detection with minimal sample preparation. *J. Chromatogr.* **427**, 229–238.

Naoi, M., Takahashi, T. and Nagatsu, T. (1988b). Effect of 1-methyl-4-phenylpyridinium ion (MPP$^+$) on catecholamine levels and activity of related enzymes in clonal rat pheochromocytoma PC12 cells. *Life Sci.* **43**, 1485–1491.

Naoi, M., Takahashi, T., Ichinose, H. and Nagatsu. T. (1988c). Inhibition of aromatic L-aminoacid decarboxylase in clonal pheochromocytoma PC12 cells by N-methyl-4-phenylpyridinium ion (MPP$^+$). *Biochem. Biophys. Res. Commun.* **152**, 15–21.

Naoi, M., Takahashi, T., Ichinose, H., Wakabayashi, K., Sugimura, T. and Nagatsu, T. (1988d). Reduction of enzyme of tyrosine hydroxylase and aromatic L-aminoacid decarboxylase in clonal pheochromocytoma PC12 cells by cartinogenic heterocyclic amines. *Biochem. Biophys. Res. Commun.* **157**, 494–499.

Naoi, M., Matsuura, S., Parvez, H., Takahashi, T, Hirata, Y., Minami, M. and Nagatsu, T. (1989a). Oxidation of N-methyl-1,2,3,4-tetrahydrosioquinoline into the N-methylisoquinolinium ion by monoamine oxidase. *J. Neurochem.* **52**, 653–655.

Naoi, M., Matsuura, S., Takahashi, T. and Nagatsu, T. (1989b). An N-methyltransferase in human brain catalyses N-methylation of 1,2,3,4-tetrahydroisoquinoline into N-methyl-1,2,3,4-tetrahydroisoquinoline, a precursor of a dopaminergic neurotoxin, N-methylisoquinolinium ion. *Biochem. Biophys. Res. Commun.* **161**, 1213–1219.

Naoi, M., Takahashi, T., Parvez, H., Kabeya, R., Taguchi, E. K., Yamaguchi, K., Hirata, Y., Minami, M. and Nagatsu, T. (1989c). N-Methylisoquinolinium ion as an inhibitor of tyrosine hydroxylase, aromatic L-amino acid decarboxylase and monoamine oxidase. *Neurochem. Int.* **15**, 315–320.

Narabayashi, H., Kondo, S., Nagatsu, T., Sugimoto, T. and Matsuura, S. (1982). Tetrahydropterinadministration for parkinsonian symptoms. *Proc. Jap. Acad.* **57–B**, 351–354.

Oka, K., Kato, T., Sugimoto, T., Matsuura, S. and Nagatsu, T. (1981). Kinetic properties of tyrosine hydroxylase with natural tetrahydrobiopterin as cofactor. *Biochim. Biophys. Acta* **661**, 45–53.

Oka, K., Ashiba, G., Sugimoto, T., Matsuura, S. and Nagatsu, T. (1982). Kinetic properties of tyrosine hydroxylase purified from bovine adrenal medulla and bovine caudate nucleus. *Biochim. Biophys. Acta* **706**, 188–196.

Origitano, T., Hannigan, J. and Collins, M. A. (1981). Rat brain salsolinol and blood-brain barrier. *Brain Res.* **224**, 446–451.

Phillips, R. S. and Kaufman, S. (1984). Ligand effects on the phosphorylation state of hepatic phenylalanine hydroxylase. *J. Biol. Chem.* **259**, 2474–2479.

Sawada, M., Sugimoto, T., Matsuura, S. and Nagatsu, T. (1986). (6R)-Tetrahydrobiopterin increases the activity of tryptophan hydroxylase in rat raphe slices. *J. Neurochem.* **47**, 1544–1547.

Sjoequist, B., Eriksson, A. and Winblad, B. (1982). Salsolinol and catecholamines in human brain and their relation to alcoholism. *Prog. Clin. Biol. Res.* **90**, 57–67.

Strolin-Benedetti, M., Dostert, P. and Carminati, P. (1989). Influence of food intake on the enantiometric composition of urinary salsolinol in man. *J. Neural Transm.* **78**, 43–51.

Teitel, S., O'Brien, J. and Brossi, A. (1972). Alkaloids in mammalian tissue. 2. Synthesis of (+) and (−) substituted-6,7-dihydroxy-1,2,3,4-tetrahydroisoquinolines. *J. Med. Chem.* **15**, 845–846.

Weiner, C. D. and Collins, M. A. (1978). Tetrahydroisoquinolines derived from catecholamines or dopa: Effects on brain tyrosine hydroxylase activity. *Biochem. Pharmac.* **27**, 2699–2703.

Zigmond, R. Z., Schwazschild, M. A. and Rittenhouse, A. R. (1989). Acute regulation of tyrosine hydroxylase by neurotransmitters via phosphorylation. *Ann. Rev. Neurosci.* **12**, 415–461.

Tyrosine Hydroxylase, pp. 37–57
M. Naoi *et al.* (Eds)
© VSP 1993

NON-CATECHOLAMINERGIC NEURONAL EXPRESSION OF HUMAN TYROSINE HYDROXYLASE IN THE BRAIN OF TRANSGENIC MICE WITH SPECIAL REFERENCE TO AROMATIC L-AMINO ACID DECARBOXYLASE

IKUKO NAGATSU,[1] KEIKI YAMADA,[1] NOBUYUKI KARASAWA,[1] NORIO KANEDA,[3] TOSHIKUNI SASAOKA,[3] KAZUTO KOBAYASHI,[2] KEISUKE FUJITA[2] and TOSHIHARU NAGATSU[2]

[1]*Department of Anatomy and* [2]*Institute for Comprehensive Medical Science, School of Medicine, Fujita Health University, Toyoake, Aichi 470-11, Japan*
[3]*Department of Biochemistry, Nagoya University School of Medicine, Nagoya, Japan*

Abstract—In the brain of transgenic (Tg) mice, carrying the entire human tyrosine hydroxylase (TH) gene, we observed TH expression not only in typical catecholaminergic (CAnergic) neurons, but also in non-CAnergic neurons. We classified atypical mouse neurons expressing human TH in the following three cell groups: 1) the cell groups with TH-like immunoreactivity (LI) alone, 2) some of the cell groups (D-type) with aromatic L-amino acid decarboxylase (AADC)-LI alone, and 3) the cell groups (B-type) with both AADC-LI and serotonin(5HT)-LI. 4) TH was expressed in CAnergic (A/C types) neurons; A16-A1 and C3-C1 neurons. 1) The cell groups with TH-LI alone are observed in the *cortex frontalis*, n *olfactorius anterior*, n *septi lateralis, cortex piriformis, cortex suprarhinalis, cortex entorhinalis*, n *caudatus putamen*, n *anterior ventralis thalami*, n *suprachiasmaticus, hippocampus, amygdaloid complex, cortex temporalis*, n *dorsalis corporis geniculati*, n *parabigeminalis, corpus callosum*, n *tractus spinalis nervi trigemini*, n *cuneatus*, caudal portion of the dorsal motor nucleus of the vagus, and n *originis nervi hypoglossi*. 2) Some of the cell groups with AADC-LI alone (D-type) are contained in the n *accumbens*, around n *suprachiasmaticus*, n *habenulae lateralis*, n *amygdaloideus centralis, tractus corticospinalis*, n *pretectalis, colliculus superior, colliculus inferior, commissura posterior*, n *mamillaris medialis, substantia grisea centralis*, n *centralis superior*, n *parabrachialis lateralis, fasciculus longitudinalis medialis*, and caudal portion of the dorsal motor nucleus of the vagus. 3) The cell groups with AADC and 5HT (B-type) are localized in B9-B1 nuclei such as n *raphe dorsalis*, n *raphe pallidus*, n *raphe magnus*, and n *raphe obscurus*.

Ultrastructurally these atypical TH expressing cells without or with AADC have neuronal characteristics and do not belong to glial cells.

Key words: human tyrosine hydroxylase expression; transgenic mice; aromatic L-amino acid decarboxylase; immunocytochemistry.

List of abbreviations

a	*nucleus accumbens*
abl	n *amygdaloideus basalis, pars lateralis*
ac	n *amygdaloideus centralis*
ap	*area postrema*
BA	*accessory olfactory bulb*
C	n *caudatus*
CA	*commissura anterior*
CCA	*corpus callosum*
CE	*cortex entorhinalis*
CI	*colliculus inferior*
co	*chiasma opticum*
cp	n *caudatus putamen*
CP	*commissura posterior*
CS	*colliculus superior*
cs	*cortex suprarhinalis*
cu	n *cuneatus*
dcgl	n *dorsalis corporis geniculati lateralis*
DP	*decussatio pyramidis*
F	*fornix*
fc	*cortex frontalis*
FLM	*fasciculus longitudinalis medialis*
FR	*fasciculus retroflexus*
gp	*globus pallidus*
gr	n *gracilis*
HI	*hippocampus*
hl	n *habenulae lateralis*
lc	*locus ceruleus*
LG	*lamina glomerulosa bulbi olfactorii*
LGR	*lamina granularis bulbi olfactorii*
lh	*area hypothalamicus lateralis*
LNO	*lamina nervi olfactorii*
LP	*lamina plexiformis externa bulbi olfactorii*
me	*median eminence*
MFB	*fasciculus medialis prosencephali* (medial forebrain bundle)
mm	n *mamillaris medialis*
ncs	n *centralis superior*
ndm	n *dorsomedialis hypothalami*
nist	n *interstitialis striae terminalis*
npd	n *parabrachialis dorsalis*
npe	n *periventricularis*

npmv	n *premamillaris ventralis*
npt	n *pretectalis*
nsc(SCN)	n *suprachiasmaticus*
ntd	n *tegmenti dorsalis (Gudden)*
nts(NTS)	n *tractus solitarii*
ntV	n *tractus spinalis nervi trigemini*
nXII	n *originis nervi hypoglossi*
oa(AON)	n *olfactorius anterior*
oc	*cortex occipitalis*
OI	*oliva inferior*
os	n *olivaris superior*
P	*tractus corticospinalis*
pbi	n *parabigeminalis*
pbc	n *parabrachialis*
pc	*cortex parietalis*
pi	*cortex piriformis*
pv	n *periventricularis thalami*
rd	n *raphe dorsalis*
rm	n *raphe magnus*
ro	n *raphe obscurus*
rp	n *raphe pallidus*
rpo	n *raphe pontis*
sc	*spinal canal*
SGC	*substantia grisea centralis*
snc	*substantia nigra, pars compacta*
snr	*substantia nigra, pars reticularis*
sl	n *septi lateralis*
ST	*stria terminalis*
tav	n *anterior ventralis thalami*
tc	*cortex temporalis*
tu	*tuberculum olfactorium*
ZI	*zona incerta*

INTRODUCTION

Recently, we have reported the production of transgenic (Tg) mice carrying the human tyrosine hydroxylase (TH) gene, and described tissue-specific expression of the transgene in catecholaminergic (CAnergic) neurons and adrenal glands with non-transgenic (nTg) mice as controls (Kaneda *et al.*, 1991). Subsequently we found TH expression in non-CAnergic (nCAnergic) neurons of Tg mouse brain, i.e. in the olfactory (typically, the anterior olfactory nucleus (AON) and piriform cortex), and visual (typically, n *suprachiasmaticus* and n *parabigeminalis*) systems (Nagatsu *et al.*, 1991).

Aromatic L-amino acid decarboxylase (AADC) (E.C.4.1.1.28) catalyzes the conversion of L-3,4-dihydroxyphenylalanine (L-DOPA) to dopamine (DA) and that of 5-hydroxytryptophan to 5-hydroxytryptamine (5HT, serotonin) (Lovenberg *et al.*, 1962). Using specific antiserum to AADC, AADC-immunoreactive (IR) cells were visualized not only in aminergic systems but also non-aminergic cells of rat (Jaeger *et al.*, 1983; 1984; Nagatsu *et al.*, 1988) and cat (Kitahama *et al.*, 1988; 1990).

The present paper aims to show the distribution of increasing numbers of TH-IR cell bodies and fibers in CAnergic neurons as well as in nCAnergic neurons of Tg mouse brains, with special reference to AADC. Specific expression of human TH was examined both by immunocytochemistry of TH protein at light and electron microscopic levels, and also by *in situ* hybridization of THmRNA.

MATERIALS AND METHODS

Construction of the transgene and production of Tg mice were described elsewhere (Kaneda *et al.*, 1991). We employed 11 kb DNA fragment of the human TH gene consisting of 2.5 kb of 5′ upstream region, the entire structure of 14 exons and 13 introns, and 0.5 kb of the 3′ downstream region, to produce Tg mice. Tg mice contained about 60–100 copies of transgene. Mice carrying a transgene that included about 2.5 kb upstream and 0.5 kb downstream of the TH transcription unit were identified by PCR screening of tail tip DNA. These Tg mice manifested specific immunocytochemical expression of TH in CA neurons.

Immunocytochemistry

For light and electron microscopic examinations, Tg mice and nTg littermate mice were anesthetized with Nembutal (50 mg/kg i.p.) and perfused intracardially with saline followed by 4% paraformaldehyde (PFA) in 0.1M phosphate buffer (PB, pH 7.4) for 6 min. Fixed tissues were dissected out and immersed overnight in the same fixative at 4 °C. After rinsing with 10–30% sucrose in the PB for 2 days, the brains were cut on a cryostat or microslicer at 40 μm section thickness. The tissue sections were processed for immunocytochemistry by the peroxidase-antiperoxidase (PAP) method (Sternberger, 1986) as described previously (Nagatsu *et al.*, 1988; 1990b). Tissue sections were incubated with TH antiserum (1:10000), AADC antiserum (1:5000), dopamine-β-hydroxylase (DBH) antiserum (1:3000), and phenylethanolamine-N-methyltransferase (PNMT) antiserum (1:5000). To allow better penetration of the antibodies, microsliced sections were treated with liquid nitrogen for 20 sec. Briefly, sections were treated as follows: 1% sodium borohydride, 30 min, or 3% hydrogen peroxide (H_2O_2), 10 min; washing with phosphate buffered saline (PBS); 5% normal swine serum (NSS), 30 min; washing; anti-TH or anti-AADC antiserum in 0.01M PB containing 0.05% NaN_3 for 2 days at 4 °C; washing; swine anti-rabbit immunoglobulins (Dakopatts, Z196, 1:100, 1h); washing; PAP complex (Dakopatts, Z113, 1:200, 1h); washing; 0.4 mg/ml, 3,3′-diaminobenzidine tetrahydrochloride (DAB, Sigma) containing 0.01% H_2O_2 in 0.05M Tris buffer (pH 7.6), 5–20 min to produce a brown reaction product. Sections were mounted on slides, dried well, and then dehydrated and coverslipped. For electron microscopy, the tissue sections were osmicated, embedded in epon, and ultrathin sectioned.

Highly specific antisera have been prepared in rabbits to adrenal TH (Nagatsu *et al.*, 1977, 1983, 1988), AADC (Nagatsu *et al.*, 1988), DBH (Nagatsu *et al.*, 1990a), PNMT (Nagatsu, 1977; Nagatsu *et al.*, 1977, 1983, 1989), in addition to DA (Nagatsu *et al.*, 1988) and 5HT (Nagatsu *et al.*, 1988), and have been characterized in our laboratory. The staining for anti-AADC was completely inhibited after preincubation with excessive antigen AADC. Neither glutamic acid decarboxylase containing reticular thalamic n, nor histidine decarboxylase containing mast cells were stained with anti-AADC antiserum.

In situ *hybridization*

Tg and nTg mice were anesthetized with Nembutal and perfused with 4% PFA. Brains were dissected out, postfixed in the same fixative for 2–4 h, washed with 20% sucrose in the PB for overnight. Frozen sections were cut on a cryostat (40 μm), floated in the buffer and processed for *in situ* hybridization. Briefly, sections were immersed in 10mM Tris-5mM EDTA buffer containing proteinase K (1 μg/ml) for 20 min at 37 °C. They were washed twice in 2 $\times$ SSC and, without further pretreatment, hybridized to a probe complementary to nucleotides of human THmRNA (hTHmRNA). The cDNA probe (probe 1; 0.25 kb SacI-EcoRI) hybridizing only with hTHmRNA, or the cDNA prove (probe 2; 1.3 kb EcoRI-SacI) hybridizing with both mouse and human THmRNA was labeled with [α-35S]dCTPαS (1000 Ci/mmol, 1Ci = 37 GBq, Amersham) by the multiprime DNA labelling systems. Following hybridization for 16 h at 37 °C with 50% formamide and 4 $\times$ SSC, 1 $\times$ Denhardt's solution, 1% sarcosyl, 0.02M sodium phosphate buffer (pH 7.0), 10% dextran sulphate, 500 μg/ml yeast tRNA, 250 μg/ml heat-denatured salmon sperm DNA and 80mM dithiothreitol. Sections were rinsed four times in 2$\times$SSC at room temperature for 2–3 min, washed twice in 0.1 $\times$ SSC at 30 °C for 1 h, and placed on glass slides. They were exposed to X-ray film (Hyperfilm-Bmax, Amersham) for a week, and thereafter, dipped into Kodak NTB2 or Konica NR-M2 liquid emulsion (Yamada *et al.*, 1992).

RESULTS

Immunocytochemistry

Specific neuronal expression of hTH and AADC in the brain of Tg mice was schematically shown in Fig. 1, using TH and AADC immunohistochemistry, and classified in the four groups.

The neurons and fibers with only TH-like immunoreactivity (LI). Tg mouse brain revealed TH-staining in a discrete population of cells and fibers not only in

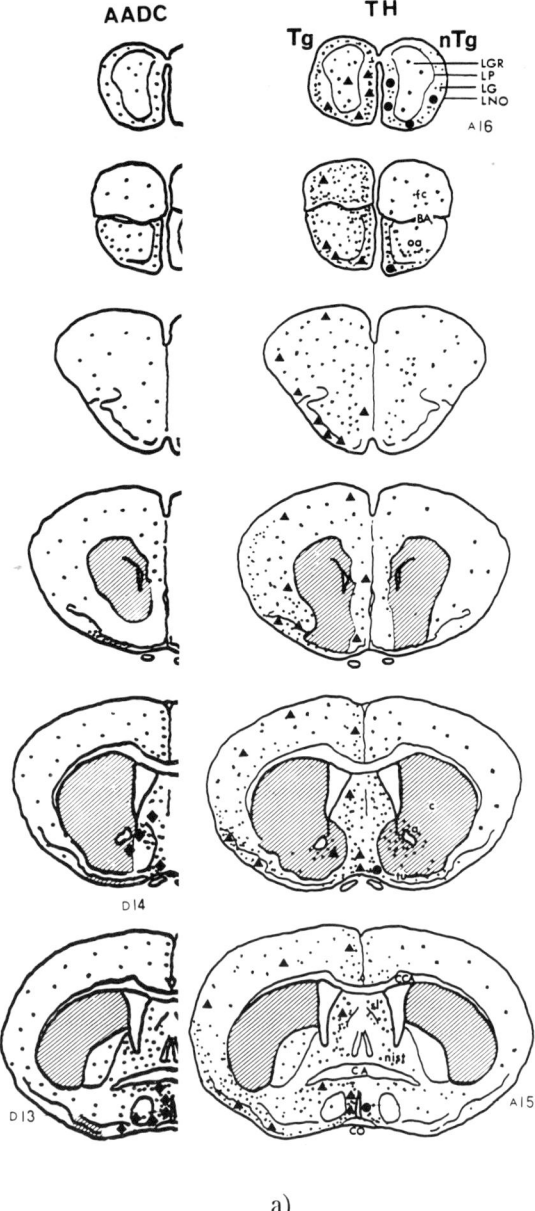

a)

Figure 1. Schematic drawings of AADC (left line, solid square) and TH (middle line, solid triangle) immunostaining perikarya in the brains of Tg mice and nTg mice (right line, solid circle) on frontal sections. Figures illustrate the distribution of AADC- and TH-LI at twenty-three representative levels. AADC- or TH-IR cell bodies show each symbol represents 3–5 cell bodies. Fine dots represent AADC- or TH-IR fibers.

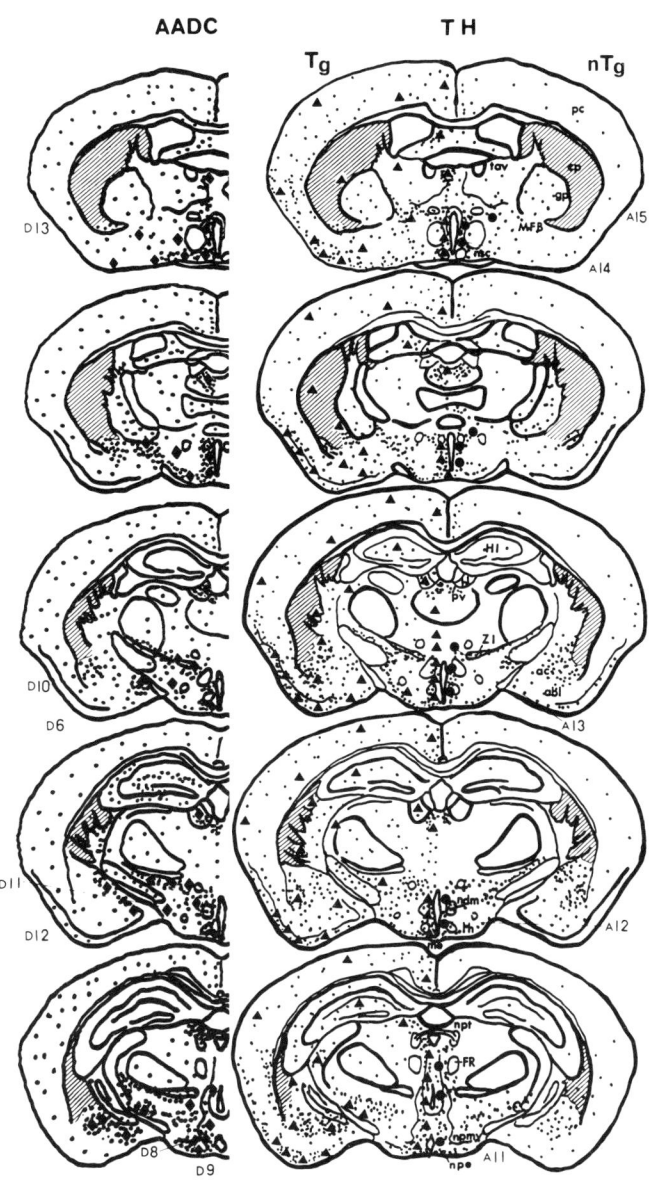

b)

Figure 1 (continued).

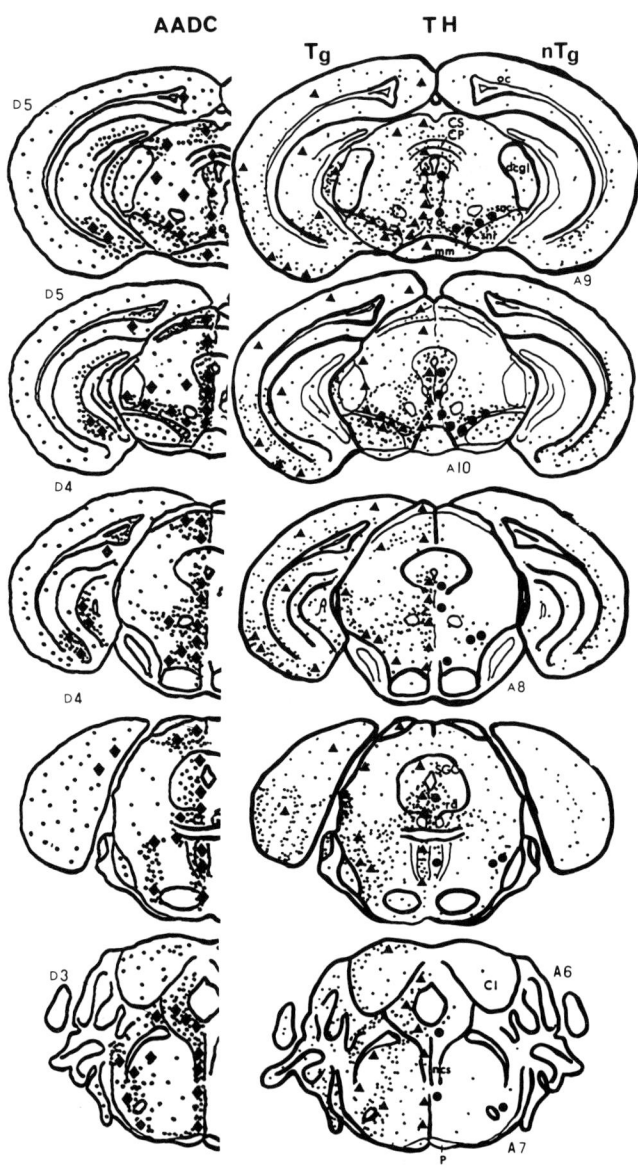

c)

Figure 1 (continued).

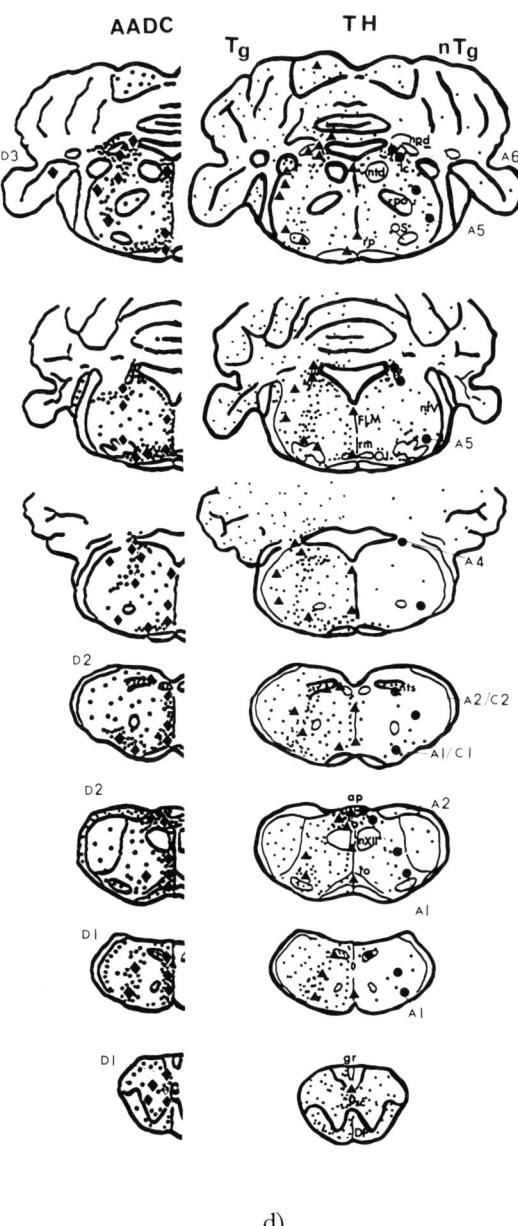

d)

Figure 1 (continued).

CAnergic neurons but also in nCAnergic neurons, for example, the n *olfactorius anterior* (AON) (Figs 2A–B (Nagatsu *et al.*, 1991), *cortex frontalis, cortex piriformis* (Fig. 3B), *cortex suprarhinalis, cortex temporalis, cortex entorhinalis, anterior cingulate cortex, septum, corporis callosi, amygdaloid complex* (Figs 3B and 3H), *hippocampus*, n *suprachiasmaticus* (SCN) (Fig. 3B) (D13), n *caudatus putamen*, n *corpus geniculati lateralis*, n *parabigeminalis* (Figs 4B and 4G), n *tractus spinalis nervi trigemini*, n *cuneatus*, and n *originis nervi hypoglossi.* Although a small number of TH-IR cells were visible in the caudal portion of the dorsal motor nucleus of the vagus (DMV) in nTg mice (Fig. 5A), a larger number of TH-IR cells were observed in cells of the entire extent of DMV (Fig. 5B) in Tg mice.

Neurons and fibers with AADC-LI (D-type) alone. The numbers of cells and fibers with specific AADC-LI were similar between Tg and nTg mouse brains (Fig. 1). The cells were widely distributed like D1–D14 cell groups in the rat (Jaeger *et al.*, 1983, 1984). In the Tg mouse brain, atypical human TH expressing cells were also observed in some of cell groups with AADC-LI alone such as n *accumbens*, around n *suprachiasmaticus*, n *habenulae lateralis*, n *amygdaloideus centralis, tractus corticospinalis*, n *pretectalis, colliculus superior, colliculus inferior, commissura posterior*, n *mamillaris medialis, substantia grisea centralis*, n *centralis superior*, n *parabrachialis lateralis, fasciculus longitudinalis medialis*, and caudal portion of the dorsal motor nucleus of the vagus. Numbers of fibers with AADC-LI in the cortices of Tg and nTg mouse are more than those of TH-IR fibers in the nTg cortex, but fewer than those of TH-IR fibers in the Tg mouse cortex (Fig. 1). In the SCN of adult Tg mice, numerous TH-IR cells (Fig. 3B) were observed in the median part of the SCN. In contrast, the neurons with AADC-LI alone (Fig. 3C) were seen around the SCN. No immunoreactivity was observed using antiserum to DBH, PNMT, or DA (data not shown). Numerous AADC-IR cells were stained in the DMV (Fig. 5C).

AADC/5HT-IR (B-type) cell groups (Dahlström and Fuxe, 1964). In the Tg mouse brain, the expression of TH-IR cells (Fig. 1, middle line, solid triangle) was observed not only in the predicted monoaminergic neurons, but also in AADC-containing neurons (Fig. 1, left line, solid square).

CAnergic (A/C-types) cell groups (Dahlström and Fuxe, 1964; *Hökfelt et al.*, 1984). TH-IR cells of the nTg mouse brain (Fig. 1, right line, solid circle) were also observed in Tg mouse brain (Fig. 1, middle line, solid triangle).

In situ *hybridization*

Neurons that showed extremely high-level expression of hTHmRNA were found in Tg mice by autoradiography not only in CAnergic neurons, but also in nCAnergic sensory neurons, such as the AON (Nagatsu *et al.*, 1991), *cortex piriformis*

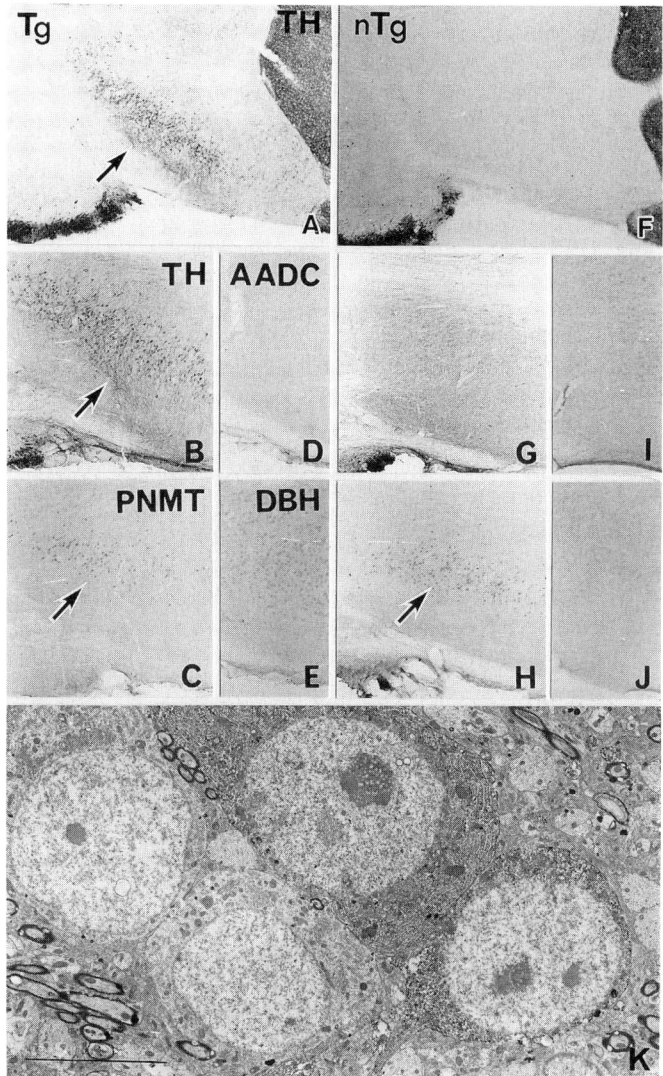

Figure 2. Immunocytochemical micrographs of the n olfactorius anterior (AON) at adult stage of Tg (A-E, K) and nTg (F-J) mice, after incubation with antiserum to TH (A, B, F, G), AADC (D, I), DBH (E, J), and PNMT (C, H). B-E, and G-J show adjacent tissue sections 40 μm thick. Numerous TH-positive cells (A, B) are seen in the median part of the AON, as well as PNMT-positive neurons (C, H). These cells are AADC-negative (D, I) and DBH-negative (E, J). A, F × 25; B-E, G-J × 65. Arrows indicate immunoreactive neurons. Ultrastructurally, TH-labeled perikarya with TH-negative nucleus, mitochondria and Golgi apparatus is observed to contact with TH-negative neurons. Bar indicates 10 μm. (K).

(Fig. 3E, using probe 2), SCN (Fig. 3E), *amygdaloid complex* (Fig. 3E), n *parabigeminalis* (Fig. 4C, using probe 1; Figs 4D-F, using probe 2), and localized similarly with neurons stained immunocytochemically for TH (Figs 4B and 4G). Cells labeled by *in situ* hybridization were morphologically comparable to cells that contain TH protein in three different tissue preparations. In contrast, hTHmRNA was not found in the AON, *cortex piriformis*, SCN, *amygdaloid complex*, and n *parabigeminalis* of nTg mouse (Fig. 3D).

Ultrastructure

TH and/or AADC labelled cells were medium-sized and multipolar-shaped neurons. Ultrastructurally, TH- and/or AADC-labelled perikarya in the above mentioned nuclei of adult Tg mice showed TH- and/or AADC-negative nuclei, Golgi apparatus, and mitochondria. One of these bipolar TH- and/or AADC-IR cells was observed to contact with TH- and/or AADC-negative neuronal processes (Figs 6B/6A). Immunonegative inclusion bodies (Katoh and Shimizu, 1982) were often observed in the *raphe dorsalis* with anti-TH antiserum (Fig. 6B), or with anti-AADC antiserum (Fig. 6A). Moreover in the NTS, TH-IR inclusion body was observed (Fig. 5E). Subsurfaced cisternae were occasionally seen in the ventral tegmental area, or D13 (data not shown). We confirmed the above results by immunoelectron microscopy. Neither TH-LI nor hybridization signal was observed in nCAnergic neurons of nTg mice. Both CAnergic and nCAnergic neurons as mentioned above were TH-labeled in brains of Tg mice. Medium-sized TH-IR cells in the *piriform cortex* (Fig. 3G, one of the centrior showed ciliar form), *amygdala* (Fig. 3I), n *parabigeminalis* (Fig. 4H), and NTS (Fig. 5E) of Tg mice were ultrastructurally observed to contact with TH-immunonegative neuronal processes. Deep nuclear infoldings were observed in the NTS (Figs 5D-E) and *raphe dorsalis* (Fig. 6A).

DISCUSSION

TH is the first enzyme (Nagatsu, T. *et al.*, 1964), and AADC is the second enzyme (Lovenberg *et al.*, 1962) in the CA biosynthesis. These enzymes are essential for the production of CAs, and TH expression is necessary for establishment of the CAnergic phenotype. Expression of this gene is restricted to adrenal chromaffin cells and a large number of CAnergic neuronal cells. Recently we (Kaneda *et al.*, 1991) have produced Tg mice carrying human TH transgene. The brain of Tg mouse revealed immunocytochemical staining in a discrete population of cells not only in CAnergic neurons, but also in nCAnergic neurons (Nagatsu *et al.*, 1991).

(1) Possible existence of L-DOPA-positive/-negative neurons with TH-LI alone.
The present study provides the first evidence of a new group of neurons composed of a large number of TH-IR neurons in the Tg mouse brain, where no

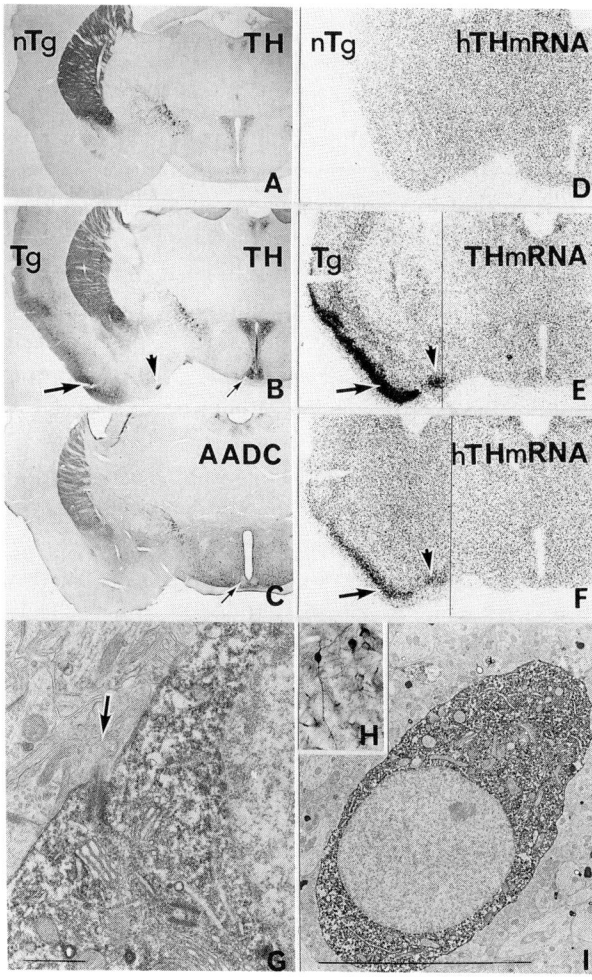

Figure 3. The micrographs display the low magnification appearance of the *cortex piriformis* (large arrow), and *amygdaloid complex* (arrow head) of nTg (A) and Tg (B, C, H) mice, demonstrated by immunocytochemistry against TH (A), or AADC (C) antiserum. *In situ* hybridization (ISH) for mouse and human THmRNA (E) and hTHmRNA (F) is shown in the *piriform cortex* (large arrow) and *amygdaloid complex* (arrow head) of Tg mice. No signal for hTHmRNA (D) was observed in nTg mice. We note medium- to large-sized TH-IR cells having very long dendritic trees in the *medial amygdaloid* n (H). In the *anterior hypothalamus*, intensely stained AADC-IR small cells were observed in the n *suprachiasmaticus* (SCN) (small arrow), clustered in the n *habenulae lateralis*, and n *paraventricularis thalami*. In the Tg mice, TH-IR cells can be detected inside the SCN (small arrow), although distinctly labeled AADC-IR cells were found around the nucleus (small arrow). A-F × 12; H × 200. Cilia (arrow) was shown in the TH-IR pyramidal cell of *piriform cortex* (G, Bar = 1μm). Very strong TH-LI was observed in the cytoplasm of the medial amygdaloid nucleus (I, Bar = 10μm).

monoamine neurons had been reported. Some of fibers in the Tg mouse cortex were immunoreactive to TH, but immunonegative to AADC. This shows an existence of fibers with TH-LI alone in the Tg mouse cortex. These TH-IR cells did not show any AADC-, DBH-, or PNMT-LI, suggesting the absence of synthesis of 5HT, DA, noradrenaline, and adrenaline. Using our highly sensitive anti-DA, or anti-5HT antiserum, we confirmed the lack of DA or 5HT. It is interesting to note that in the Tg mouse brain, a large number of increasing TH-IR neurons may contain L-DOPA possibly as an end product. The application of L-DOPA immunocytochemistry would be required to solve the problem. Moreover, recently Mons *et al.* (1991) described the existence of TH-positive/L-DOPA-negative cells besides TH-positive/L-DOPA-positive cells in the dorsal part of the hypothalamus, using their specific L-DOPA antiserum. Furthermore, some TH-IR cells of the mesencephalon might not express the AADC gene. Tison *et al.* (1991) observed TH-positive mesencephalic neurons containing a very low or no AADC mRNA signal, and suggested that monoaminergic neurons could have L-DOPA as an end product. Such neurons have been unambiguously demonstrated in the arcuate nucleus of the hypothalamus (Meister *et al.*, 1988; Okamura *et al.*, 1988; Komori *et al.*, 1991b, c). Just recently, Karasawa *et al.* (1992) produced a highly specific antibody against L-DOPA, and showed TH- and L-DOPA-positive but AADC- and DA-negative neurons in the lateral habenular nucleus of the house-shrew (*Suncus murinus*) brain, ultrastructurally.

(2) *Possible existence of D-type cells.* Besides monoaminergic cell systems, AADC-IR cell bodies that did not contain TH- or 5HT-LI were called "D1–D14 cell groups" in the rat (Jaeger *et al.*, 1983, 1984), and "D-type cells" in the cat (Kitahama *et al.*, 1990). As new additional D-type cell groups that have not been reported in the rat (Jaeger *et al.*, 1984), Kitahama *et al.* (1990) described dense or loose clusters of D-type cells, which were localized in the dorsal motor n of the vagus, n *praepositus hypoglossi*, central, pontine, periaqueductal gray, superficial layer of the *superior colliculus*, and area medial to the *retroflexus*. Nagatsu *et al.* (1988) found very small cerebrospinal fluid-contacting D-type cells, which were not only TH- or 5HT-negative, but also DBH-, PNMT-, or DA-immunonegative, within and beneath the ependymal layer of the rat and mouse 10th area of Rexed surrounding the central canal (D-1).

In the Tg mouse brain, TH-immunonegative/AADC-IR cells were widely distributed as mentioned in Results, similarly to some of the above-described regions. Moreover, we found AADC-IR cells in the caudoputamen and frontal cortex, at light and electron microscopic levels (Komori *et al.*, 1991a).

Tashiro *et al.* (1989a) found a small number of striatal neurons showing AADC-LI in the caudoputamen. The number of these neurons was increased after an electrothermic or 6-hydroxydopamine-induced lesion, injected around

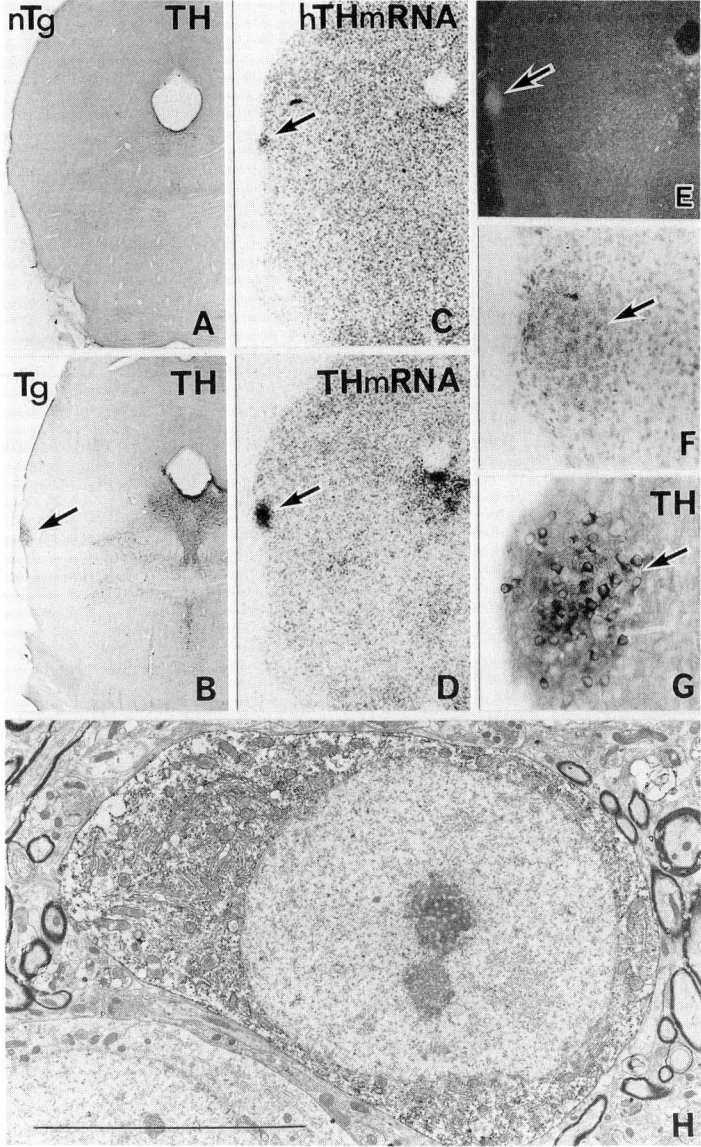

Figure 4. Immunocytochemical micrographs of the n *parabigeminalis* of nTg (A) and Tg mice (B, G), after incubation with antiserum to TH. The appearance of neurons labeled for the presence of hTHmRNA (C) and THmRNA (D, E, F) show the same pattern as TH-IR neurons (B, G). Small arrow marks THmRNA positive cell seen at dark-field (E) and at higher magnification in bright-field (F) photomicrographs of the n *parabigeminalis* demonstrated by ISH for THmRNA. A-E × 16; F × 75; G × 300. At the electron microscopic level, TH-IR cytoplasm (H) showed well developed Golgi apparatus and no glial filaments. Bar = 10μm.

the *substantia nigra pars compacta*. They (Tashiro *et al.*, 1989b) previously observed similar striatal neurons showing TH-LI, and more increased number of these neurons after deafferentation from dopaminergic neurons in the midbrain.

AADC-containing neurons in the Tg mouse brain distributed similarly to those in the nTg mouse brain. The result suggests that AADC is unaffected between Tg and nTg mouse brain. Previous studies have shown that when TH genes are introduced individually into Tg mice, they are developmentally regulated. This implies that stage-specific transacting factors interact with the TH gene itself to bring about gene switching, but not with the AADC gene.

The morphological appearance of AADC-immunoreactive structures strongly suggests that they are neuronal elements. We confirmed the above results by immunoelectron microscopy. Inclusion bodies were very often observed in the mouse brain as mentioned by Katoh and Shimizu (1982), but in the inclusion bodies TH or AADC-LI was not contained. Therefore functional significance of the inclusion bodies is not clear yet.

(3) Functional considerations. Our understanding of the possible functions of the neurons with TH-LI alone or AADC-LI alone is still limited, therefore the physiological roles of these neurons remain speculative.

Kosaka *et al.* (1987) observed numerous TH-IR cells all over the isocortex (frontal, temporal, parietal and occipital regions), and in some parts of the allocortex (the anterior cingulate cortex, the retrosplenial cortex and anterior part of the insular cortex) of the adult rat brain. In the Tg mice brain, we noted markedly increased TH-IR cells in the perirhinal cortex, posterior part of the insular cortex, piriform cortex, entorhinal cortex and hippocampal formation, where usually TH-IR cells were rare in the nTg mice. In Tg mice, nCAnergic neurons represent an induced increase in TH synthesis in neurons which may express the protein at levels normally too low to be detected.

By colchicine-treatment, Kitahama *et al.* (1990) just recently found TH-IR, but AADC- or DA-immunonegative cells in the posterior, dorsal and periventricular hypothalamic areas, and in the area ventral to the mamillothalamic tract (A13c) (Hökfelt *et al.*, 1984), medial and lateral preoptic areas, anterior commissure nucleus, basal forebrain, area closely related to the *organum vasculosum laminae terminalis*, and some in the bed nucleus of the *stria terminalis*. In these areas, TH-IR cells were clearly observed in Tg mice.

In pre- and postnatal mice, transient appearance of TH-IR neurons was reported in the AON (Nagatsu *et al.*, 1989, 1990b), caudate putamen (Komori *et al.*, 1991c), and in the deeper layers of the cerebral isocortex (Satoh and Suzuki, 1990). These neurons contained neither AADC-LI nor DA-LI. These cells are present in the subplate and marginal zones during development, when their location is correlated with the arrival and accumulation of ingrowing axonal systems and with synapses. However, as the brain matures, the cells disappear.

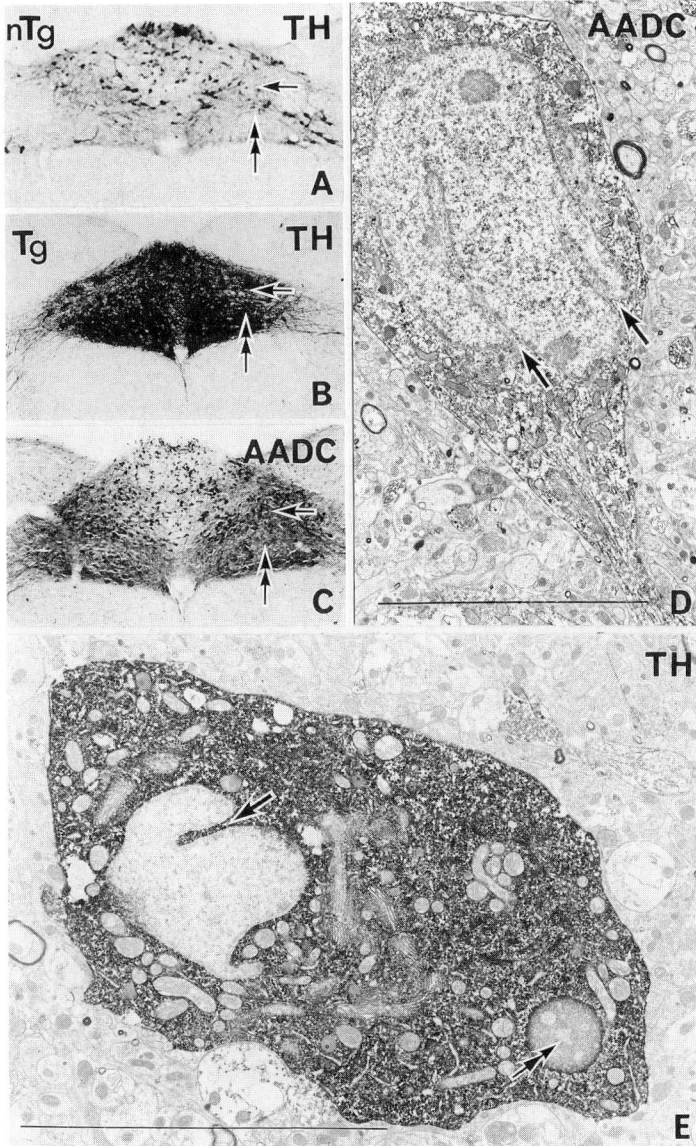

Figure 5. The area postrema and medial part of the n *tractus solitarii* (arrow) show many small TH-IR cells (A, B). The dorsal motor nucleus of the vagus (double arrow) contains medium-sized AADC-IR cells (C), whose distribution pattern is similar to that of TH-IR ones of Tg (B) mouse brains. A-C × 60. The ultrastructure of AADC(D)-, or TH(E)-positive perikarya is shown in the area postrema, or NTS of Tg mouse, respectively. The staining is not localized in the deformed nucleus (arrow), mitochondria, Golgi apparatus, and inclusion bodies (double arrow). Bar = 10μm.

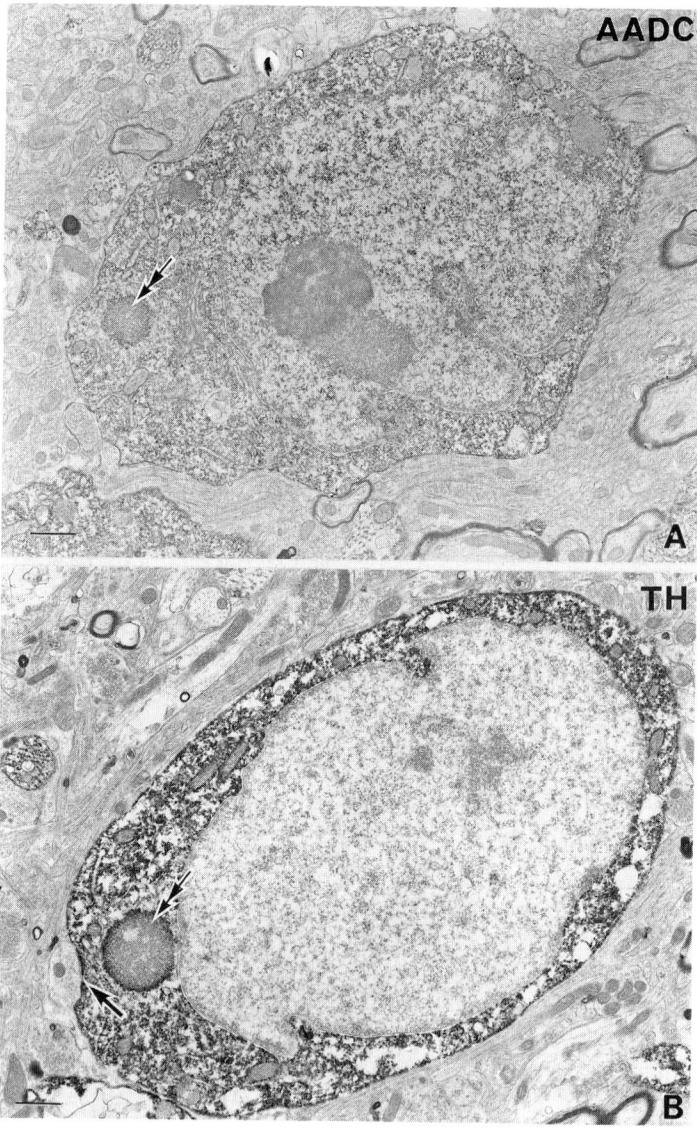

Figure 6. At electron microscopic level, the inclusion body (double arrow) is also observed in the n *raphe dorsalis* using anti-AADC(A), or anti-TH(B) antibody. One of these bipolar TH-IR cells is observed to contact with TH-negative nerve terminals. A synaptic contact (arrow) was occasionally shown. Bar = 1μm.

Chun *et al.* (1987) reported that the subplate cells have properties typical of mature neurons receiving synaptic contacts. In contrast, TH-IR neurons have the ultrastructural appearance of atypical deformed or immature neurons without synaptic contacts. These observations agree well with our previous report of transient TH-IR cells in the mouse AON regions (Nagatsu *et al.*, 1990b). During the fetal and early postnatal development of mouse telencephalon, the subplate cells may function as neurons in local circuitry that disappears by adulthood.

This unusual expression of human TH in nCAnergic neurons in the Tg brain suggests that negative regulatory elements or suppressors or silencers, which act in such regions as the AON, may be absent in the transgene. The regulatory regions in the TH gene should be analyzed using the transient expression system with neuronal cell lines. To identify the cis-elements responsible for the tissue- and stage-specific expressions of the TH gene, Tg mice carrying the chimeric genes consisted of promoter region and some reporter genes, should be constructed and analyzed.

Acknowledgments

This work was supported in part by Grants-in-Aid for Scientific Research from the Japanese Ministry of Education, Science and Culture, and by a Grant-in-Aid from Fujita Health University, Japan. The authors are very grateful to Drs T. Nomura, M. Katsuki, M. Yokoyama, Messrs T. Hasegawa, R. Takahashi, K. Nakao, K. Kaseda, K. Komori, T. Fujii, Mrs K. Katsuki, T. Takeuchi, and Misses A. Morita, M. Hatanaka for their excellent assistance.

REFERENCES

Chun, J. J. M., Nakamura, M. J. and Shatz, C. J. (1987). Transient cells of the developing mammalian telencephalon are peptide-imunoreactive neurons. *Nature* **325**, 617–620.

Dahlström, A. and Fuxe, K. (1964). Evidence for the existence of monoamine-containing neurons in the central nervous system. I. Demonstration of monoamines in the cell bodies of brain stem neurons. *Acta Physiol. Scand.* **62** (Suppl.), 232, 1–55.

Hökfelt, T., Martensson, R., Björklund, A., Kleinau, S. and Goldstein, M. (1984). Distributional maps of tyrosine-hydroxylase-immunoreactive neurons in the rat brain. In: *Classical Transmitters in the CNS*. A. Björklund and T. Hökfelt (Eds). Part I. Handbook of Chemical Neuroanatomy, vol. 2, Elsevier, Amsterdam, pp. 277–379.

Jaeger, C. B., Ruggiero, D. A., Albert, V. R., Joh, T. H. and Reis, D. J. (1984). Immunocytochemical localization of aromatic-L-aminoacid decarboxylase. In: *Classical Transmitters in the CNS*. A. Björklund and T. Hökfelt (Eds). Part I. Handbook of Chemical Neuroanatomy, vol. 2, Elsevier, Amsterdam, pp. 387–408.

Jaeger, C. B., Teitelman, G., Joh, T. H., Albert, V. R., Park, D. H. and Reis, D. J. (1983). Some neurons of the rat central nervous system contain aromatic-L-amino-acid decarboxylase but not monoamines. *Science* **219**, 1233–1235.

Kaneda, N., Sasaoka, T., Kobayashi, K., Kiuchi, K., Nagatsu, I., Kurosawa, Y., Fujita, K., Yokoyama, M., Nomura, T., Katsuki, M. and Nagatsu, T. (1991). Tissue-specific and high-level expression of the human tyrosine hydroxylase gene in transgenic mice. *Neuron* **6**, 583–594.

Karasawa, N., Isomura, G. and Nagatsu, I. (1992). Production of specific antibody against L-DOPA and its ultrastructural localization of immunoreactivity in the house-shrew (*Suncus murinus*) lateral habenular nucleus. *Neurosci. Lett.* **143**, 267–270.

Katoh, Y. and Shimizu, N. (1982). The light and electron microscopic localization of intracytoplasmic nucleolus-like bodies in the mouse brain stained by Holmes' silver method. *Arch. Histol. Jap.* **45**, 325–333.

Kitahama, K., Denoyer, M., Raynaud, B., Borri-Voltattorni, C., Weber, M. and Jouvet, M. (1988). Immunohistochemistry of aromatic L-amino acid decarboxylase in the cat forebrain. *J. Comp. Neurol.* **270**, 337–353.

Kitahama, K., Denoyer, M., Raynaud, B., Borri-Voltattorni, C., Weber, M. and Jouvet, M. (1990). Aromatic L-amino acid decarboxylase immunohistochemistry in the cat lower brainstem and midbrain. *J. Comp. Neurol.* **302**, 935–953.

Komori, K., Fujii, T., Karasawa, N., Yamada, K. and Nagatsu, I. (1991a). Some neurons of the mouse cortex and caudo-putamen contain aromatic L-amino acid decarboxylase but not monoamines. *Acta Histochem. Cytochem.* **24**, 571–577.

Komori, K., Fujii, T. and Nagatsu, I. (1991b). Do some tyrosine hydroxylase-immunoreactive neurons in the human ventrolateral arcuate nucleus and globus pallidus produce only L-DOPA? *Neurosci. Lett.* **133**, 203–206.

Komori, K., Sakai, M., Karasawa, N., Yamada, K. and Nagatsu, I. (1991c). Evidence for transient expression of tyrosine hydroxylase immunoreactivity in the mouse striatum and the effect of colchicine. *Acta Histochem. Cytochem.* **24**, 223–231.

Kosaka, T., Kosaka, K., Hataguchi, Y., Nagatsu, I., Wu, J.-Y., Ottersen, O. P., Storm-Mathisen, J. and Hama, K. (1987). Catecholaminergic neurons containing GABA-like and/or glutamic acid decarboxylase-like immunoreactivities in various brain regions of the rat. *Exp. Brain Res.* **66**, 191–210.

Lovenberg, W., Weissbach, H. and Udenfriend, S. (1962). Aromatic L-amino acid decarboxylase. *J. Biol. Chem.* **237**, 89–93.

Meister, B., Hökfelt, T., Steinbusch, H. W. M., Skagerberg, G., Lindvall, O., Geffard, M., Joh, T. H., Cuello, A. C. and Goldstein, M. (1988). Do tyrosine hydroxylase-immunoreactive neurons in the ventrolateral arcuate nucleus produce dopamine or only L-DOPA? *J. Chem. Neuroanat.* **1**, 59–64.

Mons, N., Dubourg, P. and Tramu, G. (1991). Preparation and characterization of a specific antibody for the immunohistochemical detection of L-DOPA in paraform aldehyde-fixed rodent brains. *Brain Res.* **554**, 122–129.

Nagatsu, I. (1977). Phenylethanolamine-N-methyltransferase. In: *Electron Microscopy of Enzymes. Principles and Methods.* M. A. Hayat (Ed.), Van Nostrand Reinhold, New York, pp. 52–71.

Nagatsu, I. (1983). Immunohistocytochemistry of biogenic amines and immunoenzyme-histocytochemistry of catecholamine-synthesizing enzymes: Application for axoplasmic transport and neuronal localization. In: *Methods in Biogenic Amine Research.* S. Parvez, T. Nagatsu, I. Nagatsu and H. Parvez (Eds), Elsevier, Amsterdam, pp. 873–909.

Nagatsu, I., Kobayashi, K., Fujii, T., Komori, K., Sekiguchi, K., Titani, K., Fujita, K. and Nagatsu, T. (1990a). Antibodies raised to the different oligopeptide segments of human dopamine-β-hydroxylase. *Neurosci. Lett.* **120**, 141–145.

Nagatsu, I., Komori, K., Miura, K., Sakai, M., Karasawa, N. and Yamada, K. (1989). Ontogeny of phenylethanolamine-N-methyltransferase and tyrosine hydroxylase-imunoreactive expression in the mouse anterior olfactory nucleus. *Biomed. Res.* **10** (Suppl.), 3, 277–286.

Nagatsu, I., Komori, K., Takeuchi, T., Sakai, M., Yamada, K. and Karasawa, N. (1990b). Transient tyrosine hydroxylase-immunoreactive neurons in the region of the anterior olfactory nucleus of pre- and postnatal mice do not contain dopamine. *Brain Res.* **511**, 55–62.

Nagatsu, I., Kondo, Y., Inagaki, S., Karasawa, N., Kato, T. and Nagatsu, T. (1977). Immunofluorescent studies on tyrosine hydroxylase: Application for its axoplasmic transport. *Acta Histochem. Cytochem.* **10**, 494–499.

Nagatsu, I., Sakai, M., Yoshida, M. and Nagatsu, T. (1988). Aromatic L-amino acid decarboxylase-immunoreactive neurons in and around the cerebrospinal fluid-contacting neurons of the central canal do not contain dopamine or serotonin in the mouse and rat spinal cord. *Brain Res.* **475**, 91–102.

Nagatsu, I., Yamada, K., Karasawa, N., Sakai, M., Takeuchi, T., Kaneda, N., Sasaoka, T., Kobayashi, K., Yokoyama, M., Nomura, T., Katsuki, M., Fujita, K. and Nagatsu, T. (1991). Expression in brain sensory neurons of the transgene in transgenic mice carrying human tyrosine hydroxylase gene. *Neurosci. Lett.* **127**, 91–95.

Nagatsu, T., Levitt, M. and Udenfriend, S. (1964). Tyrosine hydroxylase: The initial step in norepinephrine biosynthesis. *J. Biol. Chem.* **239**, 2910–2917.

Okamura, H., Kitahama, K., Mons, N., Ibata, Y., Jouvet, M. and Geffard, M. (1988). DOPA-immunoreactive neurons in the rat hypothalamic tuberal region. *Neurosci. Lett.* **95**, 42–46.

Satoh, J. and Suzuki, K. (1990). Tyrosine hydroxylase-immunoreactive neurons in the mouse cerebral cortex during the postnatal period. *Devel. Brain Res.* **53**, 1–5.

Sternberger, L. A. (1986). The unlabelled antibody peroxidase-antiperoxidase (PAP) methods. In: *Immunocytochemistry* (3rd edn.), Wiley, New York, pp. 90–209.

Tashiro, Y., Kaneko, T., Sugimoto, T., Nagatsu, I., Kikuchi, H. and Mizuno, N. (1989a). Striatal neurons with aromatic L-amino acid decarboxylase-like immunoreactivity in the rat. *Neurosci. Lett.* **100**, 29–34.

Tashiro, Y., Sugimoto, T., Hattori, T., Uemura, Y., Nagatsu, I., Kikuchi, H. and Mizuno, N. (1989b). Tyrosine hydroxylase-like immunoreactive neurons in the striatum of the rat. *Neurosci. Lett.* **97**, 6–10.

Tison, F., Normand, E., Jaber, M., Aubert, I. and Bloch, B. (1991). Aromatic L-amino-acid decarboxylase (DOPA decarboxylase) gene expression in dopaminergic and serotoninergic cells of the rat brainstem. *Neurosci. Lett.* **127**, 203–206.

Yamada, K., Sakai, M., Okamura, H., Ibata, Y. and Nagatsu, I. (1992). Detection of tyrosine hydroxylase and phenylethanolamine-N-methyltransferase messenger RNAs in the mouse adrenal gland and the brain by *in situ* hybridization. *Histochemistry* **97**, 201–206.

Tyrosine Hydroxylase, pp. 59–69
M. Naoi *et al.* (Eds)
© VSP 1993

REGULATION OF TYROSINE HYDROXYLASE BY ADENOSINE DERIVATIVES IN ADRENAL CHROMAFFIN CELLS; LONG-TERM INDUCTION OF GENE EXPRESSION AND SHORT-TERM MODULATION OF ENZYME ACTIVITY

YOUICHIRO KURODA,[1] KAZUYO MURAMOTO[1] and
KOUNOSUKE KUMAKURA[2]

[1]*Department of Molecular and Cellular Neurobiology, Tokyo Metropolitan Institute for Neuroscience, 2–6 Musashidai, Fuchu-city, Tokyo 183, Japan*
[2]*Laboratory Neurochemistry, Life Science Institute, Sophia University, 7–1 Kioi-cho, Chiyoda-ku, Tokyo 120, Japan*

Abstract—Extracellular addition of ATP and of 2-chloroadenosine (2-Cl-Ado, an adenosine agonist) increased the tyrosine hydroxylase (TH) activity in primary cultures of bovine adrenal cells. The TH activation was dose-dependent and was not inhibited by dipyridamole, an inhibitor of adenosine uptake into the cells. The effect of 2-Cl-Ado was antagonized by the addition of deoxyadenosine. These results suggest that the short term activation of TH by adenosine derivatives is mediated by the increase of cyclic AMP in the cells. Similar increase of both TH activity and cyclic AMP level by adenosine derivatives were observed in pheochromocytoma (PC12) cells. Moreover, exposure of the chromaffin cells to ATP or 2-Cl-Ado for 1–3 h increased both the expression of TH mRNA and TH activity later. This long-term increase can be corresponding to the cyclic AMP-medicated induction of TH gene expression which is followed by the enzyme biosynthesis.

Key words: tyrosine hydroxylase mRNA; adenosine; ATP; cyclic AMP; adrenal chromaffin cells; gene induction; pheochromocytoma; catecholamine synthesis; adenosine receptor; biopterine.

INTRODUCTION

Acute stresses, for example electrical shock or decapitation, result in a short-term activation of TH in the adrenal medulla (Weiner, 1982), while a persisting stress results in a long-term increase in TH activity which was termed trans-synaptic induction (Axelrod, 1971). There is accumulating evidence that cyclic AMP-dependent process may be involved in either short-term regulation of TH (Yamauchi and Fujisawa, 1979; Weiner, 1982) or long-term regulation (Guidotti and Costa, 1977). However, the first messenger for the TH regulation, which triggers an increase of cyclic AMP through cell surface receptors, has not yet been identified.

ATP is known to be stored in catecholaminergic synaptic vesicles as well as in chromaffin granules. This ATP is released with catecholamines concomitantly on nerve stimulation and detected as a mixture of adenosine derivatives. Several years ago, we found (Kuroda and Kobayashi, 1978a) that extracellulary applied adenosine derivatives increase the level of cyclic AMP in synaptosomes from guinea pig cerebral cortex. Changes of TH activity in catecholaminergic nerve terminals after the addition of adenosine derivatives were investigated, as one of the physiological functions of the increased cyclic AMP in nerve terminals. Addition of 2-Cl-Ado activated TH in guinea pig striatal synaptosomes (Kuroda and Kobayashi, 1978b). However, this activation of TH is not due to the cyclic AMP-dependent process, since (1) other purine derivatives (guanosine, adenine) which do not increase the level of cyclic AMP also cause the activation of TH, (2) the TH activation by 2-Cl-Ado is additive to that by the addition of dibutylyl-cyclic AMP, (3) deoxyadenosine, which antagonize the action of 2-Cl-Ado for the increase of cyclic AMP level, does not inhibit the activation of TH by 2-Cl-Ado. Furthermore, an inhibitor of adenosine uptake, dipyridamole, markedly inhibits the TH activation, indicating that the activation by purine derivatives occurs after the uptake into the synaptosomes, but not through the surface receptors (Kumakura, Kobayashi and Kuroda, 1981). Preliminary experiments suggest that the activation can be explained by the increase of biopterine levels in the synaptosomes which is normally subsaturated. Present studies were carried out to know whether a similar type of short-term activation can be observed in adrenal chromaffin cells. Long-term effects of adenosine derivatives were also investigated to show that the induction of TH gene expression is involved in the delayed increase of TH activity.

MATERIALS AND METHODS

Preparation of primary cultured chromaffin cells

Adrenal chromaffin cells were isolated from fresh bovine adrenal glands and cultured by a modification of the method described by Waymire and others (1977).

Briefly, the isolation procedure consists of: dissection of the adrenal medulla from the cortex, digestion of the interstitial tissue of adrenal medulla by retrograde perfusion with 0.25% collagenase (Sigma, Type 1) to dissociate the cells, differential centrifugation and filtration, and differential plating to remove contaminant cells. The differential plating was repeated twice. The purified chromaffin cells were cultured in minimum essential medium (GIBCO 410–1200), 10% fetal calf serum (GIBCO), penicillin (100 units/ml), streptomycin (100 μg/ml), fluorodeoxyuridine (10 μM), cytosine arabinoside (10 μM), uridine (5 μM) and mycostatin (GIBCO, 25 units/ml).

Short-term experiments

Tyrosine hydroxylase activity was assayed by measurement of the rate of $^{14}CO_2$ formation from ^{14}C-tyrosine, by a modification of the tryptophan hydroxylase assay of Ichiyama and others (1970) which utilizes the endogenous pteridine cofactor contained in the chromaffin cells. The cells cultured for 2 days were suspended (approximately 10^6 cells/ml) in Krebs-Ringer bicarbonate glucose medium which contained; 125 mM NaCl, 5 mM KCl, 1.24 mM KH_2PO_4, 1.3 mM $MgSO_4$, 0.77 mM $CaCl_2$, 26 mM $NaHCO_3$, glucose and 0.1 mM Na-ascorbate, gassed by 95% O_2 and 5% CO_2. The chromaffin cell suspension was incubated for 10 min at 37 °C, and then an aliquot (200 μl) of the cell suspension was incubated for 5 min at 37 °C following the addition of 25 μl medium with or without agents. The reaction was started by the addition of 25 μl of [l-^{14}C]-tyrosine (0.025 μCi) at a final concentration of 20 μM and then carried out for 20 min. A blank incubation contained all the ingredients except that a boiled cell preparation was employed. The amount of $^{14}CO_2$ formed was calculated by subtracting the blank d.p.m. from the experimental d.p.m., then dividing by specific activity of ^{14}C-tyrosine. In some experiments, 20 μl of [l-^{14}C]-L-dopa (0.01 μCi) was substituted for ^{14}C-tyrosine. For uptake experiments, 50 μl of the same reaction mixture was incubated in the same manner as for the assay of TH. After the incubation, the mixture was immediately added by 2 ml of Kreb's-Ringer bicarbonate glucose medium and filtered on Millipore membrane filter (25 mm diameter, 0.45 μm pore size). The filter was washed 3 times with 2 ml of the medium at room temperature within 5 sec and then dried. Uptake of tyrosine was estimated by the radioactivity remaining on the filter.

Long-term exposure of adenosine derivatives and TH Assay

Test substances were added to the cultures for the indicated time, then the cells were collected by centrifugation and resuspended in the culture medium. After 24 or 48 h, the cultured cells were harvested and approximately 5×10^7 cells were homogenized in 500 μl of 50 mM Tris-acetate buffer (pH 6.0) containing 0.2% Triton X-100 after one cycle of freeze-thawing. In an aliquot of 100 μl of the homogenates the TH activity was measured using ^{14}C-tyrosine in the presence of 2 mM of 6-methyl-tetrahydropterine according to the method of Waymire and others (1971).

Assay of cyclic AMP

The cultured cells were centrifuged and washed 3 times with phosphate buffered saline, then suspended in Kreb's-Ringer HEPES medium. The suspended cells were preincubated for 10 min at 37 °C and the test substances were added. After incubation for appropriate periods, the cells were spun down within 30 sec and extracted with 1 ml of 6% trichloroacetic acid. Cyclic AMP in the extracts was determined with radioimmunoassay as previously described (Kuroda and Kobayashi, 1978a).

Isolation of total RNA

Cultured adrenal chromaffin cell suspensions were incubated with or without 0.5 mM 2-Cl-Ado for 3 h, then total RNAs were extracted from the cells after 0, 2, 4, 8, 24 and 48 h using a method described by Chirgwin *et al.* (1977). The cell suspensions from 2 culture dishes (1.0×10^7 cells/100 mm dish) were combined, then pelleted and washed by centrifugation with phosphate buffered saline. The cell pellet was lysed in a 4 ml of buffer containing 4 M guanidine thiocyanate (Fluka), 100 mM Tris-HCl (pH 7.0), 100 mM 2-mercaptoethanol (Sigma), 25 mM sodium citrate (Sigma), 0.5% sodium N-laurylsarcosine (Sigma) and 0.1% antiform A (Sigma). The cell lysate was layered over a 2 ml cushion of 5.7 M CsCl and 100 mM EDTA (pH 7.0), and centrifuged at 220 000 g (28 000 rpm, RPS40T-858 rotor (Hitachi)) for 22 h at 20 °C. Following centrifugation, the supernatant was carefully removed and the RNA pellet was resuspended in 70% ethanol. This suspension was centrifuged at 8 000 rpm for 10 min and the RNA pellet was dissolved in a 2 ml solution consisting of 10 mM Tris-HCl (pH 7.5), 1 mM EDTA (pH 7.5), 5% sodium N-laurylsarcosine and 5% phenol. Moreover, 0.1 volume of 1 M sodium chloride and equal volume (2.2 ml) of phenol-chloroform (1:1, v/v) was added to the RNA solution. Following centrifugation, aqueous phase was transferred to another tube, then total RNA was precipitated by adding 2 volume of 99% ethanol and keeping for 8 h at −20 °C. After washing with 70% ethanol, total RNA pellet was redissolved in a 1 ml of buffer containing 10 mM Tris-HCl (pH 7.5) and 0.1% SDS. This RNA solution was scanned spectrophotometrically and RNA concentration was estimated from its maximal absorbance at 260 nm.

Hybridization probe

Plasmids with human tyrosine hydroxylase type I (hTH *cDNA*), which was inserted at EcoRI site in Bluescript, were kindly gifted from Professor T. Nagatsu. The 2kb hTH *cDNA* probe was dissected by treatment of restriction endonuclease (EcoRI:TOYOBO), isolated by agarose electrophoresis and recovered using GENECLEAN II kit (Bio 101). The *cDNA* probe was nick-translated using a Nick Translation kit (Amersham) and [α-^{32}P]dCTP as radiolabel, then used as hybridization probe.

Northern blot analysis

Total RNA was denatured at 55 °C for 15 min in a 1X MOPS buffer (consisting of 20 mM morpholinopropanesulfonic acid (MOPS: Sigma, pH 7.0), 5 mM sodium acetate and 1 mM EDTA), 50% formamide (Wako) and 7% formaldehyde (Merk) (total volume 20 μl). Then samples of denatured RNA were cooled on ice and 2 μl of gel-loading buffer (50% sucrose, 1 mM EDTA (pH 8.0), 0.25% bromophenol blue and 0.25% xylene cyanol FF (all Sigma)) was added as a marker dye. Samples together with lamda DNA/Hind III digest as molecular weight marker were fractionated by electrophoresis in a 1% agarose gel containing 1X MOPS buffer and 7% formaldehyde at 100 V for 45 min in a 1×MOPS buffer. After electrophoresis RNA was transferred to a Gene Screen Plus membrane (NEN) using vacuum blotting system (Vacuum Gene TM XL:Pharmacia) according to the instruction manual, and cross-linked by UV-cross-linker (FS-800: Funakoshi). Prehybridization of membrane-bound RNA was in 5×SSC (1×SSC: 0.15 M sodium chloride, 0.015 M sodium citrate, pH 7.4), 5×Denhardt's reagent (Sigma, 1×Denhardt's reagent: 0.02% polyvinylpyrolidone, 0.02% Ficoll 400 and 0.02% bovine serum albumin), 50% formamide, 100 μg/ml denatured salmon sperm single strand DNA (ssssDNA, Sigma) and 0.1% SDS at 42 °C for 2 h with gentle shaking. The membrane-bound RNA was hybridized to the [^{32}P]-labelled cDNA probe in a same solution of prehybridization at 42 °C for 19 h. Following hybridization the membrane was washed twice at room temperature in 2×SSC for 5 min per wash, twice at 60 °C in 2×SSC and 1% SDS for 30 min per wash and 3 times at room temperature in 0.1×SSC for 30 min per wash. The membrane was covered with polyvinyl wrap and exposed to Imaging Plate (Fuji Film) for 2 h, then the Imaging Plate was analyzed by using Image Analyzer (BAS2000: Fuji Film). Using this analysis, the relative content of mRNA cording for TH was determine as a level of radiolabel of hybridized probe.

RESULTS AND DISCUSSION

Short-term activation of TH

When 100 μM of 2-Cl-Ado was added to the chromaffin cell suspension, the formation of ^{14}CO$_2$ was increased to about 270% of the control. The formation of $^{14}CO_2$ was not changed when ^{14}C-dopa was substituted for ^{14}C-tyrosine, suggesting that the stimulatory effect of 2-Cl-Ado did not occur after the TH step, for example on the dopa decarboxylase step. Uptake of ^{14}C-tyrosine into the cells was not increased by the addition of 2-Cl-Ado (Table 1).

Table 1.
Effects of 2-Cl-Ado on catecholamine metabolism in primary cultures of adrenal chromaffin cells

Treatment	Formation of $^{14}CO_2$		Uptake of $^{14}CO_2$
	from ^{14}C-dopa	^{14}C-tyrosine	
	(nmol/hr/10^6 cells)		(μmol/hr/10^6 cells)
Control	1.50 ± 0.25	13.1 ± 0.5	33.8 ± 2.4
2-Cl-Ado (0.1 mM)	4.08 ± 0.99	12.4 ± 1.0	28.2 ± 1.7
	(271%)	(95%)	(92%)

These results indicate that the increase of $^{14}CO_2$ formation produced by 2-Cl-Ado corresponds to the activation of TH. In order to know whether 2-Cl-Ado works extracellularly or after the uptake, effect of dipyridamole was investigated. Dipyridamole (0.01 mM) itself did not affect the TH activity. When dipyridamole was added together with 2-Cl-Ado, the activation of TH by 2-Cl-Ado was not changed significantly (Table 2).

Table 2.
Effects of 2-CL-Ado and Dipyridamole on TH activity in chromaffin cells

Treatment	TH activity (% of control $\pm$ SEM)
Control	100
2-Cl-Ado (0.1 mM)	282 ± 18**
Dipyridamole (0.01 mM)	100 ± 7
2-Cl-Ado (0.1 mM) +Dipyridamole (0.01 mM)	254 ± 29*

TH activity in control: 0.88 ± 0.1 nmol CO_2/hr/10^6 cells. **$p < 0.001$; *$p < 0.01$; when compared to control.

On the other hand, the adenosine receptors, failed to produce any activation in TH activity (data not shown). These results suggest that 2-Cl-Ado causes an activation of TH through receptors located on the chromaffin cell surface membrane.

To further study the existence of adenosine receptor coupled adenylate cyclase system in the chromaffin cells, we have determined changes of cyclic AMP content after the addition of adenosine derivatives. As shown in Fig. 1, the addition of 2-Cl-Ado as well as ATP significantly increased the level of cyclic AMP in the cells.

We have also studied effects of adenosine derivatives on cyclic AMP content in pheochromocytoma cells (PC12). The addition of 0.01 mM adenosine produced about a 4-fold increase in cyclic AMP content in PC12 cells (control 23.70 ± 2.29 pmol cyclic AMP per 10^6 cells; adenosine treated 91.26 ± 14.41 pmol cyclic AMP/10^6 cells) within 5 min. In addition, the same concentration of adenosine failed to produce any significant increase in cyclic AMP content when it was added in the presence of 2 mM theophylline or 0.1 mM 2'-deoxyadenosine. Recently Erny and others (1981) have reported that 2-Cl-Ado increases both TH activity and cyclic AMP levels in intact pheochromocytoma cells. Our present observation is in good agreement with their report, and together could support the

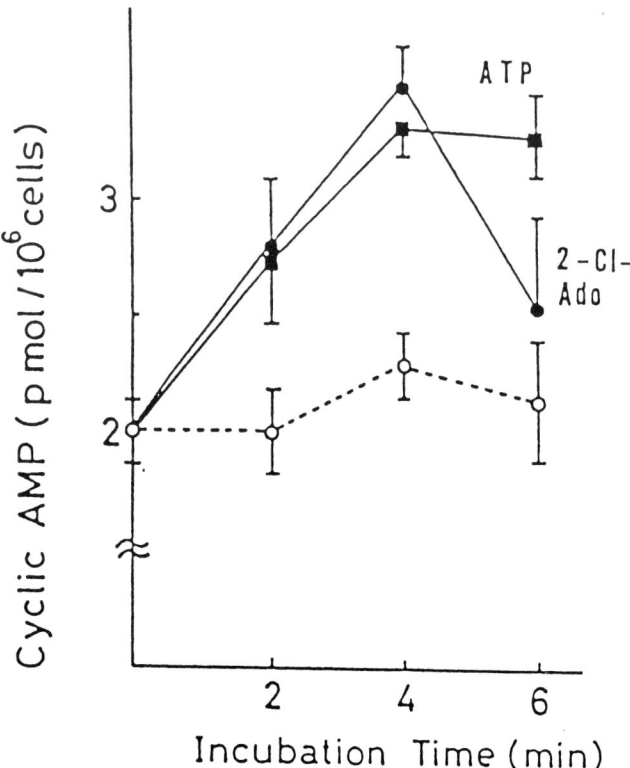

Figure 1. Effects of ATP and 2-Cl-Ado on cyclic AMP levels in primary cultures of chromaffin cells.

possible involvement of adenosine in the short-term regulation of catecholamine synthesis through cyclic AMP-dependent mechanisms in chromaffin tissue. However, when other purine derivatives were added to the cell suspension, 1 mM adenosine and ATP Inosine was rather inhibitory to the TH activity as shown in Table 3.

Table 3.
Effects of purine compounds on TH

Treatment	TH activity (% of control ± S.E.)
Control	100
ATP	229 ± 20**
Adenosine	174 ± 19*
Adenine	230 ± 11**
Guanosine	186 ± 10**
Inosine	72 ± 5

Each drug was added at the concentration of 1 mM. TH activity in control was 4.59 ± 0.32 nmol CO_2/hr/10^6 cells. $^*p < 0.01$ and $^{**}p < 0.001$ when compared to control.

Therefore at present, there seems to be two different mechanisms of the activation of TH as the short-term regulation of catecholamine synthesis. One involves a cyclic AMP-dependent process, another in non-cyclic AMP mediated process. The latter might be mediated, for example, by changes of biopterine metabolism in the cells.

Long-term regulation of TH

When the chromaffin cells were exposed to 2-Cl-Ado at the concentration of 0.1–0.5 mM for 3 h, the activity of TH was significantly increased after 48 h (Table 4). This delayed increase elicited by 2-Cl-Ado seems to be due to the de novo synthesis of enzyme molecules, since the delayed increase was completely blocked by the addition of 5 μg/ml actinomycin D (data not shown).

These results together with the results in Fig. 1 suggest that the delayed increase in TH activity observed in the present study could correspond to cyclic AMP-mediated induction of TH (Kumakura et al., 1979), although further studies must be done precisely.

The induction of TH in the adrenal medulla after the stimulation of splanchnic nerve is believed to be mediated by the increase of cyclic AMP which leads to a chain of events in the chromaffin cells (see for review, Guidotti and Costa, 1977).

Table 4.
Long-term effects of 2-Cl-Ado on TH activity in adrenal chromaffin cells

Cell exposure to 2-Cl-Ado	TH activity measured at 24 h	48 h
10 μM×3 h		103 ± 1.5
50 μl		109 ± 10.5
100 μM	116 ± 4.2	156 ± 20.3*
500 μM	148 ± 11.3*	143 ± 3.9*
500 μM×10 min	110 ± 2.4	
×30 min	114 ± 6.7	
×1 h	137 ± 10.0*	
×3 h	148 ± 11.3*	

TH activity was expressed as % of control ± S.E.; TH activity in control cells was 0.8 ± 0.01 nmol CO_2/hr/10^6 cells; *$p < 0.01$ when compared to control.

However, it seems that acetylcholine which is known to be released from splanchnic nerve terminals, neither produce a significant increase in cyclic AMP levels nor elicit an induction of TH in the isolated adrenal chromaffin cells. Therefore, the present study may allow us to speculate that ATP and its metabolites which are released from splanchnic nerve terminals, concomitantly with acetylcholine, may have a role to trigger an increase in the level of cyclic AMP in the trans-synaptic induction of TH in the adrenal medulla.

Time course of induction of TH gene expression

To study whether the delayed increase in TH activity was due to new synthesis of enzyme molecule or not, the rate of transcription of the RNA cording for tyrosine hydroxylase (mRNATH) with or without 2-Cl-Ado treatment were compared. Following addition of 2-Cl-Ado to cultured adrenal chromaffin cells, an increase in mRNATH level was detectable after 2 h of treatment (Fig. 2). Maximal level of mRNATH was reached after 4 h of 2-Cl-Ado treatment, and the mRNATH level declined rapidly until 8 h after treatment. After then, mRNATH level was maintained until 24 h and decreased gradually to no detectable level after 48 h (Fig. 2). At maximal point (after 4 h), mRNATH level of the cell treated with 2-Cl-Ado was increased about 23-fold to that of non treated cell (Fig. 2). This result indicates that prior to increase in TH activity, the transcriptional level of TH gene was induced by the treatment with 2-Cl-Ado. Therefore, the increase in TH activity appeared to be correlated with mRNATH content, suggesting that the long-term increase in the enzyme activity resulted from induction of mRNATH.

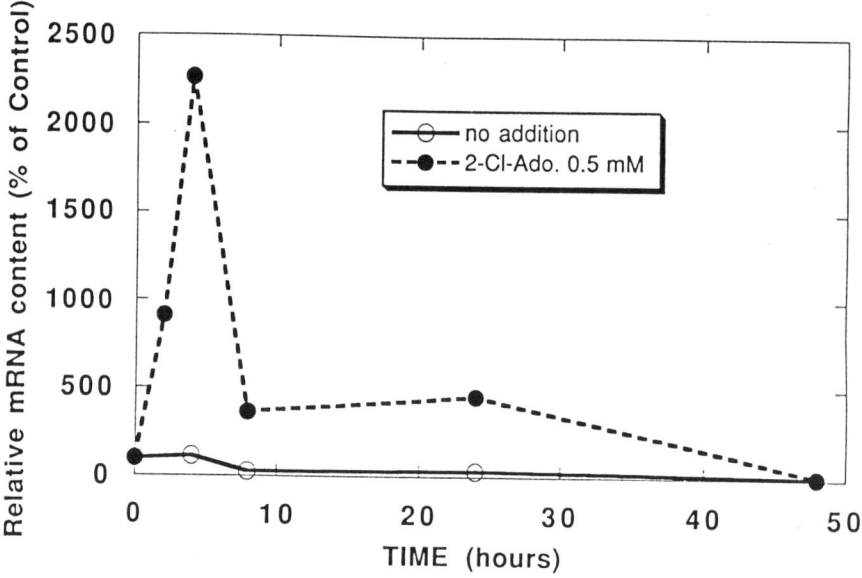

Figure 2. Time course of mRNA[TH] induction. 10^7 cells were plated in 100 mm culture dishes. Two days after plating, 2-Cl-Ado was added at the final concentration of 0.5 mM and medium was changed to agent(-)medium 3 h after the beginning of treatment. Cells were lysed at each indicated time after treatment and total RNA was extracted. The relative amount of mRNA[TH] was determined as described in MATERIALS AND METHODS. The mRNA[TH] content at time 0 is that of cells before adding 2-Cl-Ado and the open circle represents mRNA[TH] level of no-treated cells which sampled after same incubation periods to cells treated with 0.5 mM 2-Cl-Ado for 3 h (closed circle).

Around chromaffin cells, as well as synaptic junctions, ATP is released upon stimulation and adenosine derivatives are accumulated. Our data suggest that activity-dependent release of ATP induces not only short-term activation of tyrosine hydroxylase but also long-term upregulation of tyrosine hydroxylase gene expression to compensate exhausting catecholamine stores. This activity-dependent mechanism through ATP/adenosine receptors also regulates transmitter release itself (Kuroda, 1978; Kuroda *et al.*, 1991), even involved both in the induction of long-term potentiation in hippocampus (Sekino *et al.*, 1991) and in the synapse formation/maintenance between cerebral cortical neurons (Kuroda *et al.*, 1992). Since ATP is released together with other principal transmitters in most of synapses (cholinergic, catecholaminergic, glutamatergic) in an activity-dependent manner, we propose that ATP/adenosine is a general activity-dependent neuromodulator which induces many important synaptic changes with different time courses in neuronal tracing circuit for memory process (Kuroda, 1989).

Acknowledgement

The authors wish to thank Professor T. Nagatsu for kindly gifting human tyrosine hydroxylase type I $cDNA$.

REFERENCES

Axelrod, J. (1971). Noradrenaline: Fate and control of its biosynthesis. *Science* **173**, 598–606.

Chirgwin, J. M., Przybyla, A. E., MacDonald, R. J. and Rutter, W. J. (1979). Isolation of biologically active ribonucleicacid from sources enriched in ribonuclease. *Biochem.* **18**, 5294–5299.

Erny, R. E., Berezo, M. W. and Perlman R. L. (1981). Activation of tyrosine 3-mono-oxygenase in pheochromocytoma cells by adenosine. *J. Biol. Chem.* **256**, 1335–1339.

Guidotti, A. and Costa, E. (1977). Trans-synaptic regulation of tyrosine 3-mono-oxygenase biosynthesis in rat adrenal medulla. *Biochem. Pharmacol.* **26**, 817–823.

Ichiyama, A., Nakamura, S., Nishizuka, Y. and Hayaishi, O. (1970). Enzymatic studies on the biosynthesis of serotonin in mammalian brain. *J. Biol. Chem.* **245**, 1699–1709.

Kumakura, K., Guidotti, A. and Costa, E. (1797). Primary cultures of chromaffin cells; Molecular mechanisms for the induction of tyrosine hydroxylase by 8-Br-cyclic AMP. *Mol. Pharmacol.* **16**, 865–876.

Kumakura, K., Kobayashi, K. and Kuroda, Y. (1981). Regulation of catecholamine synthesis by adenosine derivatives in nerve terminals and adrenal chromaffin cells. In: *Physiology and Pharmacology of Adenosine Derivatives.* J. Daly, Y. Kuroda, J. Phillis, H. Shimizu, and M. Ui (Eds).Raven Press, New York.

Kuroda, Y. (1978). Physiological roles of adenosine derivatives which are released during synaptic transmission in mammalian brain. *J. Physiol. (Paris)* **74**, 463–470.

Kuroda, Y. (1989). Tracing circuit model for the memory process in human brain: Roles of ATP and adenosine derivatives for dynaumi change of synaptic connections. *Neurochem. Intern.* **14**, 309–319.

Kuroda, Y. (1991). Activity-dependent release of ATP and adenosine derivatives can trigger molecular cascades for the memory process in human brain. In: *Role of Adenosine and Adenine Nucleotides in the Biological System.* S. Imai and M. Nakazawa (Eds). Elsevier, The Netherlands.

Kuroda, Y. and Kobayashi. K. (1978a). Increase of cyclic AMP levels by adenosine derivatives in mammalian brain synaptosomes. *Proc. Jap. Acad.* **54-B**, 243–247.

Kuroda, Y. and Kobayashi, K. (1978b). Activation of tyrosine hydroxylase in a synaptosomal preparation from guinea pig striatum by an adenosine derivative. *Proc. Jap. Acad.* **54-B**, 640–644.

Kuroda, Y., Ichikawa, M., Muramoto, K., Kobayashi, K., Matsuda, Y., Ogura, A. and Kudo, Y. (1992). Block of synapse formation between cerebral cortical neurons by a protein kinase inhibitor. *Neurosci. Lett.* **135**, 255–258.

Sekino, Y., Ito, K., Miyakawa, H., Kato, H. and Kuroda, Y. (1991). Adenosine (A2) antagonist inhibits induction of long-term potentiation of evoked synaptic potentials but not of the population spike in hippocampal CA1 nuerons. *Biochem. Biophys. Res. Commun.* **181**, 1010–1014.

Waymire, J. C., Bjur, R. and Weiner, N. (1971). Assay of tyrosine hydroxylase by coupled decarboxylation of Dopa formed from $1\text{-}^{14}C$-tyrosine. *Anal. Biochem.* **43**, 588–600.

Waymire, J. C., Waymire, K. G., Boehme, R., Noritake, D. and Wardell, J. (1977). Regulation of tyrosine hydroxylase by cyclic 3′, 5′-adenosine monophosphate in cultured neuroblastoma and cultured dissociated bovine adrenal chromaffin cells. In: *Structure and Function of Monoamine Enzymes*. E. Usdin, N. Weiner, and B. H. Youdim (Eds), Marcel Dekker, New York, pp. 327–363.

Weiner, N., Masserano, J. M., Meligeni, J. and Tank, A. W. (1982). Activation of adrenal tyrosine hydroxylase following acute and chronic stress. In: *Advances in the Biosciences*. F. Izumi, M. Oka, K. Kumakura (Eds), Vol. 36, Pergamon Press, Oxford, pp. 37–46.

Yamauchi, T. and Fujisawa, H. (1979). *In vitro* phosphorylation of bovine adrenal tyrosine hydroxylase by adenosine 3′, 5′-monophosphate-dependent protein kinase. *J. Biol. Chem.* **254**, 503–507.

Tyrosine Hydroxylase, pp. 71–89
M. Naoi *et al.* (Eds)
© VSP 1993

AMINE SYNTHESIZING ENZYMES, PEPTIDES, GROWTH FACTORS AND ONCOGENE PRODUCTS IN MIDGUT CARCINOID TUMOURS AND PHEOCHROMOCYTOMAS IN TISSUE CULTURE

HÅKAN AHLMAN,[1] ANNICA DAHLSTRÖM, MENEK GOLDSTEIN,[2]
SVANTE JANSSON,[1] AMANDA McRAE, OLA NILSSON and
BO WÄNGBERG[1]

Department of Histology, University of Göteborg and [1]Department of Surgery I, Sahlgren's Hospital, University of Göteborg, P.O. Box 33031, S-400 33 Göteborg, Sweden
[2]Department of Psychiatry, New York University, 10024 NY, USA

INTRODUCTION

Catecholamines (CA) and tryptamines are normally secreted from adrenal medullary chromaffin cells and gut enterochromaffin cells, respectively. Both cell types may have a common embryonic origin in the neural crest and are sometimes referred to as paraneurons, since they share functional and structural features with neurons (Wängberg *et al.*, 1990). In fact, the chromaffin cells in the adrenal medulla can be regarded as modified postganglionic neurons, not growing long neurites. Neurite formation in adrenal medullary cells is normally suppressed by high glucocorticoid levels from the adjacent cortical cells (Jones *et al.*, 1977; Unsicker *et al.*, 1978). Transplantation of adrenal medullary cells into the anterior eye chamber (Olson, 1970; Strömberg *et al.*, 1985) or into the brain (Strömberg *et al.*, 1984) induces the extension of vast neurite outgrowth, due *inter alia* to the fact that the cells no longer are exposed to high corticoid levels. Thus, their true nature of adrenergic neurons is disclosed.

Enterochromaffin cells and adrenal medullary cells share the chromaffin staining properties (Bolande, 1974; Schimke, 1977; Ayer-Le Lievre and Fontaine-Perus, 1982). The soluble carrier proteins, the chromogranins, were first isolated from the secretory granules of adrenal chromaffin cells (Blaschko *et al.*, 1967). These proteins are present in several paraneurons and endocrine cell types (O'Connor *et al.*, 1983; Winkler *et al.*, 1986).

Pheochromocytomas and midgut carcinoid tumours are derived from two different types of chromaffin cells and secrete CA and tryptamines, respectively. Determination of the urinary excretion of monoamine metabolites is important for the diagnosis of these tumours. There are several experimental studies on CA secretion and growth patterns of rat adrenal pheochromocytoma cells *in vitro*, but only few studies have been carried out on human cells (Greene and Tischler, 1976; Jones *et al.*, 1977; Unsicker *et al.*, 1978; Livett *et al.*, 1981; Wilson *et al.*, 1981; Yoffe *et al.*, 1982). Pheochromocytomas, as well as midgut carcinoid tumour cells, can express neuronal phenotypes in cell culture, especially when exposed to growth factors (Edgar *et al.*, 1979; Tsugawa *et al.*, 1987; Ahlman *et al.*, 1989a). This is comparable with the phenotypic expression of neurons seen in adrenal medullary cells after transplantation. Thus it seems as if the phenotypic expression of chromaffin stem cells is determined after migration into the target organ, i.e. adrenal glands, paraganglia or intestinal mucosa.

Under more primitive conditions, e.g. neoplasia, several common traits can be seen in the two tumour types, i.e. synthesis and storage of certain peptides like enkephalin, substance P (SP) (Polak *et al.*, 1976; Alumets *et al.*, 1978; Nilsson *et al.*, 1986a) and a novel peptide, Delta Sleep-Inducing Peptide (DSIP) (Ahlman *et al.*, 1989b; Nilsson *et al.*, 1991). Serotonin (5-HT)-immunoreactive cells constitute the main part of midgut carcinoid tumours, but may occur also in pheochromocytomas (Nilsson *et al.*, 1986a). Recently, biochemical analyses of midgut carcinoid tumour tissue have shown small amounts of CA, especially dopamine (DA), in addition to large amounts of tryptamines (Feldman and Moore, 1989).

IN OCULO TRANSPLANTATION

Small pieces ($1 \times 1 \times 1$ mm) of human pheochromocytomas (7 patients) and midgut carcinoid tumours (4 patients) were successfully transplanted into the anterior eye-chamber of rats immunosuppressed by treatment with cyclosporin A (Nilsson *et al.*, 1986a, b). The tumour transplants were vascularized by vessels from the host iris within 2 days. No increase in size of the transplants was noted during 1–4 weeks.

Interestingly, however, individual animals with ocular transplants of pheochromocytoma tumour, displayed ataxia and piloerection, most probably due to secretion of tumour products into the host circulation (Theodorsson *et al.*, 1989; Ahlman *et al.*, 1990a). Within the first week after transplantation the tumour cells sent out long delicate neurite-like processes on the host iris (Fig. 1). An ingrowth of iris nerves (immunopositive for the synaptic vesicle protein SV 10) into the tumour transplants was also observed 7–10 days after inoculation (Theodorsson *et al.*, 1989).

All pheochromocytoma tumour cells had a bluish-green formaldehyde-induced fluorescence reaction (Hillarp-Falck technique; cf. Corrodi and Jonsson, 1968)

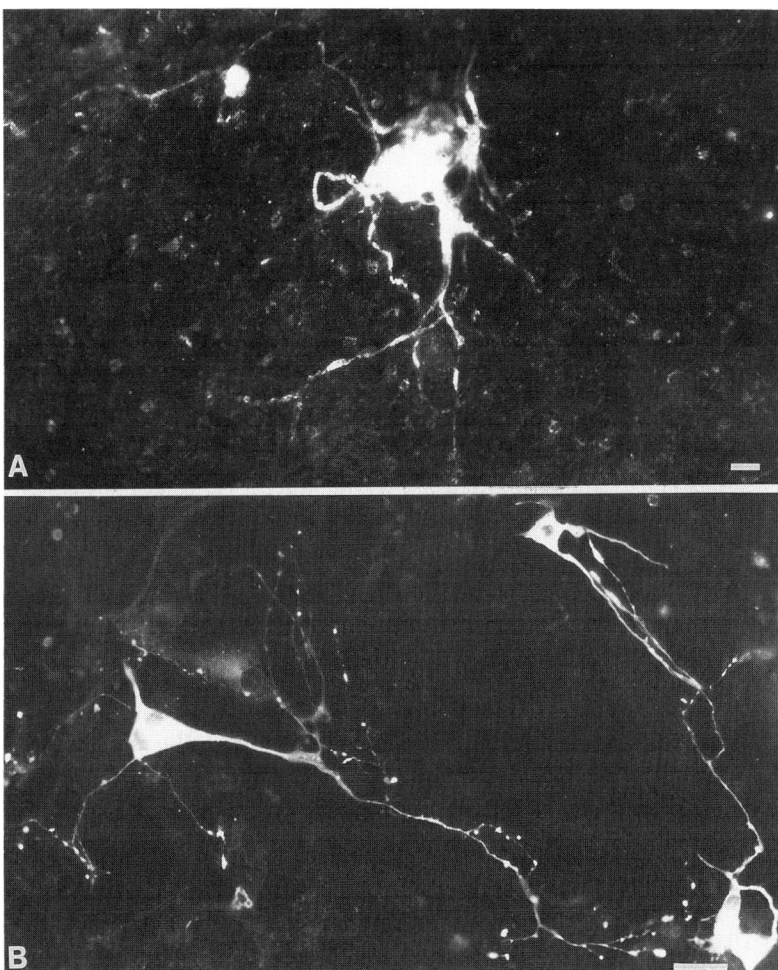

Figure 1. A: Fluorescence micrograph of a human pheochromocytoma tumour transplant grown for 2 weeks in the anterior eye-chamber of an immunosuppressed rat. The host iris with attached transplant was treated according to the Hillarp-Falck technique to visualize CA. The eye was sympathetically denervated prior to transplantation to remove adrenergic nerve fibres from the host iris. Densely packed tumour cells with CA fluorescence were observed extending neurite-like cell-processes on the host iris. B: Fluorescence micrograph demonstrating TH-LI in pheochromocytoma cells in tissue culture. The cells display long cell-processes, similar to those seen in A) with varicose densities. Bars indicate 25 μm.

characteristic of the CA fluorophore. Immunocytochemically the tumour cells displayed a positive reaction against the CA-synthesizing enzymes tyrosine hy-droxylase, TH (Nagatsu *et al.*, 1964) and dopamine-β-hydroxylase, DBH (Friedman and Kaufman, 1965). Large cell populations also showed positive immunoreactions with antisera against neuropeptide Y (NPY) and enkephalins (ENK). Only few tumours displayed minor populations with serotonin (5-HT)-immunopositive cells. Occasionally cell populations with positive immunoreactions with antisera against vasoactive intestinal peptide (VIP), somatostatin (SOM) and substance P (SP) were seen. The immunoreactive pattern of the primary tumour and the heterotransplants was identical in all cases. However, it appeared as if VIP-like immunoreactivity (LI), SOM-LI and SP-LI were only present in the largest pheochromocytomas (75–115 g), while the smaller tumours appeared to lack these peptides. However, in each tumour, where these peptides were observed, they occurred in only a minority of the cells in the population (Nilsson *et al.*, 1986a).

All midgut carcinoid tumour cells had a strong yellow formaldehyde-induced fluorescence reaction (Hillarp-Falck technique, cf. Corrodi and Jonsson, 1968). Immunocytochemically all tumour cells displayed a positive reaction with antibodies against 5-HT and with antisera against tachykinin peptides, like SP and neuropeptide K (NPK) (Nilsson *et al.*, 1986b; Ahlman *et al.*, 1988).

AMINE HANDLING PROPERTIES

The spontaneous and drug-induced release of 5-HT from midgut carcinoid tumour cells in tissue culture has been studied in detail (Wängberg *et al.*, 1990). Reserpine caused an increase in the 5-HT levels in the culture medium, probably due to an escape of the amine from the storage granules, whose amine storage capacity is irreversibly blocked by reserpine (cf. Häggendal and Dahlström, 1971). This effect was further enhanced after blockade of the membrane pump with imipramine, or after inhibition of monoamine oxidase (MAO) with nialamide. After 24 h of reserpine treatment (10^{-7}–10^{-9}M) individual midgut carcinoid tumour cells still contained 5-HT immunoreactivity located in granules, as well as in the cytoplasm, when studied immunocytochemically. The 5-HT synthesis capacity in the tumour cells was pronounced, as judged by a very high cumulative release of 5-HT found in media samples after hourly changes of media. If media were changed only every 4 days the 5-HT levels in the medium soon reached a saturation level, strongly indicating the presence of a control mechanism of the type "end-product inhibition" in the carcinoid tumour cells (Wängberg *et al.*, 1990).

In a comparative study pheochromocytomas and midgut carcinoid tumours were treated with reserpine (10^{-7}–10^{-11}M) present in the medium during 4 days in culture. Biochemical estimations, using HPLC with electrochemical detection, of amines in sonicated cultured cells, showed that reserpine in high doses reduced or depleted the 5-HT and CA contents, respectively, from both types of tumours studied (Fig. 2). The levels of TH- and DBH-LI appeared to be

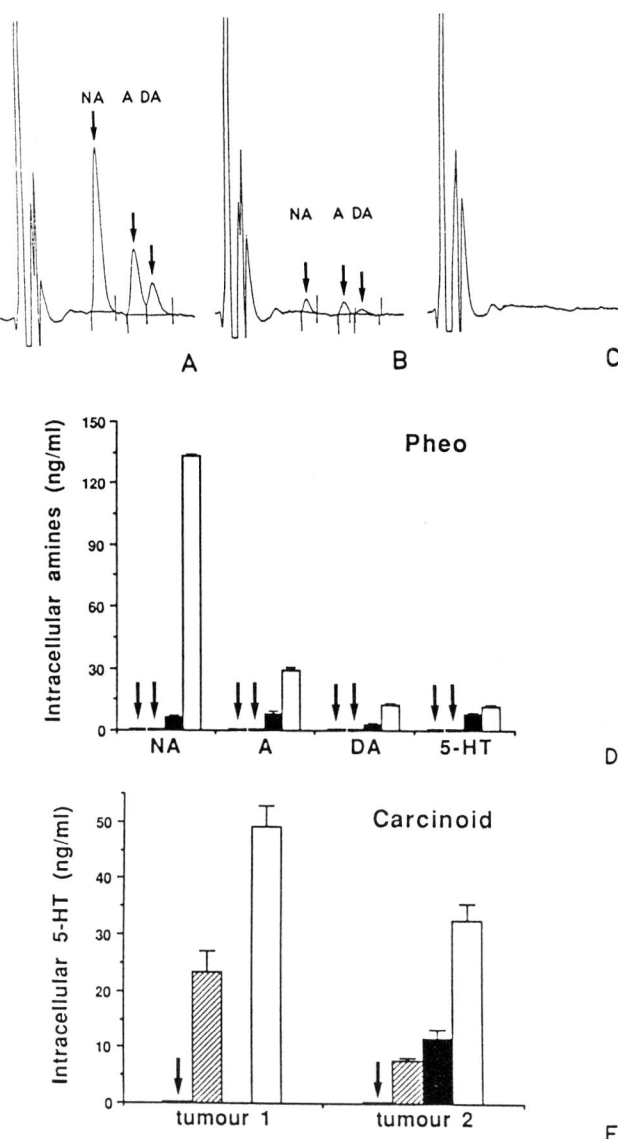

Figure 2. A–C: Chromatographic detection of CA in cultured pheochromocytoma cells under normal conditions (A), after 4 days of treatment with reserpine 10^{-11}M (B) or 10^{-7}M (C) respectively. (D) Dose-dependent reduction of intracellular CA and 5-HT in one pheochromocytoma tumour after reserpine treatment. (Control = white bar; reserpine 10^{-11}M = black bar; reserpine 10^{-9}M and 10^{-7}M indicated by arrows, $n = 8$). (E) Dose-dependent reduction of intracellular 5-HT in two midgut carcinoid tumours after reserpine treatment. (Control = white bar, reserpine 10^{-11}M = black bar; reserpine 10^{-9}M = hatched bar; and reserpine 10^{-7}M indicated by arrows; $n = 8$).

unchanged after such treatment. In pheochromocytoma cultures treated with reserpine (10^{-7}–10^{-9}M) DA- and 5-HT-LI were almost totally abolished, while small amounts of amines were present after treatment with reserpine 10^{-11}M (Fig. 3). In cell cultures from different pheochromocytoma tumours some variation in the occurrence of TH-LI and DBH-LI between individual tumour cells was observed (Fig. 4). Biochemically, when the pheochromocytoma cell cultures had been sonicated, a clear dose-dependent reduction of the intracellular levels of both CA and 5-HT was verified (Fig. 2).

For the midgut carcinoid tumours there appeared to be a slight discrepancy between the biochemical and immunocytochemical findings; thus, high-dose treat-

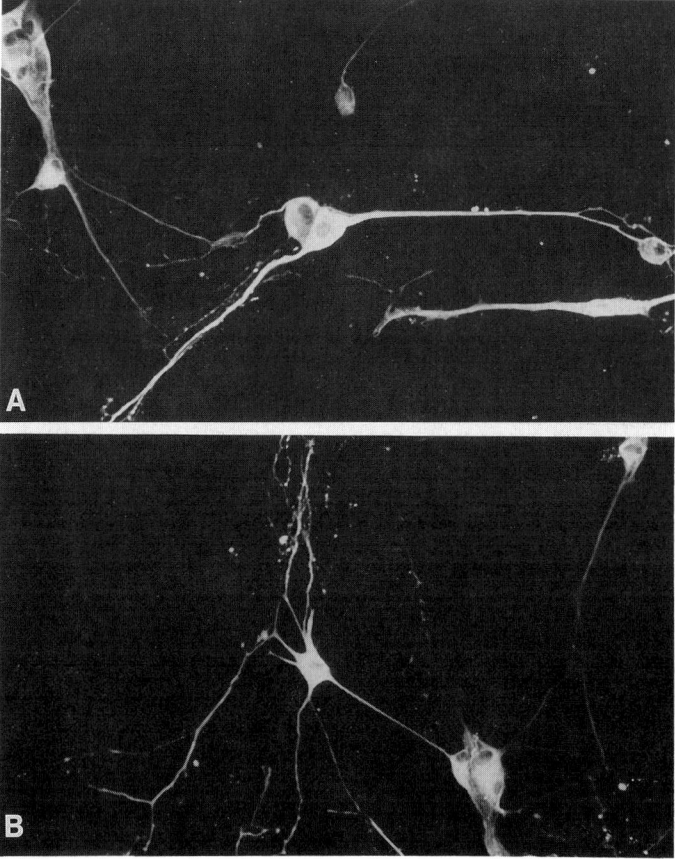

Figure 3. TH-LI in cultured pheochromocytoma cells, studied in a confocal laser scanning microscope (Biorad MRC 600). The photographs show contacts between cell processes from different tumour cells. Note the variable contents of TH-LI in different tumour cells (A). Bar indicates 75 µm.

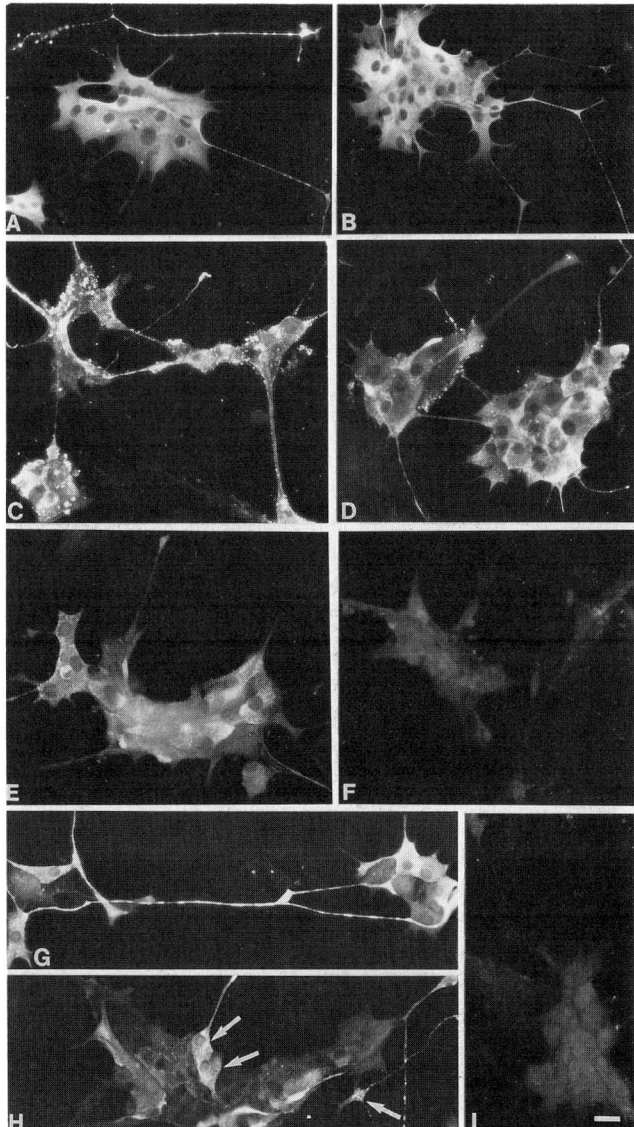

Figure 4. Fluorescence micrographs, showing immunofluorescence studies of the same pheochromocytoma cells as studied biochemically in Fig. 2, before (A ,C, G, E) and after (B, D, F, H, I) reserpine treatment for 4 days. No evident changes in the appearance of TH-LI (A–B) or DBH-LI (C–D) could be observed after reserpine 10^{-7}M as compared to control. In contrast, a depletion of DA was observed after reserpine treatment (F) as compared to untreated (E). This pheochromocytoma tumour also contained 5-HT-LI with mainly granular location (G). A few tumour cells appeared resistant to treatment with the low concentration of reserpine, 10^{-11}M (H, arrows), while all tumour cells were depleted of 5-HT-LI after 4 days in the high concentration of reserpine 10^{-7}M (I). Bar indicates 25 μm.

ment with reserpine depleted the tumour cells completely, studied biochemically in sonicated cells, while individual tumour cells still contained fair amounts of 5-HT-LI observed in the fluorescence microscope. Cross-reactivity with 5-HTP and 5-HT-metabolites present within the granules and cytoplasm of the cells cannot be excluded. Individual cells may, on the other hand, be more resistant to reserpine than the majority of cultured cells, since, under physiological conditions enterochromaffin cells in general are more resistant to the monoamine depleting effect of reserpine than adrenergic nerve endings and adrenal medullary cells (Ahlman and Enerbäck, 1974; Nilsson *et al.*, 1981).

STUDIES ON MIDGUT CARCINOID TUMOURS AND
PHEOCHROMOCYTOMA IN MONOCULTURES

Immunocytochemically, TH-LI as well as DBH-LI have been demonstrated in cultured tumour cells from 3 patients with midgut carcinoid tumours, in analogy with the findings in human pheochromocytoma cell cultures (Ahlman *et al.*, 1990b). The tumour cell cultures were studied over 3-6 weeks, and we observed that the majority of cells formed strands and clusters of spherical cells filled with cytoplasmic granules (the endocrine phenotype). Regularly, however, also tumour cells with neuritic extensions were observed (the neuronal phenotype). Such elongating cells were often located in the peripheral zone of a tumour cell colony. Both phenotypes had similar immunocytochemical properties, with positive immunoreactions against TH (Figs 3, 4), DBH and DA (Fig. 4). The immunoreactive material was present in granules within the cellbody and in the neuritic processes. Three CA (noradrenaline (NA), adrenaline (A) and dopamine (DA)) were identified by HPLC with electrochemical detection in media from all 3 midgut carcinoid tumours. The major peaks represented NA and DA, but minor peaks of A were also detected. Sneddon (1973) first suggested that midgut carcinoid tumours could extract DA from the blood by means of a pump mechanism, rather than by genuine synthesis. However, Feldman (1985) reported that one out of five patients with carcinoid tumour had elevated urinary excretion of the main DA metabolite, homovanillic acid. Feldman and Moore (1989) recently analyzed CA from tumour tissue of 12 midgut carcinoid tumours, which all contained DA (mean 9.3 μmol/kg) and small amounts of A and NA (means 0.3 and 0.5 μmol/kg, respectively). The CA content was, however, small in comparison with the average content of 5-HT (6 569 μmol/kg) in these tumours.

Our own observations of all three CA detected in culture media and intracellulary, together with the presence of both TH- and DBH-LI in tumour cells, indicate that midgut carcinoid tumours are able to synthesize not only DA, but also NA. The fact that A was present in measurable amounts indicates that the carcinoid tumour cells probably also may express phenylethanolamine-N-methyl transferase (PNMT).

The above mentioned observations have had clinical implications already; for instance, tumour biopsy specimens from patients with ganglioneuromas and some DA-producing pheochromocytomas, exhibit positive immunoreactivity after incubation with TH-antisera, but are negative with DBH-antisera. This is related to selective hypersecretion of DA as the sole CA, and lack of hypertensive symptoms clinically (Jansson *et al.*, 1989).

Based on observations in tissue cultures, it is difficult to evaluate to what extent CA production occurs *in situ*, in the patient. In cell cultures, CA synthesis is facilitated due to the presence in the culture medium of the precursor amino acid tyrosine. It is tempting to speculate that the genes coding for CA-synthesizing enzymes in a chromaffin progenitor cell are suppressed in the enterochromaffin cell by an intestinal factor, in analogy with the suppression of the neural phenotype of adrenal chromaffin cells by glucocorticoids from the adrenal cortex. In neoplasia such inhibitory mechanisms may be partly lost. And in tissue culture, where neoplastic cells are removed from the surrounding tissue, possible inhibitory influence of such factors is further reduced.

In Fig. 5 the immunohistochemical properties of tissues cultured pheochromocytoma cells and midgut carcinoid tumour cells are presented. Both types of tumour cells express TH-LI, DBH-LI as well as DA-LI in their cytoplasm.

MIDGUT CARCINOID TUMOURS IN CO-CULTURE WITH NEURONS

For co-culture experiments rat fetal DA-neurons were harvested from the mesencephalon of fetuses (day E 15) and were grown initially as monoculture under optimal conditions (for neurons) in MEM (1:1) with addition of 8% inactived fetal calf serum, 5% glucose, insulin (5 mg/ml), L-glutamine (5mM) and PEST (Ahlman *et al.*, 1990b). Cholinergic neurons were harvested from the medial septum of fetuses (day E 18) and were characterized with acetylcholinesterase (AChE) staining and immunocytochemically by the presence of NGF-receptor-LI (Wigander *et al.*, 1991). If monocultures of neurons were transferred to serum-free medium, the neurons did not survive more than 4–8 days.

However, in co-cultures with midgut carcinoid tumour cells good survival under serum-free conditions was observed up to 28 days. The neurons frequently exhibited very long neurites, indicating a neuronotrophic action exerted by the tumour cells. In separate studies cholinergic neurons thrived in serum-free medium, if this was supplemented with conditioned medium from tumour cell cultures, grown serum-free (Fig. 6). This indicates that the tumour cells produce transferable neuronotrophic growth factor(s). Immunocytochemical studies have demonstrated the presence of NGF-like material in the tumour cells, which also exhibit immunoreactive NGF-receptor material on their cell membrane. Thus, an autocrine production of a NGF-like factor with action on intrinsic NGF-receptors on the tumour cells has been suggested (Ahlman *et al.*, 1989a). DA-neurons were

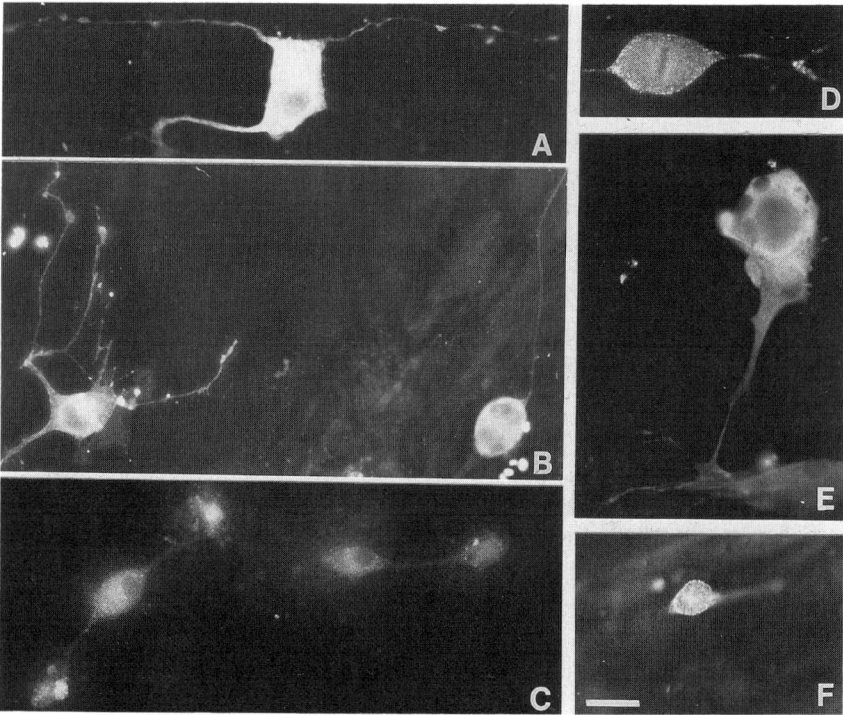

Figure 5. Fluorescence micrographs showing immunofluorescence investigations of cultured pheochromocytoma cells (left column, A–C) in comparison with cultured midgut carcinoid tumour cells (right column, D–F). TH-LI (A and D), DBH-LI (B and E) and DA (C and F) are demonstrated. Note that also carcinoid cells can express a neuronal phenotype with neurite extensions, like the pheochromocytoma cells. Bar indicates 25 μm.

often observed to make contacts with co-cultured tumour cells. Ultrastructurally, bouton-like formations at the terminal end of the DA-immunoreactive varicosities were found. These were filled with clear vesicles and occasional large dense core vesicles, and were observed to establish direct contacts with the cell membrane of tumour cells (Fig. 7). This morphological observation may indicate that there is probably some kind of influence of the nerve cells on the cultured tumour cells. This suggestion has obtained support recently, since pilot studies indicate that the release of 5-HT from cultured tumour cells into media was higher in co-cultures with DA-neurons than in co-cultures with cholinergic neurons (Fig. 8). This may indicate the existence of functional synapses between DA neurons and carcinoid tumour cells.

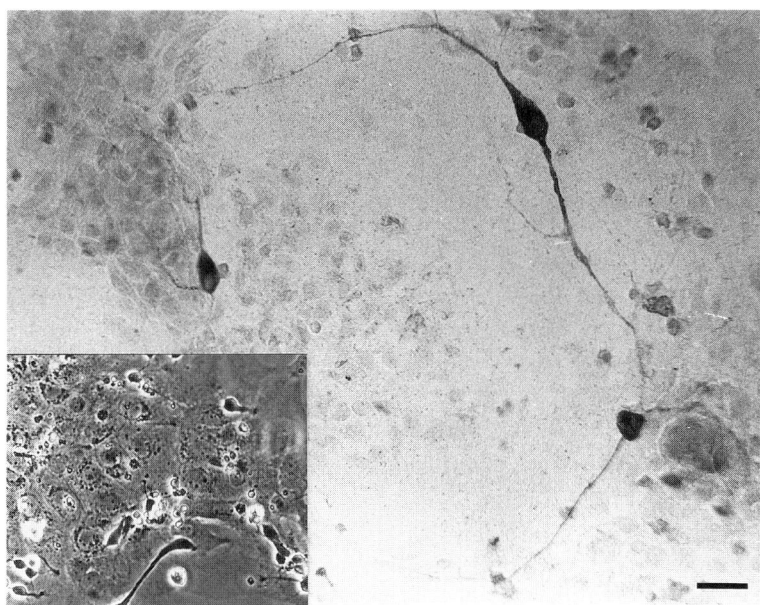

Figure 6. Rat fetal cholinergic neurons in culture, visualized with AChE staining, show good survival and outgrowth of neuritic processes in long-term culture (10–28 days) under serum-free conditions, if supplemented with conditioned media from midgut carcinoid tumour cells grown in serum-free medium. Inset shows serum-free control cultures of cholinergic neurons, which deteriorate within 4–6 days without supplementation of tumour conditioned media. Bar indicates 25 μm.

DSIP

DSIP is a nonapeptide, first isolated from cerebral venous blood in rabbits during thalamic stimulation to induce delta-sleep (Schoenenberger *et al.*, 1977). The presence of DSIP-LI has been demonstrated by radioimmunoassay, not only within the CNS, but also in peripheral organs, e.g. stomach, small intestine and kidney (Graf and Kastin, 1986). Using immunohistochemistry, DSIP-LI has also been shown to occur in chromaffin tissues, e.g. gut enterochromaffin cells and adrenal medulla (Ekman *et al.*, 1987). In midgut carcinoid tumours and pheochromocytoma DSIP-LI is co-localized with biogenic amines in secretory granules (Fig. 9). Chromatographic analyses of conditioned culture media revealed the secretion into the media of DSIP-like peptides, with hydrophobicity and size different from that of synthetic DSIP (Ahlman *et al.*, 1989b; Nilsson *et al.*, 1991). The function of DSIP is not yet elucidated, and there is so far no evidence for the existence of a specific DSIP receptor. However, DSIP has been shown to modulate adrenoceptors post-synaptically (Graf and Kastin,

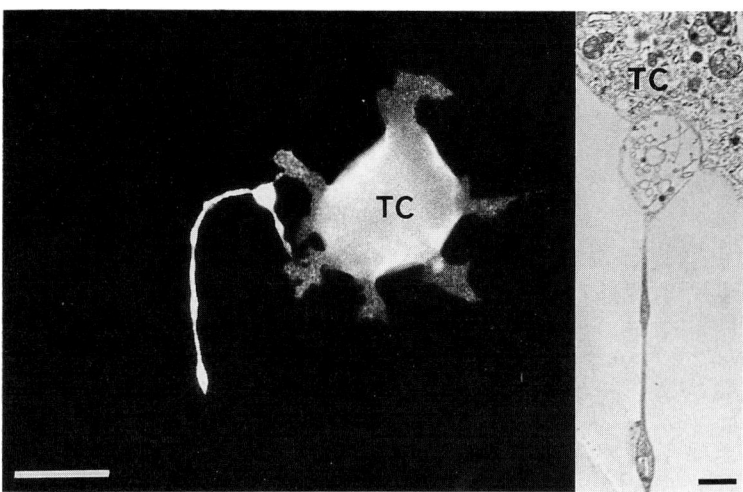

Figure 7. Fluorescence micrograph (left) and electronmicrograph (right) of cells in a co-culture of rat fetal DA neurons and human midgut carcinoid tumour cells (TC). Synapse-like contacts between neurons and TC can be seen in both figures. The DA-nerve cells as well as the TC are immunoreactive after incubation with TH-antiserum. Ultrastructurally the bouton-like formations appeared to be in direct membrane contact with the TC, but without membrane specializations typical for a true synapse. White bar indicates 25 μm and black bar 2 nm.

1986). DSIP may also increase the local metabolism of 5-HT via induction of N-acetyltransferase, as demonstrated in rat pineal gland (Graf and Schoenenberger, 1986). Our demonstration of DSIP-LI in pheochromocytomas, as well as in midgut carcinoid tumours, further emphasizes the phenotypic similarities between these tumours.

ONCOPROTEINS AND ENDOCRINE TUMOURS

Oncogenes are derived from genetically damaged normal genes (protooncogenes) and expression of such oncogenes is part of the neoplastic transformation, with uncontrolled cell growth as a consequence. Human tumours may thus express activated oncogenes, some of which have been related to prognosis (Nishimura and Sekiya, 1987). Increasing knowledge about specific oncoproteins and endocrine tumours is presently being assembled. To date the oncogene expression has been investigated in some detail for endocrine pancreatic tumours, medullary thyroid carcinoma, bronchial carcinoids and pheochromocytomas (Lee *et al.*, 1987; Höfler *et al.*, 1988; Klimpfinger *et al.*, 1988; Boultwood *et al.*, 1989; Goto *et al.*, 1990; Roncalli *et al.*, 1991).

c-erbB-2 is a transmembrane receptor oncoprotein with structural similarities to the epidermal growth factor (EGF) receptor, which is overexpressed in sev-

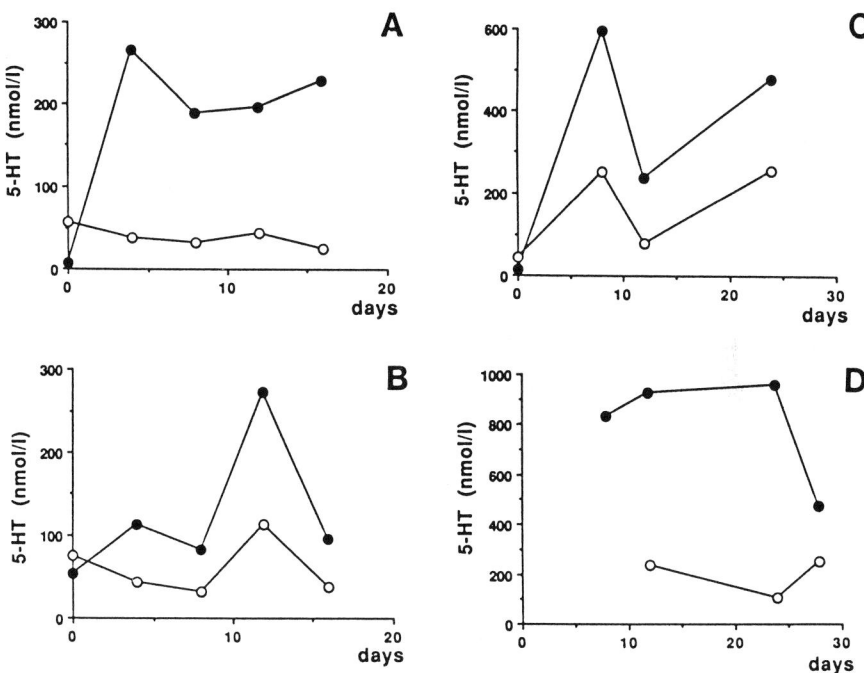

Figure 8. Concentrations of 5-HT in culture media over time studied in 4 co-culture experiments between human midgut carcinoid tumour cells (2 different tumours, A–B and C–D respectively) and rat fetal cholinergic (o——o) or dopaminergic (•——•) neurons, respectively. In co-cultures with DA-neurons there was always higher 5-HT levels than in co-cultures with cholinergic neurons, under otherwise identical conditions.

eral tumours (ovarian, breast, and gastric carcinomas) (cf. Maguire and Greene, 1989). This oncoprotein is present in one out of five pulmonary atypical carcinoids or pheochromocytomas (Roncalli *et al.*, 1991).

The large family of myc-oncoproteins represents nuclear-associated phosphoproteins, regulating cell cycling, which can be overexpressed in pulmonary and gastrointestinal carcinomas (Kaczmarek, 1986). c-myc can be demonstrated in one out of three pulmonary atypical carcinoids, while L-myc is present in 15 % of pheochromocytomas. N-myc is expressed in medullary thyroid carcinomas (Roncalli *et al.*, 1991). It thus seems, as if some oncoproteins are restricted to specific tumour types despite a certain intra- and intertumoral heterogeneity. Expression of such oncoproteins has no obvious relationship to growth pattern, mitotic index, or prognosis of the tumours studied (Roncalli *et al.*, 1991). In future, oncoprotein detection may prove to be of assistance in the classification of neuroendocrine tumours.

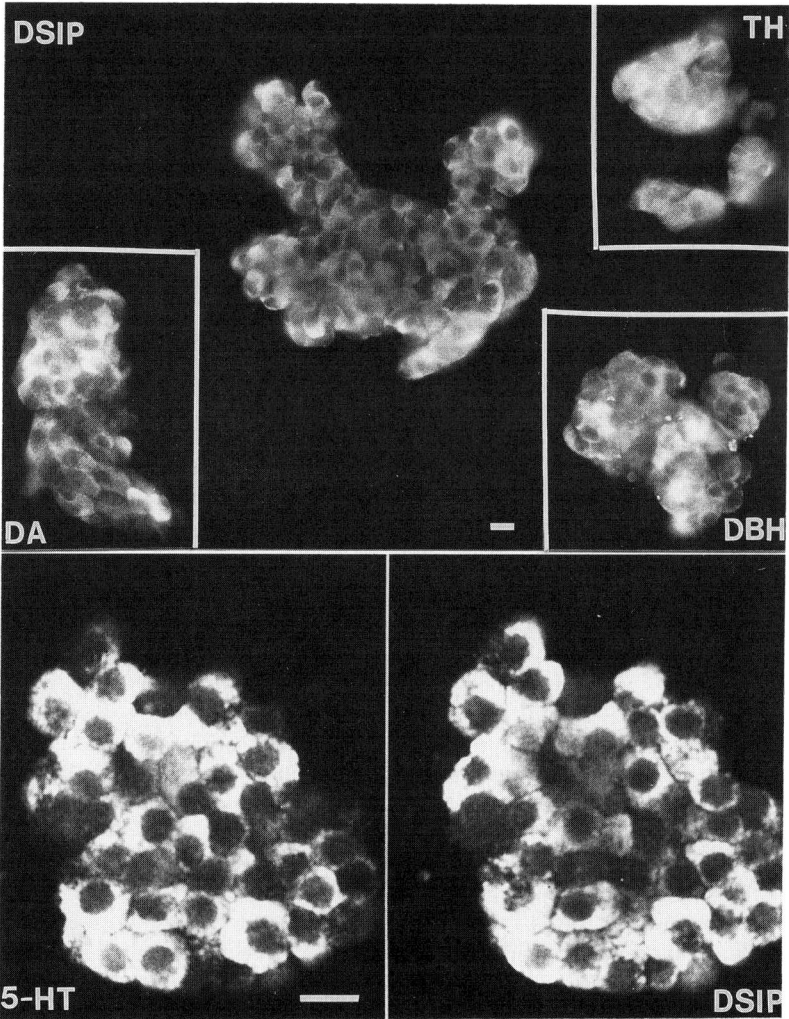

Figure 9. Upper half: Fluorescence microphotographs showing the presence of immunoreactive DSIP, TH, DBH and DA in tumour cells of pheochromocytoma cell colonies in culture. In this colony-forming tumour the endocrine phenotype clearly dominates. The lower half of the figure shows co-existence of 5-HT-LI and DSIP-LI, visualized by confocal laser scanning (Biorad MRC 500), in cells of a midgut carcinoid colony. The tumour cultures were double-incubated with rat monoclonal anti-5-HT antibodies and polyclonal anti-DSIP anti-serum. Bars indicate 25 μm.

The influence of specific oncogenes, e.g. c-erbB-2 and L-myc, on CA handling properties (e.g. synthesis of enzymes, growth factors, and adrenoceptors) of pheochromocytomas has not been investigated. However, introduction of the ras p21 oncogene into the PC12 pheochromocytoma tumour cell line, by microinjection or triturization, induced the expression of a neuronal phenotype in these cells (Bar-Sagi and Feramisco, 1985). The outgrowth of neurites from these tumour cells had a strong resemblance to that seen in some of our monocultures of pheochromocytomas. Overexpression of ras p21 in such cultures remains to be investigated.

Recent studies performed in our laboratory (Nilsson *et al.*, 1992) has demonstrated the presence of myc- and ras-oncoproteins in cultured midgut carcinoid tumours. All tumours investigated so far ($n = 5$) contained both oncoproteins in a variable number of cells. The c-myc monoclonal antibodies gave a strong labelling over the cytoplasm of tumour cells, while nuclei were weakly labelled (Fig. 10a). The ras monoclonal antibody, detects all three families of ras-proteins, and reacts with both normal and activated forms. This antibody gave a distinct labelling over tumour cell membranes, with little or no labelling of the cytoplasm (Fig. 10b). No obvious relationship between clinical characteristics and oncoprotein pattern could be detected.

The importance of c-myc- or ras-oncogene activation in midgut carcinoid tissue remains to be clarified. However, it is noteworthy that c-myc amplification has been recognized for Colo 320 HSR (Quinn *et al.*, 1979), a colonic tumour cell line. These colonic tumour cells, originally of non-endocrine origin, in culture display several endocrine characteristics with secretory granules and monoamine production.

In neural tissues the expression of c-fos genes is coupled to neuronal activity or to trauma and regeneration (Sheng and Greenberg, 1990). The expression of these "early immediate genes" in the nervous system has in fact been used extensively to map neuronal connections and to study drug targets in the CNS (e.g. Graybiel *et al.*, 1991). Thus, the expression of such genes in endocrine tumour tissue, most probably derived ontogenetically from neuroblasts, may be related to stimulus-induced alterations of intracellular messenger systems.

Acknowledgements

Studies performed in the authors' laboratories were supported by the Swedish Medical Research Council (grants no. 2207 and 5220), the Swedish Cancer Foundation (grant no. 2998), Ingabritt och Arne Lundberg's Foundation, T. and R. Söderberg's Foundations, Assar Gabrielsson's Foundation, the Swedish Medical Society, the Medical Association in Göteborg, and Funds for Medical Research at the Medical Faculty, University of Göteborg. The skilful technical assistance of Ms Kerstin Lundmark, Ms Annelie Wigander, Ms Ann-Christine Illerskog, and Ms Lena Johansson is gratefully acknowledged.

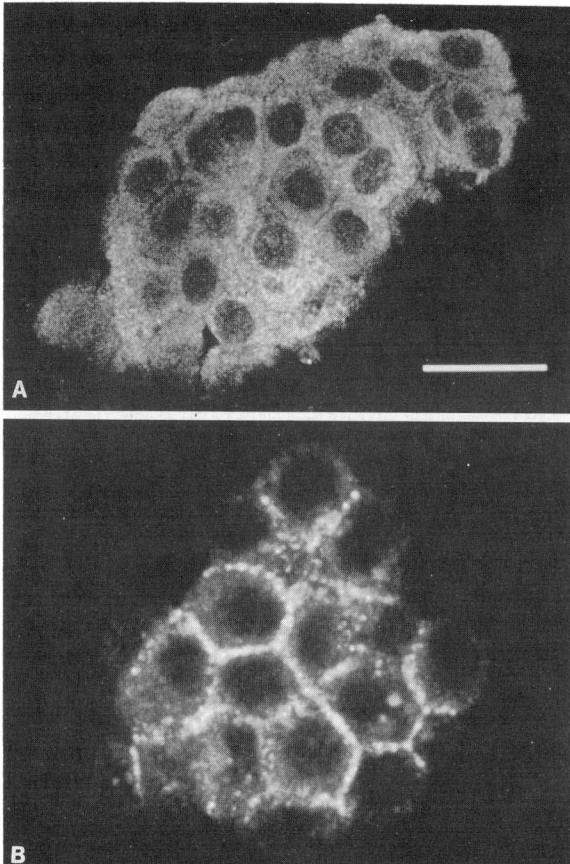

Figure 10. Photographs from the screen of a confocal laser scanning equipment (Biorad MRC 500 attached to a NIKON Optiphot FX epifluorescence microscope) showing the distribution of immunoreactive material after incubation with antibodies against two oncoproteins, present in midgut carcinoid tumour cells: (A). After incubation with anti-human c-myc protooncogene product, the immunoreactive material can be observed to be present mainly in the cytoplasm. (B). After incubation with anti-ras p21 the immunoreactive product is present in patches in the cell membrane. Bar indicates 25 μm.

REFERENCES

Ahlman, H. and Enerbäck, L. (1974). A cytofluorometric study of the myenteric plexus in the guinea-pig. *Cell Tiss. Res.* **153**, 419–434.

Ahlman, H., Åhlund, L., Nilsson, O., Theodorsson, E. and Dahlström, A. (1988). Carcinoid tumour cells in long-term culture. Release of serotonin, but not of tachykinins, at stimulation with adrenoceptor agonists. *Int. J. Cancer* **42**, 506–510.

Ahlman, H., Wigander A., Mölne, J., Nilsson, O., Karlsson, J. E. , Theodorsson, E. and Dahlström, A. (1989a). Presence of nerve growth factor-like immunoreactivity in carcinoid tumour cells and induction of a neuronal phenotype in long-term culture. *Int. J. Cancer* **43**, 949–955.

Ahlman, H., Åhlund, L., Nilsson, O., Dahlström, A., Bjartell, A. and Ekman, R. (1989b). Coexistence of delta sleep-inducing peptide and serotonin in midgut carcinoid tumour cells *in vivo* and *in vitro*. *Int. J. Cancer* **44**, 641–647.

Ahlman, H., Åhlund, L., Dahlström, A., Grimelius, L. and Theodorsson, E. (1990a). Somatostatin analogue and tissue cultures in the study of a human malignant glucagonoma. *J. Surg. Oncol.* **44**, 191–199.

Ahlman, A., McRae, A., Jansson, S., Mölne, J. and Dahlström, A. (1990b). Immunocytochemical and biochemical evidence of catecholamine secretion from human midgut carcinoid tumour cells in culture. *Biogenic Amines* **5**, 399–408.

Alumets, J., Håkansson, R., Sundler, F. and Chang, K. J. (1978). Leu-enkephalin-like material in nerves and enterochromaffin cells in the gut. An immunocytochemical study. *Histochemistry* **56**, 187–196.

Ayer-Le Lievre, C. and Fontaine-Perus, J. (1982). The neural crest: its relations with APUD and paraneuron concepts. *Arch. Histol. Jap.* **5**, 409–429.

Bar-Sagi, D. and Feramisco, J. R. (1985). Microinjection of the ras oncogene protein into PC12 cells induces morphological differentiation. *Cell* **42**, 841–848.

Blaschko, H., Comline, R. S., Schneider, F. H., Silver, M. and Smith, A. D. (1967). Secretion of a chromaffin granule protein, chromogranin, from the adrenal gland after splanchnic stimulation. *Nature* **215**, 58–59.

Bolande, R. P. (1974). The neurocristopathies: a unifying concept of disease in neural crest maldevelopment. *Human Pathol.* **5**, 409–429.

Boultwood, J., Wyllie, F. S., Williams, E. D. and Wynford-Thomas, D. (1989). N-myc expression in neoplasia of human thyroid C-cells. *Cancer Res.* **48**, 4073–4077.

O'Connor, D. T., Burton, D. and Deftos, L. J. (1983). Chromogranin A: immunohstology reveals its universal occurrence in normal polypeptide hormone producing endocrine glands. *Life Sci.* **33**, 1657–1663.

Corrodi, H. and Jonsson, G. (1967). The formaldehyde fluorescence method for the histochemical demonstration of biogenic amines. A review on the methodology. *J. Histochem. Cytochem.* **15**, 65–78.

Edgar, D., Barde, Y. A. and Thoenen, H. (1979). Induction of fibre outgrowth and cholineacetyltransferase in PC12 pheochromocytoma cells by conditioned media from glial cells and organ extracts. *Exp. Cell. Res.* **121**, 353–361.

Ekman, R., Bjartell, A., Ekblad, E. and Sundler, F. (1987). Immunoreactive delta sleep-inducing peptide in pituitary adrenocorticotrophin/alpha-melanotropin cells and adrenal medullary cells of the pig. *Neuroendocrinology* **45**, 298–304.

Feldman, J. M. (1985). Increased dopamine production in patients with carcinoid tumours. *Metabolism* **34**, 255–260.

Feldman, J. M. and Moore, J. O. (1989). Biogenic amines in carcinoid tumours. *Biogenic Amines* **6**, 247–252.

Friedman, S. and Kaufman, S. (1965). 3,4-Dihydroxyphenylethylamine β-hydroxylase. Physical properties, copper content, and role of copper in the catalytic activity. *J. Biol. Chem.* **240**, 4763.

Goto, K., Ogo, A., Yanase, T., Haji, M., Ohashi, M. and Nawata, H. (1990). Expression of c-fos and c-myc protooncogenes in human adrenal phaeochromocytomas. *J. Clin. Endocrinol. Metab.* **70**, 353–357.

Graf, M. V. and Kastin, A. J. (1986). Delta sleep-inducing peptide (DSIP). An update. *Peptides* 7, 1165–1187.

Graf, M. V. and Shoenenberger, G. A. (1986). DSIP affects adrenergic stimulation of rat pineal N-acetyltransferase *in vivo* and *in vitro*. *Peptides* 7, 1001–1006.

Graybiel, A. M., Moratalla, R. and Robertson, H. A. (1991). Amphetamine and cocaine induce drug-specific activation of the c-fos gene in striosome-matrix and limbic subdivisions of the striatum. *Proc. Natl. Acad. Sci. USA* **87**, 6912–6916.

Greene, L. A. and Tischler, A. S. (1976). Establishment of a noradrenergic clonal line of rat adrenal pheochromocytoma cells which respond to nerve growth factor. *Proc. Natl. Acad. Sci. USA* **73**, 2424–2428.

Häggendal, J. and Dahlström, A. (1971). Recovery of noradrenaline in adrenergic nerve terminals after reserpine treatment. *J. Pharm. Pharmacol.* **23**, 81–89.

Höfler, H., Ruhri, C., Putz, B., Wirnsberger, G. and Hauser, H. (1988). Oncogene expression in endocrine pancreatic tumours. *Virchows. Arch.* B **55**, 355–361.

Jansson, S., Dahlström, A., Hansson, G., Tisell, L. E. and Ahlman, H. (1989). Concomitant occurrence of an adrenal ganglioneurona and a contra-lateral pheochromocytoma in a patient with von Recklinghausen's neurofibromatosis. An immunocytochemical study. *Cancer* **63**, 324–329.

Jones, M. T., Hillhouse, E. W. and Burden, J. L. (1977). Dynamics and mechanisms of corticosteroid feed-back at the hypothalamus and anterior pituitary gland. *J. Endocr.* **73**, 405–417.

Kaczmarek, L. (1986). Protooncogene expression during the cell-cycle. *Lab Invest.* **54**, 365–376.

Klimpfinger, M., Ruhri, C., Putz, B., Pfragner, R., Wirnsberger, G. and Höfler, H. (1988). Oncogene expression in a medullary thyroid carcinoma. *Virchows. Arch.* B **54**, 256–259.

Lee, I., Gould, V. and Radosevich, J. A. (1987). Immunohistochemical evaluation of ras oncogene expression in pulmonary and pleural neoplasms. *Virchows. Arch.* B **53**, 146–152.

Livett, B. G., Dean, D. M., Whelan, L. G., Udenfriend, S. and Rossier, J. (1981). Co-release of enkephalin and catecholamines from cultured adrenal chromaffin cells. *Nature (London)* **289**, 317–319.

Maguire, H. C. and Greene, M. I. (1989). The neu (c-erbB-2) oncogene. *Semin. Oncol.* **16**, 148–155.

Nagatsu, T., Levitt, M. and Udenfriend, S. (1964). Tyrosine hydroxylase. The initial step in norepinephrine biosynthesis. *J. Biol. Chem.* **239**, 2910–2917.

Nilsson, O., Dahlström, A., Tisell, L. E. and Ahlman, H. (1986a). Growth of human pheochromocytomas in the anterior eye-chamber of the rat. A histochemical study of amine and peptide content of pheochromocytoma tumour cells. *Regul. Peptides* **15**, 9–141.

Nilsson, O., Ahlman, H., Dahlström, A., Ericsson, L. E. and Skolnik, G. (1986b). Release of serotonin from human carcinoid tumour cells *in vitro* and grown in the anterior eye-chamber of the rat. *Cancer* **58**, 676–684.

Nilsson, O., Wängberg, B., Wigander, A., Lundmark, K., Dahlström, A., Ahlman, H., Bjartell, A. and Ekman, R. (1991). Human pheochromocytoma cells studied in culture contain large amounts of DSIP-like material. *Peptides* **12**, 1077–1083.

Nilsson, O., Ahlman, H., Wängberg, B. (1992). The expression of oncogene products in midgut carcinoid tumour cells in tissue culture. *Int. J. Cancer* (submitted).

Nishimura, S. and Sekiya, T. (1987). Human cancer and cellular oncogenes. *Biochem. J.* **243**, 313–327.

Olson, L. (1970). Fluorescence histochemical evidence for axonal growth and secretion from transplanted adrenal medullary tissue. *Histochemie* **22**, 1–7.

Polak, J. M., Heitz, P. and Pearse, A. G. E. (1976). Differential localization of substance P and motilin. *Scand. J. Gastroenterol.* **39**, 39–42.

Quinn, L. A., Moore, G. E., Morgan, R. T. and Woods, L. K. (1989). Cell lines from human colon carcinoma with unusual cell products, double minutes and homogeneously staining regions. *Cancer Research* **39**, 4914–4924.

Roncalli, M., Springall, D. R., Varndell, I. M., Gaitonde, V. V., Hamid, Q., Ibrahim, N. B. N., Grimelius, L., Wilander, E., Polak, J. M. and Coggi, G. (1991). Oncoprotein immunoreactivity in human endocrine tumours. *J. Pathol.* **163**, 117–127.

Schimke, R. N. (1977). Tumours of the neural crest system. In: *Genetics of Human Cancer.* J. J. Mulvihill, R. W. Miller and J. F. Fraumeni Jr (Eds). Raven Press, New York, pp. 179–198.

Schoenenberger, G. A., Maier, P. F., Tobier, K. J. and Monnier, M. (1977). A naturally occurring delta-EEG-enhancing nonapeptide in rabbits. Final isolation, characterization and activity test. *Pflüg. Arch. Ges. Physiol.* **369**, 99–109.

Sheng, M. and Greenberg, M. E. (1990). The regulation and function of c-fos and other immediate early genes in the nervous system. *Neuron* **4**, 477–485.

Sneddon, J. M. (1973). Blood platelets as a model for monoamine-containing neurons. *Progr. Neurobiol.* **2**, 151–198.

Strömberg, I., Ebendal, T., Seiger, Å. and Olson, L. (1985). Nerve fibre production by intraocular adrenal medullary grafts; Stimulation by nerve growth factor or by sympathetic denervation of the host iris. *Cell Tissue Res.* **241**, 241–249.

Strömberg, I., Herrera-Marschitz, M., Hultgren, L., Ungerstedt, U. and Olson, L. (1984). Adrenal medullary implants in the dopamine denervated striatum. I. Acute catecholamine levels in grafts and host caudate as determined by HPLC-electrochemistry and fluorescence histochemical image analyses. *Brain Res.* **297**, 41–51.

Theodorsson, E., Ryberg, B., Nilsson, O., Ericson, L. E., Dahlström, A. and Ahlman, H. (1989). Intraocular transplants of a human gastrinoma in immunosuppressed rats: morphological, chromatographic and functional studies. *Regul. Peptides* **24**, 97–110.

Tsugawa, M., Morikawi, K., Miyagawa, J., Gomi, M., Fujii, H., Iida, S. and Tarvi, S. (1987). Induction of differentiation of human pheochromocytoma cells in culture by epidermal growth factor and insulin. *Anticancer Res.* **7**, 1161–1164.

Unsicker, K., Kirsch, B., Otten, U. and Thoenen, H. (1978). Nerve growth factor-induced fiber outgrowth from isolated rat adrenal chromaffin cells: impairment by glucocorticoids. *Proc. Natl. Acad. Sci. (Washington)* **75**, 3498–3502.

Wängberg, B., Ahlman, H., Nilsson, O., Haglid, K., Denney, R. M. and Dahlström, A. (1990). Amine handling properties of human carcinoid tumour cells in tissue culture. *Neurochem. Int.* **17**, 331–341.

Wigander, A., Lundmark, K., McRae, A., Mölne, J., Nilsson, O., Haglid, K., Dahlström, A. and Ahlman, A. (1991). Production of transferable neuronotrophic factor(s) by human midgut carcinoid tumour cells; studies using cultures of rat fetal cholinergic neurons. *Acta Physiol. Scand.* **141**, 107–117.

Wilson, S. P., Abou-Donia, M. M., Chang, K. J. and Viveros, O. H. (1981). Reserpine increases opiate like peptide content and tyrosin-hydroxylase activity in adrenal medullary chromaffin cells in culture. *Neuroscience* **6**, 71–79.

Winkler, H., Apps, D. K. and Fischer-Colbrie, R. (1986). The molecular function of adrenal chromaffin granules; established facts and unresolved topics. *Neuroscience* **18**, 261–290.

Yoffe, J. R. and Borchardt, R. T. (1982). Characterization of serotonin uptake in cultured pheochromocytoma cells. Comparison with norepinephrine uptake. *Mol. Pharmacol.* **21**, 386–373.

Tyrosine Hydroxylase, pp. 91–105
M. Naoi *et al.* (Eds)
© VSP 1993

DEVELOPMENTAL CHANGES IN TYROSINE HYDROXYLASE IN PERIPHERY

HIROSHI KUZUYA

Division of Molecular Biology, Institute for Comprehensive Medical Science, Fujita Health University, Toyoake, Aichi, Japan

INTRODUCTION

Catecholamines are synthesized in the biosynthetic pathway in the following order: tyrosine dihydroxyphenylalanine (DOPA) → dopamine → norepinephrine → epinephrine. Each reaction is catalyzed by tyrosine hydroxylase (TH) (Nagatsu *et al.*, 1964), which is the rate-limiting enzyme regulating the synthesis of catecholamines, DOPA decarboxylase (DDC) (Christenson *et al.*, 1970), dopamine-β-hydroxylase (DBH) (Levin *et al.*, 1960), and phenylethanol-amine-N-methyltransferase (PNMT) (Kirshner and Goodall, 1957; Axelrod, 1962), respectively. Among these enzymes, TH and DBH are specific enzymes for adrenergic neurons, and are markers (Christenson *et al.*, 1970) of differentiation and development of adrenergic neurons. The distribution of PNMT, thought to be a marker enzyme of adrenals, is highly localized in the adrenal gland. DDC is present in most organs. In contrast, choline acetyltransferase generally is a marker of cholinergic neurons.

This paper mainly discusses the superior cervical ganglion and adrenal gland. The sympathetic ganglion has been frequently used in studies of the effects of decentralization by axotomy of preganglionic cholinergic nerves, denervation by axotomy of post-ganglionic adrenergic nerves, and excision of target organs such as salivary gland and iris. The adrenal medulla is also an organ derived from the neural crest as well as the sympathetic neuron and contains all the enzymes involved in the biosynthesis of catecholamines. Consequently, this organ, considered to be an integral part of the sympathetic nervous system, is often used as a model of adrenergic neurons.

Relationships between developments of catecholamine-synthesizing enzymes in the superior cervical ganglion and salivary gland, as its target organ, will also be discussed.

Abbreviations are as follows: TH, tyrosine hydroxylase; DBH, dopamine-β-hydroxylase; DDC, DOPA decarboxylase; PNMT, phenylethanolamine-N-methyltransferase; MAO, monoamine oxidase; ChAT, choline acetyltransferase; NGF, nerve growth factor.

DEVELOPMENTAL CHANGES IN CATECHOLAMINE-SYNTHESIZING ENZYMES IN THE SUPERIOR CERVICAL GANGLION

Enzyme activities involved in catecholamine synthesis were studied in the superior cervical ganglion of young growing rats (Thoenen *et al.*, 1972), mice (Black *et al.*, 1971; Gaetani *et al.*, 1975), and hamsters (Jonas *et al.*, 1979).

Rats were studied from birth to 75 days of age (Thoenen *et al.*, 1972). TH activity progressively increased and reached the approximate adult levels (about 3-fold) 14–16 days after birth. Thereafter, it continued to increase slightly up to 75 days, the longest period examined. The specific activity of TH also reached maximal levels (1.4-fold) 14–16 days after birth. After that, it remains about the same. DBH exhibited developmental changes different from those of TH. The total activity of DBH did not increase for a few days after birth, but thereafter, rose fairly steadily up to 40 days of age to about 8-fold. Then, it remained at about the same level. The maximal increase in the specific activity of the enzyme amounted to 2.5-fold and was also reached around 40 days of age. The specific activity of DBH did not change thereafter. The profile of developmental changes in the total activity of DDC resembled that of protein content. The total activity increased slowly until 30 days of age to about 2-fold. Thereafter, it rose rapidly up to 40 days of age. Then, the increase became gradual again. The specific activity of the enzyme at birth was equal to adult levels, thereafter, it rose slightly up to 8 days of age, then at 15 days, decreased to about 90% of the rate of activity present at birth. After that, it persisted at about the same levels.

Mice were examined from birth to 50 or 60 days of age (Black *et al.*, 1971; Gaetani *et al.*, 1975). Total activity of TH temporarily ceased to increase during 3 to 7 days of age. After that, it increased rapidly reaching adult and maximal levels at 12 days old. Although development of the enzyme activity to adult levels essentially appeared to be similar between rats and mice, the increase is about 7-fold greater in mice than in rats.

The developmental increase in TH activity in the superior cervical ganglion of rats and mice was not attributed to depletion of inhibitors or activation and increase in the active sites of the enzyme. It also was not attributable to decrease in proteolytic activity. It was dependant on the increase in the enzyme molecule numbers. This conclusion was derived from experiments including the inhibition of protein synthesis by actinomycin D and cycloheximide (Black *et al.*, 1972; Thoenen, 1972) and immunotitration by anti-NGF antibody (Black *et al.*, 1974).

Activities of TH and ChAT in the 6th lumbar sympathetic ganglion were examined in newborn rats. The pattern of maturation was similar to that for the superior cervical ganglion, but was delayed, suggesting a rostrocaudal gradient of sympathetic development (Hamill *et al.*, 1977).

The superior cervical ganglion of hamsters was examined from birth to adulthood (Jonas *et al.*, 1979). The total activity of TH changed little during 3 postnatal days. Thereafter, it rose rapidly up to about 10 days of age, then slowly for several days. After that, it rapidly increased until about 20 days of age when it reached maximal levels. TH activity at adulthood was 15-fold higher in total activity and 5-fold higher in specific activity compared to those at birth. The total activity of DBH rose rapidly between 4 and 14 days of age and reached a level slightly lower than that at adulthood. Thereafter it stopped increasing for several days, and then reached a maximum at 35 days of age; after that it slightly decreased and became a plateau. The enzyme activity at adulthood increased to about 10-fold in total activity and about 3-fold in specific activity compared to those at birth. The total activity of DDC steadily rose up to 50 days of age to reach mature levels, paralleling developmental change in protein contents. The activity of DDC at mature levels was about 10-fold higher in total activity, and about 3-fold higher in specific activity compared to those at birth.

TH activity, a specific enzyme for adrenergic neurons, rapidly increased to almost adult levels by 15 days of age in rats, 12 in mice, and 20 in hamsters, respectively. DBH activity, another marker enzyme for adrenergic neurons, reached adult levels a little behind the development of TH activity; 40 days of age in rats, 21 in mice, and 30 in hamsters. Mature levels of DDC activity, an enzyme distributed to most organs, were not reached before 50 days of age in either rats or hamsters.

The rates of increase in activities of TH and DBH were summarized as follows comparing rats, mice and hamsters at birth and adulthood. The total activity of TH, whose differences among the three animal species were the biggest among the three enzymes, increased to about 3-fold in rats, about 7-fold in mice, and about 15-fold in hamsters. The specific activity of TH rose to about 1.4-fold in rats, and about 5-fold in hamsters. The total activity of DBH increased to about 8-fold in rats and about 10-fold in hamsters, and the specific activity of the enzyme rose to about 2.5-fold in rats and about 3-fold in hamsters.

Effects of preganglionic cholinergic nerves on the development of postganglionic adrenergic neurons

Prolonged increase in the activity of the peripheral sympathetic nervous system, for instance by reserpine treatment, stimulates synthesis of TH in the terminal adrenergic neuron (Mueller *et al.*, 1969a). The induction of TH is mediated by increased activity of the preganglionic cholinergic nerves (Thoenen *et al.*, 1969a). On the other hand, developmental increase in TH activity in the superior cervical ganglion of newborn mice occurs subsequent to the early phase of preganglionic synapse formation, indicating that the development of this enzyme may depend on the innervation of the post-synaptic cells (Black *et al.*, 1971). To investigate this possibility, superior cervical ganglia were decentralized by surgical section of the preganglionic nerves. Decentralization in 4-day-old mice reduced the total activity of TH to 50%, and the total activity of MAO to about 60% 11 days after surgery compared to nonresected controls (Black *et al.*, 1972). The same surgical resection in 3-day-old rats caused decreases in protein content to 68%, in the specific activity of both TH and DBH to about 80% two weeks after the operation (Thoenen *et al.*, 1972). Thus, surgical decentralization prevented the developmental increases in protein and MAO activity, which are distributed to most organs. This indicates that presynaptic nerves regulate general growth of superior cervical ganglia, while TH and DBH activities, which specifically distribute to adrenergic neurons, were more strongly affected by the decentralization than protein and MAO. This suggests that trans-synaptic regulation may influence the maturation of adrenergic neurons (Black *et al.*, 1971). It is suggested that there is a critical period during the first few postnatal weeks when trans-synaptic influences from afferents are necessary for the induction of TH in the sympathetic neurons (Smolen *et al.*, 1985). Consequently, acetylcholine and ChAT are considered to take part in the regulation. Chlorisodamine and pempidine, which are compounds preventing depolarization of postsynaptic neurons in sympathetic ganglia by competing with acetylcholine for receptor sites, were given to 2-day-old mice, resulting in reduced activity of TH by 50% at 14 days of age (Thoenen, 1972; Black, 1973). Treatment of rats with reserpine (Mueller *et al.*, 1969a), which leads to a reflex increase in the activity of the preganglionic nerves and augmented synthesis of TH, also stimulated synthesis of ChAT in the soluble fraction (Oesch and Thoenen, 1973), but not in the whole homogenate in superior cervical ganglia of rats (Hendry, 1976). However, treatment of neonate mice with carbacol and physostigmine, an agonist of cholinergic agents and an inhibitor of choline esterase, did not influence the normal development of TH activity in the superior cervical ganglion (Black *et al.*, 1972). Although all of these findings are not explained by the same development of the postganglionic neuron is suggested to be regulated by the presynaptic cholinergic nerves. The presynaptic regulation was well reviewed (Thoenen, 1972; Black, 1977), and is discussed in another section.

Decentralization of presynaptic nerves of the superior cervical ganglion also prevented the normal maturation of adrenergic nerve terminals, and the development of innervation of the iris and pineal gland, target organs of the ganglion (Black, 1977).

Effects of descending central pathways on the development of adrenergic neurons

The 6th lumbar ganglion was examined. The presynaptic cholinergic neurons innervating this ganglia are located in the intermediolateral column of the lower thoracic and upper lumbar spinal cord. Although midthoracic spinal transection interrupts descending central pathway to the intermediolateral column cells, this transection does not directly injure the preganglionic neurons which lie caudal to the lesion. Transection of the spinal cord at the 5th thoracic segment prevented the normal developmental increase in TH activity of the 6th ganglion in neonate rats. However, this transection did not change the development of TH activity in the superior cervical ganglion in neonate rats, which derives its innervation from spinal segments rostral to the surgical lesion (Black *et al.*, 1976). Nor did such surgery affect the 6th lumbar ganglion in 30-day-old rats (Hamill *et al.*, 1983). Treatment of rats with reserpine induces TH biosynthesis (Mueller *et al.*, 1969a) through reflex sympathetic activation. However, the onset and ontogenic pattern of TH induction by reserpine are different among all three areas of rats tested, the superior cervical ganglion, adrenal gland and nucleus locus coeruleus (Black and Reis, 1975). Midthoracic spinal cord transection in newborn rats induced TH induction in the 6th lumbar ganglion by treatment with reserpine (Hamill and Cochard, 1984). However, in adult rats receiving lesions, the increase in TH activity did not occur by treatment with reserpine after motor and autonomic spinal reflexes returned in paraplegic animals (Hamill *et al.*, 1983). These observations indicate that during the first month of life central pathways exert critical facilitatory influences on sympathetic ganglion maturation (Hamill *et al.*, 1983) and that trans-synaptic regulation of adrenergic maturation in the periphery is governed by suprasegmental mechanisms in the central nervous system (Black *et al.*, 1976) rather than spinal mechanisms (Hamill *et al.*, 1983).

Effects of postganglionic adrenergic nerves on the development of adrenergic neurons

Effects of axotomy of the postganglionic nerves of the adrenergic neurons in the superior ganglion of rats were examined (Hendry, 1975). When the axotomy was done after 21 days of age, the total protein content and TH and DDC activities in the ganglion hardly differed from those of controls. However, if the procedure was performed prior to 12 days of age, a marked atrophy of the ganglion and a reduction in the total content of both enzymes occurred. When the axotomy was done between 12 and 21 days of age, intermediate values for all three enzymes. Thus, a critical period during the development of the superior cervical ganglion is suggested during which the adrenergic neurons undergo a maturation governed by their contact with the peripheral target cell via axons (Hendry, 1975).

Selective destruction of adrenergic neurons in neonatal mice with either 6-hydroxydopamine or antiserum NGF prevents the normal maturation of TH and ChAT activity in presynaptic terminals of the superior cervical ganglion. This suggests that the postsynaptic adrenergic neuron contributes to the development of presynaptic cholinergic fibers in the ganglion (Black, 1977).

The trans-synaptic induction of TH by reserpine was not affected by postganglionic axotomy (Hendry, 1976).

Effects of target organs on the development of adrenergic neurons

The salivary gland and iris innervated from the superior cervical ganglion were employed as a model system (Dibner and Black, 1976). Unilateral sialectomy and iridectomy in 3-day-old rats prevented the normal development of ganglion TH and DDC activities, resulting in TH levels of 61% at 6 weeks of age and 64% in DDC levels at 16 days postoperatively compared to controls at 6 weeks of age. In contrast, total ganglion protein initially developed normally until 40 days after surgery. However, 6 weeks after surgery, protein content was significantly lower in ganglia deprived of the normal field of innervation. The same surgical excision in adult rats did not significantly alter TH activity or protein in rats followed up to one month after surgery (Dibner and Black, 1976). These results coincide with findings obtained from axotomy of postganglionic nerves of the superior cervical ganglion. The end organ may have a role providing trophic factors at a critical period during the development of the ganglion (Hendry, 1975).

The submaxilla gland contains large amounts of NGF, especially in the mature male mouse, although the biological purpose of this concentration is obscure. In addition, since the induction of TH and DBH by treatment with NGF was recognized in the superior cervical ganglion of rats (Thoenen *et al.*, 1971) and mice (Hendry and Iversen, 1971), effects of NGF on the development of sympathetic neurons were studied intensively. Administration of NGF to newborn and adult rats produces selective increases in TH and DBH both in superior cervical ganglion and adrenal medulla. However, the rate of increase in the activities are higher in the superior cervical ganglion in newborn rats and in the adrenal medulla in adult rats (Otten *et al.*, 1977). Relationships between TH and NGF are discussed in another section.

Effects of other factors on the development of adrenergic neurons

Adrenal medulla. Adrenoectomy in rats at 12 days of age prevented further developmental increase in TH activity in the superior cervical ganglion. Adrenoectomy at 7 days of age did not affect normal development of enzyme activity when examined at 12 days of age. Daily injections of epinephrine reversed the effect of adrenoectomy, suggesting that further developmental increase in TH activity after about 2 weeks of age is an epinephrine-dependent process (Markey and Sze, 1981).

Male sex organ. Both neonatal (10–11 days of age) (Hamill and Guernsey, 1983) and adult castration (Hamill *et al.*, 1984) in rats prevented the normal development of TH activity in the hypogastric ganglion, but there was no change in the enzyme activity in the superior cervical ganglion. Enzyme activity was restored following replacement therapy with testosterone, suggesting that hormonal factors modulate adrenergic development in peripheral sympathetic ganglia. These effects appear restricted to ganglia whose targets include hormonally dependent sex organs. Neonatal castration at 10–11 days of age also prevented normal development of ChAT activity. Testosterone restored the activity of the enzyme, indicating that testosterone regulates the postorganizational maturation of post-synaptic adrenergic and presynaptic ChAT activity (Melvin and Hamill, 1986).

Adrenal cortex. Treatment of newborn rats with hydrocortisone injected daily for 7 days after birth caused a great and temporary increase in the number of the small, intensely fluorescent (SIF) cells in the superior cervical ganglion. These SIF cells disappeared 7–21 days after the treatment. It was suggested that these were different from normal SIF cells of the same ganglion, most of which appear at a later stage of postnatal development. These normal SIF cells contain TH, but not DBH or PNMT, and they remain in the ganglion permanently (Eranko *et al.*, 1982).

Appearance of tyrosine hydroxylase immunoreactivity in developing sympathetic tissues of rat fetus.

Initial development of the adrenergic phenotypes was studied in rat embryos by immunocytochemical detection of the catecholamine-synthesizing enzymes. TH and DBH, not detected in the neural crest (Teitelman *et al.*, 1979), were initially detected in trigeminal ganglion anlages as early as 10.5 days of embryonic development (Ionakait *et al.*, 1984). At 11 or 11.5 days, the two enzymes appear almost simultaneously in neuroblasts of the sympathetic ganglion primordia (Cochard *et al.*, 1979; Teitelman *et al.*, 1979), in neuroblasts of primordia of the sensory ganglion serving the glossopharyngeal and vagal cranial nerves (Ionakait *et al.*, 1979), and in neuroblasts of the gut wall (Cochard *et al.*, 1979; Teitelman *et al.*, 1979). In abdominal and lumbar ganglia, TH and DBH immunoreactivity appeared 1–2 days after the appearance in neuroblasts of the thoracic sympathetic ganglia, exhibiting a rostral-caudal gradient of differentiation (Teitelman *et al.*, 1979). The cells stained for TH and DBH in the gut wall disappeared by 13.5–14 days of development (Cochard *et al.*, 1979; Teitelman *et al.*, 1979). At 15 days, cells stained for TH and DBH appeared in the adrenal anlage. However, PNMT, a marker enzyme of adrenal medullary cells, was not detected in the adrenal until 17 or 18 days of development (Teitelman *et al.*, 1979; Verhofstad *et al.*, 1979). TH immunoreactivity in the glossopharyngeal petrosal ganglion was investigated as a model for sensory neurons innervating a single target, the carotid body (Katz and Erb, 1990). Two temporally distinct waves of TH expression were detected during the development of the embryo. TH immunoreactivity was initially detectable at 11.5 days of development as well; the number of cells stained for TH increased markedly by 12.5 days and then fell off sharply to near 0 by 15.5 days of development. A second, sustained phase of TH expression began on 16.5 days, and by birth, the number of TH immunoreactive cells increased to adult levels, indicating that TH expression in the glossopharyngeal petrosal ganglion begins around 16.5 days of embryonic development. The possibility that neuron-target interactions regulate biochemical differentiation of these catecholamine sensory neurons was suggested by the fact that TH expression begins after the start of peripheral target organ innervation (Katz and Erb, 1990).

DEVELOPMENTAL CHANGES IN THE CATECHOLAMINE-SYNTHESIZING ENZYMES IN THE SALIVARY GLAND OF YOUNG GROWING RATS

The salivary gland is innervated by the superior cervical ganglion as described above. Since TH and DBH are the enzymes highly localized to adrenergic neurons, these enzymes contained in the salivary gland originate in the nerve terminals of the superior cervical ganglion. Comparing the superior cervical ganglion and an innervating neuron, the developmental changes in the activities of TH and DBH in the salivary gland of rats are discussed.

The total activity of TH reaches mature levels by about 2 weeks of age in the superior cervical ganglion (Thoenen *et al.*, 1972), whereas in the submaxilla gland, it increased at about a constant rate until about 8 weeks of age (Kuzuya *et al.*, 1980). These results may not indicate a decrease in synthesis or an acceleration of degradation of TH molecules in the ganglion, but may indicate increases in the number of nerve terminals or synapses and in the axonal transport during development. The activity of TH per unit weight of glands reached the maximum as early as 2 weeks of age, indicating maturation of the nerve terminals in the early days of life.

Developmental change in the total activity of DBH also differs between salivary gland and superior cervical ganglion. The enzyme activity in the ganglion steadily increases until 40 days of age, and thereafter becomes level (Thoenen *et al.*, 1972), while in the submaxilla gland it continued to increase up to 70 days of age, the last age tested (Kuzuya *et al.*, 1980). The developmental changes in DBH activity were similar to those in TH activity with respect to the fact that DBH activity continues to increase in the salivary gland after ceasing to increase in the ganglion. However, the time course of the development of DBH activity in the salivary gland differed from that in the ganglion; the enzyme activity in the latter gradually increased until 2 weeks of age, rapidly from 2 to 3 weeks, then slightly to 4 weeks, and thereafter rose steadily to 70 days of age. DBH activity per unit weight in the salivary gland reached adult levels by 3 weeks of age, 1 week less than the development of TH. This delay of the developmental increase in DBH activity might be associated with nerve activities in the nerve terminals. Catecholamine content, norepinephrine in major, per unit weight in the salivary gland reached adult levels by 2 weeks of age as did TH activity (Iversen *et al.*, 1967; Kuzuya *et al.*, 1980). The ability of the salivary gland to take up tritiated-norepinephrine is already fully developed at birth. The uptake increases to 1.5-fold until 12 days of age, then, by 3 weeks of age, decreases to the level at birth (Iversen *et al.*, 1967). We speculate from these results that exocytosis of the DBH molecule (Axelrod, 1972) from the nerve terminals may cause the delay in accumulation of DBH in the salivary gland.

The activity of DDC is thought not to be assayed exactly in the nerve terminals distributing to the salivary gland, since parenchymal cells of the gland also contain a considerable amount of DDC. Moreover, proteolytic activity in the homogenate of the salivary glands, especially potent for DDC degradation, is also very high. Therefore, DDC activity in the gland cannot be assayed precisely. However, the proteolytic activity does not affect the other enzymes examined, TH, DBH and MAO (Kuzuya *et al.*, 1980).

DEVELOPMENTAL CHANGES IN TYROSINE HYDROXYLASE IN THE ADRENAL GLAND

Male Wistar rats were examined for catecholamines and their synthesizing enzymes from neonate to 600 days of age (Kvetňanský *et al.*, 1978). The total activity of TH per gland gradually increased with age until 360 days, then decreased slightly. The specific activity per unit weight of protein of the enzyme changed almost in the same way the total enzyme activity did, except for a temporary increase in the activity at 17 days of age. The epinephrine levels increased progressively until 600 days of age, whereas norepinephrine levels showed a marked increase around 17 days of age, decreased at 30 days, remained at about the same levels up to 120 days, and thereafter gradually increased until 600 days of age. The developmental patterns of the catecholamines were similar to each other in contents per gland and per unit weight of protein. DBH was also assayed; the enzyme activity gradually increased in a manner similar to TH activity up to 200 days of age, not 360 days as shown in TH. Thereafter, it remained about the same.

Remarkable increases in TH activity around 2 or 3 weeks after birth and somewhat delayed increases in DBH activity appear a common phenomenon in the superior cervical ganglion, salivary glands, and adrenal of rodents (Kvetňanský *et al.*, 1970; Black *et al.*, 1971; Thoenen *et al.*, 1972; Jonas *et al.*, 1979; Kuzuya *et al.*, 1980).

Factors influencing the development of tyrosine hydroxylase in the adrenal gland

Splanchnic nerve activity. Adrenal TH activity increases following experimental conditions such as insulin-induced hypoglycemia (Patrick and Kirshner, 1971), treatment with reserpine (Mueller *et al.*, 1969a), 6-hydroxydopamine (Thoenen *et al.*, 1969b), phenoxybenzamine (Mueller *et al.*, 1969b) and immobilization (Kvetňanský *et al.*, 1970) or cold stress (Thoenen, 1971). These produce an increase in splanchnic nerve activity. Effects of denervation on the developmental changes of adrenal TH of male Sprague-Dawley rats were investigated (Patrick *et al.*, 1972). In younger animals, during growth from 4 to 12 weeks of age, TH activity rose 2- to 3-fold in intact adrenal glands. Unilateral denervation of the gland at 23 days of age markedly prevented the increase in the enzyme activity to approximately one half that seen in innervated glands. During growth from 8 to 15 weeks of age, TH activity increased by 51% in intact adrenal glands, whereas unilateral denervation of the gland at 8 weeks of age completely blocked the increase seen in the innervated gland. DBH activity was also substantially in a manner similar to TH activity (Patrick and Kirshner, 1972).

Nerve growth factor. The induction of TH by nerve growth factor between the adrenal medulla and sympathetic ganglia of adult and new born rats was compared (Otten *et al.*, 1977). Although both adrenal medulla and sympathetic ganglion originate from the neural crest, adrenal chromaffin cells from rodents continue to exhibit significant mitotic activity after birth, but sympathetic neurons do not. Treatment with NGF increased TH activity both in adrenal medulla and superior cervical ganglion both of adult and new born Sprague-Dawley rats. The pattern of the enzyme change was similar in both organs. However, both in adult and new born rats the response of the adrenal medulla was less than that of the superior cervical ganglion. A single injection of NGF increased the enzyme activity in the adrenal medulla by 32% and 49% in new born and adult rats, respectively, whereas in the superior cervical ganglion by 116% and 89%, respectively. When NGF was repeatedly injected in new born rats on 10 consecutive days, a markedly greater difference between the adrenal medulla and superior cervical ganglion was observed, i.e. the enzyme activity increased by 69% in the adrenal medulla and by 700% in the superior cervical ganglion. These effects of NGF were not affected by decentralization of the superior cervical ganglion or denervation of the adrenal gland (Otten *et al.*, 1977).

Thyroid. TH activity in the adrenal gland of neonatal hypothyroid and hyperthyroid rats (Gripois *et al.*, 1980; Lau *et al.*, 1988). Thyroid function has been shown to be associated with catecholamine metabolism in the adrenal gland (Roy *et al.*, 1977; Tu *et al.*, 1975; Zenker *et al.*, 1976). There is evidence that thyroid hormones may serve as primary regulators of the development of the peripheral autonomic nervous system (Slotkin, 1985). Neonatal Sherman male hypothyroid rats were studied from birth to 20 days of age. The activity of TH was enhanced in hypothyroid rats. This enhancement was clearer when the activity comparison was based on unit weight of the adrenals because of impairment in adrenal growth in the hypothyroid rats (Gripois *et al.*, 1980). Neonatal hyperthyroid rats were also studied; Sprague-Dawley rats of both sexes were given triiodothyronin daily for 9 days beginning 1 day after birth. The neonatal hyperthyroidism accelerated synaptic development in the sympatho-adreno axis, whereas the TH activity per pair of adrenal gland decreased by about 70% on 9 days of age, indicating suppression of maturation of chromaffin cell growth (Lau *et al.*, 1988).

Hypoxia. Developmental profiles of TH in an hypoxic environment at high altitude were investigated on perinatal (Garvey *et al.*, 1980) and young growing rats (Vaccari *et al.*, 1977). Exposure to an hypoxic environment may be regarded as a form of stress. Adult Long-Evans rats were acclimatized for 1 month before breeding at high altitude. The TH activity per unit weight of adrenal gland was significantly higher in rats born and raised at high altitude during experimental periods from birth to adulthood (60 days of age), meaning long-lasting increase. The increase is thought to be specific for the enzyme, since MAO and catechol-O-methyltransferase activities exhibited no differences as compared to controls. The increase in TH is known to be induced by hypoxia in adult rats. However, the increase is usually temporary (Lau and Timiras, 1972). Plasma levels of corticosterone measured simultaneously from birth to 60 days of age were also higher in the rats born and raised at high altitude. Thus, the increase in tyrosine hydroxylase activity could be interpreted as a compensatory response to prenatal alterations of an adaptive response to postnatal hypoxia as a form of stress. The activity of adrenal TH of perinatal (19 days of gestation to one day after birth) Long-Evans rats was also investigated. In the controls, the enzyme activity temporarily decreased a little on day 20, markedly increased on day 21 of gestation and again decreased one day after birth. The activity in the rats exposed to an hypoxic environment progressively increased during the same periods as controls; a little lower on day 19, about the same on 20 and 21, and a little higher one day after birth in the hypoxia group compared to controls. Profiles of the developmental changes of the enzyme activity were similar to each other between the activity per gland and per unit weight of protein.

Nicotinic response. The implication of nicotine response to the induction of adrenal TH in rats was studied at various ages from one to 50 days of age (Rosenthal and Stolkin, 1977). Adrenomedullary nicotinic receptors are present in neonatal rats before development of functional splanic innervation. However, the ability to induce TH was lower than that in adults, evidencing the trans-synaptic regulation of TH.

Appearance of catecholamine-synthesizing enzyme immunoreactivities in the adrenal gland of fetuses

In rats, TH and DBH immunoreactivity appeared in the cells in the adrenal anlage on 14 or 15 days of embryonic development. PNMT, a marker enzyme of adrenal medullary cells, became immunoreactive positive on 17 or 18 days (Teitelman *et al.*, 1979; Verhofstad *et al.*, 1979). PNMT could only be demonstrated inside the adrenal gland and the capacity to synthesize epinephrine seems to be acquired only after the medullary cells have reached the cortical anlage, suggesting that the induction of PNMT is initiated by glucocorticoids secreted by the fetal cortex (Verhofstad *et al.*, 1979). Another study showed that *in vivo* differentiation of rat chromaffin precursors commences between 16.3 and 17.3 days of embryonic development. While epinephrine and PNMT were present at 17.3 days of development, they were not detected at 16.3 days. Small amounts of corticosterone were present in adrenals and plasma at 16.3 days of embryonic development, and marked increase in organ and plasma glucocorticoid levels occurred until 17.3 days. This suggests that glucocorticoids trigger the differentiation of noradrenergic sympathoadrenal precursors to adrenergic chromaffin cells (Seidl and Unsicker, 1989).

Normal human fetuses were investigated from gestational ages of 6 to 34 weeks (Molenaar *et al.*, 1990). Morphologically, two major cell types could be distinguished, i.e. "large" cells, which were present from 9 weeks of gestation on, and "small", primitive appearing cells, present from 14 weeks of gestation on. The large cells were already immunoreactive as early as 9 weeks of gestation for TH, chromogranin A and synaptosin, similar to adult chromaffin cells. In contrast, the small cells expressed TH, but not chromogranin A or synaptosin, more closely resembling ganglion cells in the adult adrenal medulla.

REFERENCES

Axelrod, J. (1962). Purification and properties of phenylethanoamine N-methyltransferase. *J. Biol. Chem.* **237**, 1657–1660.

Axelrod, J. (1972). Dopamin-β-hydroxylase: Regulation of its synthesis and release from nerve terminals. *Pharmacol. Rev.* **24**, 233–243.

Black, I. B. (1973). Development of adrenergic neurons *in vivo*: inhibition by ganglionic blockade. *J. Neurochem.* **20**, 1265–1267.

Black, I. B. (1977). Regulation of the growth and development of sympathetic neurons *in vivo*. *Prog. Clin. Biol. Res.* **15**, 61–71.

Black, I. B., Bloom, E. M. and Hamill, R. W. (1976). Central regulation of sympathetic neuron development. *Proc. Natl. Acad. Sci. USA* **73**, 3575–3578.

Black, I. B., Hendry, I. A. and Iversen, L. L. (1971). Trans-synaptic regulation of growth and development of adrenergic neurones in a mouse sympathetic ganglion. *Brain Res.* **34**, 229–240.

Black, I. B., Hendry, I. A. and Iversen, L. L. (1972). Effects of surgical decentralization and nerve growth factor on the maturation of adrenergic neurons in a mouse sympathetic ganglion. *J. Neurochem.* **19**, 1367–1377.

Black, I. B., Joh, T. H. and Reis, D. J. (1975). Accumulation of tyrosine hydroxylase molecules during growth and development of the superior cervical ganglion. *Brain Res.* **75**, 133–144.

Black, I. B. and Reis, D. J. (1975). Ontogeny of the induction of tyrosine hydroxylase by reserpine in the superior cervical ganglion, nucleus locus coeruleus and adrenal gland. *Brain Res.* **84**, 269–278.

Christenson, J. G., Dairman, W. and Udenfriend, S. (1970). Preparation and properties of a homogeneous aromatic L-amino acid decarboxylase. *Archs. Biochem. Biophys.* **141**, 356–367.

Cochard, P., Goldstein, M. and Black, I. B. (1979). Initial Development of the Noradrenergic phenotype in autonomic neuroblasts of the rat embryo *in vivo. Dev. Biol.* **71**, 100–114.

Dibner, M. D. and Black, I. B. (1976). The effect of target organ removal on the development of sympathetic neurons. *Brain Res.* **103**, 93–102.

Eranko, O., Pickel, V. M., Harkonen, M., Eranko, L., Joh, T. H. and Reis, D. J. (1982). Effect of hydrocortisone on catecholamine and the enzymes synthesizing them in the developing sympathetic ganglion. *Histochem. J.* **14**, 461–478.

Gaetani, S., Mengferi, E., Spadoni, M. A., Rossi, A. and Toschi, G. (1975). Effects of litter size on protein, choline acetyltransferase (CAT), and dopamine-β-hydroxylase (DBH) of a mouse sympathetic ganglion. *Brain Res.* **86**, 75–84.

Garvey, D. J., Vaccari, A. and Timiras, P. S. (1980). Developmental profiles of catecholaminergic enzymes in adrenals of perinatal rats: Effect of an hypoxic environment. *Horm. Metab. Res.* **12**, 318–322.

Gripois, D., Klein, C. and Valens, M. (1980). Tyrosine hydroxylase and catecholamine content in the adrenals of young hypo- and hyperthyroid rats. *Biol. Neonate.* **37**, 165–171.

Hamil, R. W., Bloom, E. M. and Black, I. B. (1977). The effect of spinal cord transection on the development of cholinergic and adrenergic sympathetic neurons. *Brain Res.* **134**, 269–278.

Hamill, R. W. and Cochard, P. (1984). Reserpine induction of tyrosine hydroxylase in paraplegia. *Exp. Neurol.* **84**, 241–248.

Hamill, R. W., Cochard, P. and Black, I. B. (1983). Long-term effects of spinal transection on the development and function of sympathetic ganglia. *Brain Res.* **266**, 21–27.

Hamill, R. W., Earley, C. J. and Guernsey, L. A. (1984). Hormonal regulation of adult sympathetic neurons: The effects of castration on tyrosine hydroxylase activity. *Brain Res.* **299**, 331–337.

Hamill, R. W. and Guernsey, L. A. (1983). Hormonal Regulation of sympathetic neuron development, The effect of neonatal castration. *Brain Res.* **313**, 303–307.

Hendry, I. A. (1975). The effects of axotomy on the development of the rat superior cervical ganglion. *Brain Res.* **90**, 235–244.

Hendry, I. A. (1976). Effects on the trans-synaptic regulation of enzyme activity in adult rat superior cervical ganglia. *Brain Res.* **107**, 105–116.

Hendry, I. A. and Iversen, L. L. (1971). Effect of nerve growth factor and its antiserum on tyrosine hydroxylase activity in mouse superior cervical ganglion. *Brain Res.* **29**, 159–161.

Ionakait, G. M., Markey, K. A., Goldstein, M. and Black, I. B. (1984). Transient expression of selected catecholaminergic traits in cranial sensory and dorsal root ganglia of the embryonic rat. *Dev. Biol.* **101**, 51–60.

Iversen, L. L., Champlain, J. D., Glowinski, J. and Axelrod, J. (1967). Uptake, storage and metabolism of norepinephrine in tissues of the developing rat. *J. Pharmacol. Exp. Therap.* **157**, 509–516.

Jonas, P., Macia, R., Oldham, S. and Johnson, E. (1979). Time course of the development of neurotransmitter synthesizing enzymes in the superior cervical ganglion and adrenal gland of the hamster. *J. Neurochem.* **32**, 241–243.

Katz, D. M. and Erb, M. J. (1990). Developmental regulation of tyrosine hydroxylase in primary sensory neurons of the rat. *Dev. Biol.* **137**, 233–242.

Kuzuya, H., Ikeno, T., Nemoto, K. and Hashimoto, S. (1980). Catecholamine contents and activities of catecholamine synthesizing and inactivating enzymes in the salivary glands of young growing rats. *Arcs. Oral Biol.* **25**, 31–36.

Kvetňanský, R., Gewirtz, G. P., Weise, V. K. and Kopin, I. J. (1970). Effect of hypophysectomy on immobilization-induced elevation of tyrosine hydroxylase and phenylethanolamine-N-methyltransferase in the rat adrenal. *Endocrinol.* **87**, 1323–1329.

Kvetňanský, R., Jahnova, E., Torda, T., Strbak, V., Balaz, and Macho, L. (1978). Changes of adrenal catecholamines and their synthesizing enzymes during ontogenesis and aging in rats. *Mech. Ageing Dev.* **7**, 209–216.

Lau, C., Franklin, M., MaCarthy, Pylypiw, A. and Ross, L. L. (1980). Thyroid hormone control of preganglionic innervation of the adrenal medulla and chromaffin cell development in the rat. An ultrastructural, morphometric and biochemical evaluation. *Dev. Brain Res.* **44**, 109–117.

Lau, C. and Timiras, P. S. (1972). Adrenocortical function in hypothalamic deafferent rats maintained at high altitude. *Am. J. Physiol.* **222**, 1040–1042.

Levin, E. Y., Levenberg, B. and Kaufman, S. (1960). The enzymatic conversion of 3,4-dihydroxyphenylethylamine to norepinephrine. *J. Biol. Chem.* **235**, 2080–2086.

Markey, K. A. and Sze, P. Y. (1981). Influence of adrenal epinephrine on postnatal development of tyrosine hydroxylase activity in the superior cervical ganglion. *Dev. Neurosci.* **4**, 267–272.

Melvin, J. E. and Hamill, R. W. (1986). Gonadal hormone regulation of neurotransmitter synthesizing enzymes in the developing hypogastric ganglion. *Brain Res.* **383**, 38–46.

Molenaar, W. M., Lee, V. M. and Trojanowski, J. Q. (1990). Early fetal acquisition of the chromaffin and neuronal immunophenotype of human adrenal medullary cells. An immunohistological study using monoclonal antibodies to chromogranin A. synaptophysin, tyrosine hydroxylase, and neuronal cytoskeletal proteins. *Exp. Neurol.* **108**, 1–9.

Mueller, R. A., Thoenen, H. and Axelrod, J. (1969a). Increase in tyrosine hydroxylase activity after reserpine administration. *J. Pharmacol. Exp. Therap.* **169**, 74–79.

Mueller, R. A., Thoenen, H. and Axelrod, J. (1969b). Adrenal tyrosine hydroxylase: Compensatory increase inactivity after chemical sympathectomy. *Science* **163**, 468–469.

Mueller, R. A., Thoenen, H. and Axelrod, J. (1969c). Increase in tyrosine hydroxylase after reserpine administration. *J. Pharmacol. Exp. Therap.* **169**, 74–79.

Nagatsu, T., Levitt, M. and Udenfriend, S. (1964). The initial step in norepinephrine biosynthesis. *J. Biol. Chem.* **239**, 2910–2917.

Otten, U., Schwab, M., Gagnon, G. and Thoenen, H. (1977). Selective induction of tyrosine hydroxylase and dopamine-β-hydroxylase by nerve growth factor: Comparison between adrenal medulla and sympathetic ganglia of adult and newborn rats. *Brain Res.* **133**, 291–303.

Patrik, R. L. and Kirshner, N. (1972). Developmental changes in rat adrenal tyrosine hydroxylase, dopamine-β-hydroxylase and catecholamine levels: Effect of denervation. *Dev. Biol.* **29**, 204–213.

Rosenthal, R. N. and Slotkin, T. A. (1977). Development of nicotinic responses in the rat adrenal medulla and long-term effects of neonatal nicotine administration. *Br. J. Pharmacol.* **60**, 59–64.

Roy, M. L., Sellers, E. M., Flattery, K. V. and Sellers, E. A. (1977). Influence of cold exposure and thyroid hormones on regulation of adrenal catecholamines. *Can. J. Physiol. Pharmacol.* **55**, 804–812.

Seidl, K. and Unsicker, K. (1989). The determination of the adrenal medullary cell fate during embryogenesis. *Dev. Biol.* **136**, 481–490.

Smolen, A. J., Beaston-Wimmer, P., Wright, L. L., Lindley, T. and Cader, C. Neurotransmitter synthesis, storage, and turnover in neonatally deafferented sympathetic neurons. *Brain Res.* **355**, 211–218.

Teitelman, G., Baker, H., Joh, T. H. and Reis, D. L. (1979). Appearance of catecholamine-synthe-sizing enzymes during development of rat sympathetic nervous system: Possible role of tissue environment. *Proc. Natl. Acad. Sci. USA* **76**, 509–513.

Thoenen, H. (1971). Induction of tyrosine hydroxylase in peripheral and central adrenergic neurons by cold-exposure of rats. *Nature* **228**, 861–862.

Thoenen, H. (1972). Neuronally mediated enzyme induction in adrenergic neurons and adrenal chromaffin cells. In: *Biochem. Soc. Sympo. 36: Neurotransmitters and Metabolic Regulation.* William Clowes and Sons, London, pp. 3–15.

Thoenen, H. (1973). Increased activity of the peripheral sympathetic nervous system: Induction of choline acetyltransferase in the preganglionic cholinergic neurone. *Nature* **242**, 536–537.

Thoenen, H., Angeletti, P. U., Levi-Montalcini, R. and Kettler, R. (1971). Selective induction by nerve growth factor of tyrosine hydroxylase and dopamine-β-hydroxylase in the rat superior cervical ganglia. *Proc. Nat. Acad. Sci. USA* **68**, 1598–1602.

Thoenen, H., Kettler, R. and Saner, A. (1972). Time course of the development of enzymes involved in the synthesis of norepinephrine in the superior cervical ganglion of the rats birth to adult life. *Brain Res.* **40**, 459–468.

Thoenen, H., Mueller, R. A. and Axelrod, J. (1969a). Trans-synaptic induction of adrenal tyrosine hydroxylase. *J. Pharmacol. Exp. Therap.* **169**, 244–254.

Thoenen, H., Mueller, R. A. and Axelrod, J. (1969b). Increased tyrosine hydroxylase activity after drug-induced alteration of sympathetic transmission. *Nature* **221**, 1264.

Tu, T. and Nash, C. W. The influence of prolonged hyper- and hypothyroid states on the noradrenaline content of rat tissues and on the accumulation and efflux rates of tritiated noradrenaline *Can. J. Physiol. Pharmacol.* **54**, 74–80.

Vaccari, A., Cimino, J., Brotman, S. and Timiras, P. S. (1977). High altitude hypoxia and adrenal development in the rat: Enzymes for biogenic amines. In: *Environmental Endocrinology.* I. Assenmacher and D. S. Farner (Eds). Springer-Verlag, Berlin, pp. 283–289.

Verhofstad, A. A., Hokfelt, T., Goldstein, M., Steinbusch, H. W. and Joosten, H. W. (1979). Appearance of tyrosine hydroxylase, aromatic amino-acid decarboxylase, dopamine-β-hydroxylase and phenyl-ethanolamine N-methyltransferase during the ontogenesis of the adrenal medulla: an immunohistochemical study in the rat. *Cell Tissue Res.* **200**, 1–13.

Zenker, N., Goudonnet, H. and Truchot, R. (1976). Effect of thyroid status and cold stress on tyrosine hydroxylase activity in adrenal gland and brown adipose tissue. *Life Sci.* **18**, 183–188.

Tyrosine Hydroxylase, pp. 107–123
M. Naoi *et al.* (Eds)
© VSP 1993

ENZYME IMMUNOASSAY OF TYROSINE HYDROXYLASE

MAKIO MOGI and MINORU HARADA

Department of Oral Biochemistry, Matsumoto Dental College, Shiojiri, Nagano, 399-07 Japan

INTRODUCTION

In general, detection and characterization of an enzyme can be performed by determination of the activity, since enzyme has its substrate specificity. However, an enzyme *in vivo* is not always an active form such as a proenzyme. If the enzyme present as a complex form with an endogenous regulator (inhibitor or activator), the activity determined by catalytic assay is not reflective of the actual enzyme activity. Radio immuno-assay (RIA) is an excellent method for quantitative analysis of protein. However, application of RIA for tyrosine hydroxylase (TH) has not been publish and the only technique for quantitative assay for TH was immunoprecipitation as far as we know.

Therefore, we think an enzyme immunoassay (EIA) introduced by Engvall and Perlmann (1971) is very useful for TH. In addition, when EIA is combined with a catalytic assay, it then becomes possible to detect immunoreactive, but catalytically inactive or more active forms of an enzyme in tissues.

In this point of view we introduced the method for immunoassay of tyrosine hydroxylase (Mogi *et al.*, 1984) and applied the method for elucidation of reductions or activations of the enzyme activity in various physiological and pathological conditions (Mogi *et al.*, 1986, 1988).

DEVELOPMENT OF ENZYME IMMUNOASSAY METHOD FOR TYROSINE HYDROXYLASE

Purification of tyrosine hydroxylase from bovine adrenal medulla for the preparation of anti-TH antibody

Purification procedure of tyrosine hydroxylase (TH) used for the antigen was referred to the method of Oka *et al.* (1982). Bovine adrenal glands were freshly obtained at the slaughterhouse. All subsequent procedures were performed at 4 °C. TH activity (V_{max}) was measured by high-performance liquid chromatography with electrochemical detection (Nagatsu *et al.*, 1979). In a typical purification procedure from bovine adrenal glands, 25 g of the adrenal medulla were dissected and homogenized in 5 vol. 50 mM Tris-HCl buffer (pH 7.3) containing 0.25 M sucrose. The homogenate was centrifuged at 100 000 g for 60 min, and the supernatant was applied to a DEAE-Sephacel column (30 × 2.6 cm internal diameter), which had been equilibrated with 20 mM Tris-HCl buffer (pH 7.3)/8% sucrose/1 mM dithiothreitol (buffer A) containing 0.05 M NaCl.

The column was washed with the same buffer and elution was carried out with 800 ml of a linear gradient of 0.05–0.5 M NaCl. Tyrosine hydroxylase activity was separated into two fractions, the first a sharp peak (Fraction I) and the next a broad peak (Fraction II). The active fraction I was collected and brought to 40% saturation with solid ammonium sulfate. The precipitate was collected by centrifugation at 25 000 g for 15 min, dissolved in a minimum volume of 20 mM potassium phosphate buffer (pH 7.3)/8% sucrose/1 mM dithiothreitol (buffer B), and centrifuged at 25 000 g for 15 min. The resulting solution was passed through a Bio-Gel A-5 m column (97 × 2.6 cm internal diameter) equilibrated with buffer B, and eluted with buffer B. The active fraction was collected and put onto a Heparin-Sepharose CL-6B column (4.3 × 2.6 cm internal diameter) equilibrated with buffer B, as described by Yamauchi and Fujisawa (1979), washed with buffer B containing 0.1 M KCl and eluted with 200 ml of a linear gradient of 0.1–0.5 M KCl. The active fraction was collected and brought to 50% saturation with solid ammonium sulfate, and after centrifugation at 25 000 g for 15 min the resulting precipitate was dissolved in buffer A and the solution was passed through a Bio-Gel A-1.5 m column (98 × 1.6 cm internal diameter) equilibrated with buffer A. Purification steps are summarized in Table 1. The active fraction was collected, stored at −80 °C and used as enzyme for the preparation of antibody.

Table 1.
Purification of tyrosine hydroxylase from bovine adrenal medulla

Fraction	Total protein (mg)	Total activity (nmol/min)	Spec. Act. (nmol/min/mg protein)	Recovery (%)
100 000 g Supernatant	2314	4717	2.0	100
DEAE-Sephacel Fraction	356	3457	9.7	73
Bio-Gel A-5m	52	2087	40	44
Heparin-Sepharose CL-6B	5.5	1010	184	21
Bio-Gel A-1.5m	1.0	409	409	8.6

Sodium dodecyl sulfate (SDS) slab gel electrophoresis showed one major protein band from Fraction I, and molecular weight (Mr) of the subunit was estimated to be 60 000. This highly purified Fraction I was stable without any loss of the activity at −80 °C for 2 months.

Preparations of immunoglobulins from anti-TH antisera

Anti-bovine adrenal TH antisera were raised in rabbits by 2 subcutaneous injections of TH in phosphate buffered saline (PBS) emulsified with equal volumes of complete Freund's adjuvant (100 μg/rabbit on day 1 and 100 μg/rabbit on day 27). Blood was collected 28 days after the second injection. Preparation of immunological reagents was carried out according to Ishikawa and Kato (1978). Anti-TH IgG fraction was isolated from anti-TH antisera by Na_2SO_4 fractionation and then by chromatography on DEAE-cellulose. Half of the IgG fractions were digested with pepsin at pH 4.5 to obtain the $F(ab')_2$ fragments were used for preparing the solid phase of immobilized antibody, and the remaining was conjugated with β-D-galactosidase.

Labeling of the antibody with β-D-galactosidase

Anti-TH IgG was coupled with β-D-galactosidase by use of a bifunctional coupling reagent, N, N'-o-phenylenedimaleimide (Yoshitake *et al.*, 1979). In brief, anti-TH IgG (15 mg/2 ml of 0.1 M sodium acetate buffer, pH 5.0) was reduced with 10 mM 2-mercaptoethylamine at 37°C for 90 min, and resulting IgG was separated by a Sephadex G-25 column, and was treated with excess amount of N, N'-o-phenylenedimaleimide to introduce maleimide residues on anti-TH IgG. The maleimide-IgG (2 mg/ml) was reacted with β-D-galactosidase to produce the IgG-enzyme conjugate. The anti-TH IgG labeled with β-D-galactosidase was separated by a Sepharose 6B column, and stored at 4°C in 0.01 M sodium phosphate buffer (pH 7.0)/0.1 M NaCl/1 mM $MgCl_2$/0.1% BSA (buffer 1). Amounts of the labeled antibody were expressed as units of enzyme activity (one unit = 1 μmol product/min).

Immobilization of antibody $F(ab')_2$ fragments on polystyrene beads as solid-phase

The antibody $F(ab')_2$ fragments were immobilized noncovalently on polystyrene beads as solid-phase. Beads were immersed in the solution of pepsin-treated immunoglobulin G ($F(ab')_2$) (100 μg/ml in sodium phosphate buffer, pH 7.0, containing 0.1% NaN_3), kept at 4°C overnight under gentle stirring, washed with buffer 1, and stored in buffer 1 for at least 3 days before use for immunoassay. The solid-phase with immobilized antibody was stable at 4°C for at least 6 months. The $F(ab')_2$ solution could be used repeatedly.

Sandwich enzyme immunoassay procedure

The assay system was essentially similar to that described by Ishikawa *et al.* (1982). Beads of the solid-phase with immobilized antibody were incubated with various amounts of standards TH from bovine and human adrenals at

4 °C with vigorous shaking in final 150 μl of 0.01 M sodium phosphate buffer
(pH 7.0)/0.3 M NaCl/1 mM MgCl$_2$/0.5% gelatin/0.1% BSA/0.1% NaN$_3$ (buffer 2).
After 5 h the reaction medium was removed by aspiration, and the beads were
washed twice with 1 ml of chilled buffer 1 in each tube. The beads were incubated at 4 °C overnight under shaking with 1 unit of the anti-TH IgG conjugated
with β-D-galactosidase in 200 μl of buffer 1 and washed with buffer 1, and
bound enzyme protein was assayed fluorometrically with 4-methylumbelliferyl-
β-D-galactosidase as a substrate. Beads were incubated with 0.1 mM substrate
in a final 150 μl of buffer 1 at 30 °C for 20 min. The reaction was terminated by
adding 0.75 ml of 0.5 M glycine-NaOH buffer (pH 10.3), and the fluorescence
intensity of the 4-methyl-umbelliferon released was measured against a freshly
prepared standard solution at 450 nm with excitation wavelength at 360 nm. One
unit of the β-D-galactosidase activity is defined as that which hydrolyzes 1 μmol
of substrate/min.

Evaluation of enzyme immunoassay

Figure 1 shows a standard curve in a log-log scale measured by the EIA for
bovine adrenal TH using anti-bovine adrenal TH antibody (homologous assay).
TH protein concentration versus bound β-D-galactosidase activity was linear
between 3 and 300 ng of TH protein/tube. Limit of sensitivity, defined as

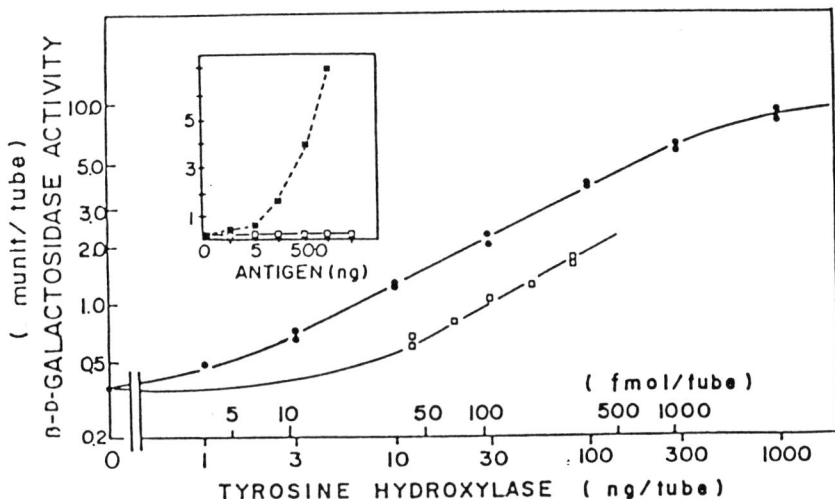

Figure 1.   Standard curves of purified TH by a sandwich EIA. Bovine adrenal TH (•) or human
adrenal TH (□) was incubated in duplicate with the solid-phase (polystyrene beads) with immobilized
anti-bovine adrenal TH F(ab')$_2$. TH molar concentrations are calculated based on Mr (280 000)
estimated by Bio-Gel A-1.5 m chromatography. Inner paragraph: ■, purified bovine adrenal TH;
□, catecholamine-relating enzymes except for TH.

the zero antigen level plus twice the standard deviation, was 2 ng/tube. The anti-TH antibody showed no cross-reactivity with dihydropteridine reductase, DBH and phenylethanolamine N-methyltransferase from bovine adrenal medulla, and phenylalanine hydroxylase from rat liver, indicating that this EIA is specific for TH in Fig. 1 (inner paragraph).

Purified TH from human adrenals (Kojima *et al.*, 1984) was determined by this EIA utilizing the cross-reactivity of anti-bovine adrenal TH antibody with human TH (heterologous assay). As shown in Fig. 1, the cross-reactivity of human TH was approximately 25% of the bovine enzyme. The standard curves were parallel, suggesting the presence of common antigenic sites in human and bovine adrenal TH. The limit of sensitivity for human TH was 10 ng/tube.

The precision of the EIA was tested with purified bovine or human adrenal TH and 100 000 g supernatant of bovine or human adrenals as samples, in ten simultaneous assays (within assay) or in five consecutive assays (between assay). All the coefficients of variations were less than 10% (Table 2).

Table 2.
Precision of the sandwich enzyme immunoassay of TH

	Sample	No. of determination	TH level (ng/ml, mean ± SD)	Coefficient of variations (%)
Within-assay	(1) Purified TH from bovine adrenals	10	183.5 ± 12.7	6.9
	(2) 100 000 g supernatant of bovine adrenals	10	2560 ± 130	5.1
	(3) Purified TH from human adrenals	10	1103 ± 107	9.7
	(4) 100 000 g supernatant of human adrenals	10	2154 ± 186	8.6
Between-assay	(1) Purified TH from bovine adrenals	5	186.5 ± 15.4	8.3
	(2) 100 000 g supernatant of bovine adrenals	5	2602 ± 187	7.2
	(3) Purified TH from human adrenals	5	1043 ± 104	10.0
	(4) 100 000 g supernatant of human adrenals	5	2113 ± 197	9.7

APPLICATION OF THE ENZYME IMMUNOASSAY FOR THE STUDIES OF IMMUNOCHEMICAL PROPERTIES OF TYROSINE HYDROXYLASE

Purification and immunochemical properties of tyrosine hydroxylase in human brain

As an extension of the purification of the human enzyme previously reported (Kojima *et al.*, 1984), the authors succeeded in homogeneously purifying TH from human brain (caudate + putamen) for the first time. In addition, the presence of multiple, inactive forms of TH in human brain was reported using EIA and by Western blot after two-dimensional electrophoresis.

MATERIALS AND METHODS

Human brain (caudate + putamen) was obtained at autopsies within 10 h after death. Ampholines (pH 3.5–10.0 and pH 4.0–6.5) were obtained from LKB (Sweden). A pI-marker (acetylated cytochrome c) was from Oriental Co. (Japan). All other reagents were of the highest purity commercially available.

All the purification steps were performed at 4°C. Human brain TH was purified by slight modification of a procedure previously described for bovine adrenal TH (Oka *et al.*, 1982) and human adrenal TH (Kojima *et al.*, 1984). Protein concentrations were determined by a protein-dye binding assay with BSA as a standard using a Bio-Rad protein assay Kit (Bradford, 1976).

Two-dimensional (2-D) electrophoresis was carried out according to the method of O'Farrell (1975). After isoelectric focusing (IEF), the gel for IEF was subjected to the second electrophoresis in 4–15% gradient gel containing 0.1% SDS and then stained either for proteins by the silver staining (Oakley *et al.*, 1980) or for protein by the Western blot technique (Towbin *et al.*, 1979). pH gradient across the gel was measured by comparing the migration of a standard pI-marker (acetylated cytochrome c).

As immunochemical detection of TH proteins on nitrocellulose sheets by Western blot, after 2-D electrophoresis, proteins in the gel were electrophoretically transferred to a nitrocellulose sheet, as described in the previous report (Mogi *et al.*, 1984).

RESULTS

Purification of active TH from human brain

Active TH in human brain (P1 enzyme with activity, as described below) was purified approximately 2 200-fold with a yield of 1.3%, as shown in Table 3.

Table 3.
Purification of TH from human brain (caudate + putamen)

Purification steps	Total protein (mg)	Total activity (nmol/min)	Specific activity (nmol/min/mg)	Yield (%)	Purification (-fold)
100 000 g supernatant	420	2.4	0.0058	100	1
DEAE-Sephacel eluate	90	1.4	0.015	56	3
Ammonium sulfate (0–40%)	19	1.3	0.066	52	11
Bio-Gel A-1.5 m eluate	3.3	0.7	0.21	28	36
Heparin-Sepharose CL-6B eluate	0.0025	0.032	12.8	1.3	2200

Starting material : Human brain (caudate nucleus + putamen), 13.6 g.

Assay condition: Tyrosine, 0.2 mM; incubation at 37 °C for 10 min.

A 2-D electrophoresis pattern of the final enzyme preparation showed a single spot by silver staining, indicating that the enzyme was approximately homogeneous (Mogi *et al.*, 1986).

Purified TH gave a subunit of $Mr = 60\ 000$ with an isoelectric point of 6.0, these properties of the subunit of human brain TH were similar to human and bovine adrenal enzyme (Kojima *et al.*, 1984; Oka *et al.*, 1982). The Mr of TH was estimated to be 280 000 by Bio-Gel A-1.5 m chromatography. Thus, the enzyme is considered to be composed of four identical subunits. The specific activity of final preparation was approximately 13 nmol DOPA formed/min/mg protein, which was only 4% of the specific activity of TH purified from human adrenals (Kojima *et al.*, 1984). This purified enzyme was unstable, and 95% of activity was lost at 4 °C for 30 min.

Immunoreactivity of purified TH by EIA

Purified TH from human brain was determined by EIA utilizing the cross-reactivity of anti-bovine adrenal TH antibody.

Figure 2 shows standard curves of bovine adrenal TH and the cross-reactivities of TH from human brain or adrenal gland in the EIA system. The standard curves were parallel, and the cross-reactivities of human brain TH were similar to those of purified human adrenal TH and was approximately 25% of the bovine adrenal enzyme. This indicates that common antigenic sites are present in human brain

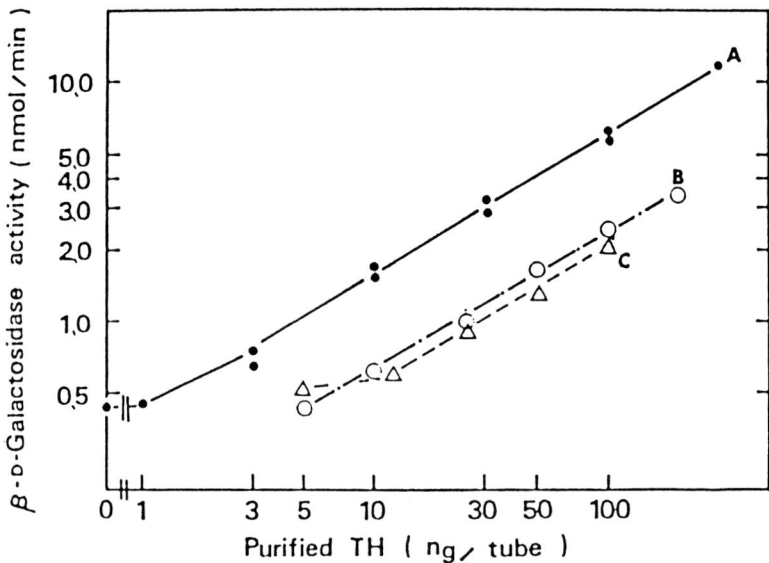

Figure 2. Assay curves obtained by a sandwich EIA with purified bovine adrenal TH and human adrenal and brain enzymes: (A) purified bovine adrenal TH; (B) purified human adrenal TH; (C) purified human brain TH.

and bovine adrenal TH, and that this EIA system can equally measure TH from different species and from different organs.

Assay of homospecific activity by EIA during purification of human brain TH

We attempted to monitor both protein content of TH with the EIA and the enzyme activity during purification from the human brain.

As shown in Fig. 3, a nearly symmetrical single peak of TH activity was found in a chromatogram on a DEAE-Sephacel column, whereas two peaks of TH protein were observed by EIA. The first small peak of TH protein (P1 enzyme, fraction No. 75–79) nearly coincided with TH activity, but the second large TH peak (P2 enzyme, fraction No. 80–93) did not coincide with TH activity. In the procedure of purification of TH, only the P1 enzyme with higher homospecific activity was collected and subjected to further purification (Table 4).

The immunoreactivity of P1 and P2 enzymes in DEAE-Sephacel chromatography with that of purified TH of the human brain was compared. When the three samples were analyzed by EIA, immunoreactivity of the enzyme in each sample was approximately parallel with each other. This indicates that this EIA system equally measure the two forms of TH with different homospecific activity

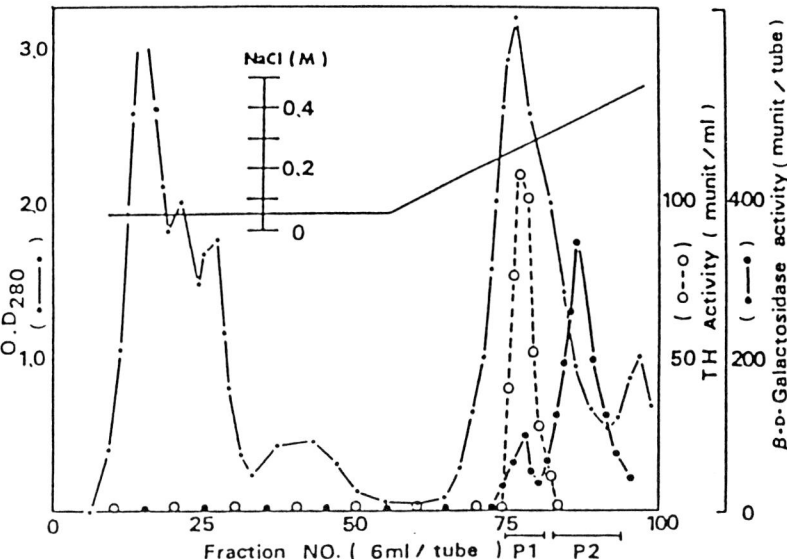

Figure 3. Chromatography of human brain TH on a DEAE-Sephacel column. o, TH activity; •, absorbance at 280 nm; •, human brain TH protein determined by EIA and expressed as β-D-galactosidase activity bound to the solid-phase.

and that the immunoreactivity of the enzyme in each peak is comparable (data not shown). When the samples were subjected to SDS-PAGE following Western blot analysis, all samples had same mobility on a nitrocellulose sheet, suggesting that P1 and P2 enzymes were immunologically similar and were composed of a same subunit with Mr = 60 000. When P1 and P2 enzymes were applied to a Bio-Gel A-1.5 m column to determine TH protein by EIA, each immunoreactive TH peak had same retention time with Mr = 280 000, which was similar to Mr of the purified TH (data not shown). Thus, P2 enzyme with low homospecific activity may be neither a degradation product of TH nor an aggregated form.

Table 4 summarizes the changes observed in specific activity and homospecific activity at DEAE-Sephacel and heparin-Sepharose CL-6B steps in the isolation of TH from human brain. While the specific activity of P1 enzyme in DEAE-Sephacel chromatography dramatically increased at each purification step, the homospecific activity was approximately constant. Although P2 enzyme was not purified further, homospecific activity of P2 enzyme was not detectable. Thus, assay of homospecific activity demonstrated the presence of two forms of TH with high and low homospecific activity in the 100 000 g supernatant of human brain. When we compared the total protein content of TH measured by EIA, the content of P2 enzyme (inactive form of TH) is approximately 8 times higher than that of P1 enzyme (active form of TH) (Table 4). It is not clear whether this inactive form in human brain exists *in vivo* or is produced in postmortem period.

Table 4.
Specific activity and homospecific activity of TH from human brain

Purification procedure or fractions	Specific activity (nmol/min/mg protein)	Homospecific activity (nmol/min/mg TH)	Total TH content by EIA (mg)
DEAE-Sephacel eluate			
(1) P1 enzyme (fraction No. 75–79)	0.015	15.0	0.11
(2) P2 enzyme (fraction No. 80–93)	n.d.	n.d.	0.92
Heparin-Sepharose CL-6B eluate (P1 enzyme)	12.8	12.8	0.0025

n.d., not detectable.

Immunochemical detection of TH on nitrocellulose sheet after 2-D separation by Western blot

When TH in the 100 000 g supernatant of human brain and the purified TH were subjected to 2-D separation followed by detection with Western blot, TH in the crude fraction of human brain was found to consist of multiple forms with different pI-values, but with same Mr (60 000) (Mogi *et al.*, 1986), whereas purified TH gave a single spot. The pI-values of the major spots and that of the minor spot ranged from 5.3 to 5.8 and 6.0, respectively. Because the pI of pure enzyme was 6.0, this protein which possesses pI at 6.0 is considered to be the active form of TH.

DISCUSSION

We have succeeded in homogeneously purifying TH from human brain (caudate + putamen) for the first time. The specific activity of the final preparation was very low as compared with human adrenal TH (Kojima *et al.*, 1984) while the Mr of the subunit, and immunoreactivity with anti-bovine adrenal TH antibody (Fig. 2) were similar to human or bovine adrenal enzyme (Kojima *et al.*, 1984; Oka *et al.*, 1982). Brain TH in mice was found to be fairly stable postmortem, and approx. 90% of the activity remained 10 h postmortem (Nagatsu *et al.*, 1977). Therefore, the low activity may not be due to postmortem changes.

Park and Goldstein (1975) reported that TH was purified from human pheochromocytoma to raise the antibody and that enzyme inhibition studies revealed extensive cross-reactivity between the antisera against human TH and the TH from bovine and rat adrenals and from rat striatum. The cross-reactivity among TH from human, bovine and rat tissues with the antibody against human enzyme indicates a structural similarity of the enzyme in these widely divergent species. The data in this study (Fig. 2) also confirmed the cross-reactivity among TH from human and bovine tissues with the antibody against bovine enzyme.

In the previous report, a sensitive and specific EIA for TH has developed (Mogi *et al.*, 1984). Sandwich EIA for TH provides an excellent means for assaying levels of TH in a manner that does not assume any *a priori* relation between catalytic efficiency and TH protein concentration. As shown in Fig. 3, the elution profile in DEAE-Sephacel chromatography revealed the presence of two forms of enzymes (P1 and P2 enzymes with different homospecific activity), as reported previously with human adrenal TH (Mogi *et al.*, 1984). These results indicate that both active and inactive forms of TH are present in 100 000 g supernatant of human brain and that the two forms of TH with high and low homospecific activity are separated by DEAE-Sephacel chromatography. Aldeson and Hedrick (1983) reported that TH in freshly prepared supernatant from rat striatum was separated in two distinct peaks by DEAE-cellulose chromatography. Their interpretation was that the two peaks from DEAE-cellulose represent the non-phosphorylated and phosphorylated forms of enzyme. We did not check TH activity after exposure to phosphorylation conditions. But the elution profile of TH activity which was fractionated by DEAE-cellulose chromatography are similar to the elution pattern of TH protein measured by EIA in DEAE-Sephacel chromatography (Fig. 3).

Tyrosine, phenylalanine and tryptophan hydroxylase are pterin-dependent mono-oxygenases and represent a family of enzymes with many common characteristics as well as distinctive properties. Smith *et al.* (1984) reported that two apparent molecular weight (Mr) forms with different pI-values represent the phosphorylated and dephosphorylated forms of phenylalanine hydroxylase, respectively. They examined the influence of phosphorylation and dephosphorylation on the 2-D electrophoretic patterns, and demonstrated that the higher apparent Mr forms are phosphorylated and that lower apparent Mr forms could be converted to the higher apparent Mr form by phosphorylation. We demonstrated that native TH from human brain was isolated in several forms of charge heterogeneity with same Mr by Western blot technique. When we compared the inactive forms of TH with the active form, each form was indistinguishable in immunological properties, except in pI-values.

Okuno and Fujisawa (1984) reported that the activity of TH was suppressed in a normal state in rat brain, and that there was an endogenous factor inhibiting the activity of TH in rat brain. In the present study, the specific activity of final preparation of brain enzyme was very low, which had only 5% of the specific activity of the purified TH from human adrenals, and this data could coincide with the report that the activity of TH was suppressed in a normal state in the brain. Inactive form of TH in human brain possessed a wide range of pI (5.3– 5.8) but the active form of TH had a pI-value of 6.0. If an endogenous factor inhibiting the activity of TH is present in human brain, the inactive form may be a complex of TH and the inhibiting factor resulting in shifts to acidic pI.

TH activity is stimulated *in vitro* by a variety of seemingly unrelated compounds. These compounds include: (1) anion and polyanions such as heparin, (2) phospholipids, (3) ATP, Mg, and cyclic AMP dependent protein kinase (Markey *et al.*, 1980; Vulliet *et al.*, 1980; Edelman *et al.*, 1981; Okuno and Fujisawa, 1982), and (4) limited proteolysis. The activation exhibited by these modulators is surprisingly similar, suggesting that a common mechanism might be involved. For example, exposure to phosphorylating conditions results in a form of TH that has a decreased K_m for pterin cofactor with no change in V_{max}. However Joh *et al.* (1978) proposed that phosphorylation of TH increased its activity by 2-fold, and was associated with an increase in V_{max} without any change in K_m for either substrate or cofactor. Their interpretation was that the pool of native TH is composed of a mixture of enzyme molecules in both active and probably inactive forms, that the active form is phosphorylated, and that phosphorylation produces an active form of the enzyme at the expense of the inactive one.

The properties of the inactive form of TH remains to be further characterized especially in the degrees of phosphorylation.

Homospecific activity (activity per enzyme protein) of tyrosine hydroxylase in Parkinsonian brain

Tyrosine hydroxylase (TH) activity in Parkinsonian brains decrease in nigrostriatal dopaminergic regions (Lloyd *et al.*, 1975; McGreer and McGreer, 1976; Nagatsu, *et al.*, 1977; Riederer *et al.*, 1978). The decrease in TH activity is thought to be due to decreased TH protein following the cell loss of the nigrostriatal dopaminergic neuron. On the other hand, the presence of inactive forms of TH was observed by an immunoassay of TH protein in control human brains (Mogi *et al.*, 1986). Although the physiological significance of the inactive TH protein remains to be further elucidated, increase in inactive forms of TH could also explain the decrease in TH activity in Parkinsonian brains.

There is also indirect evidence indicating some alternation of the property of TH in Parkinson's disease. The decreases in striatal homovanillic acid (HVA) concentration were less severe than the corresponding loss in dopamine (DA) concentrations, suggesting a compensatory increase in transmitter turnover in the surviving dopaminergic neurons (Hornykiewicz, 1966). Similar increase in the ration of HVA/DA was also reported in Parkinsonian monkeys 2 months after administration of 1-methyl-4-phenyl-1,2,3,6-tetrahydropyridine (MPTP) (Elsworth *et al.*, 1987).

In this section, we have measured both TH activity and TH protein using an EIA (Mogi *et al.*, 1984, 1986) and have found increase "homospecific activity" (Rush *et al.*, 1974; enzyme activity per enzyme protein) of TH in the nigro-striatal region of Parkinsonian brains.

MATERIALS AND METHODS

Control human brains were obtained from 9 patients without neurological diseases. Parkinsonian brains (7 cases) were obtained at autopsy in Juntendo University Hospital (Tokyo) and Ludwing Boltzman Institute for Clinical Neurobiology (Vienna). The controls and Parkinsonian patients were age matched (from 75 to 91 years). Postmortem times were from 4 h to 10 h. Caudate nucleus, putamen, and substantia nigra were dissected and stored frozen at $-80\,°C$. TH activity (V_{max}) and TH protein content were stable at $-80\,°C$.

Brain tissues were homogenized with 0.32 M sucrose. The homogenate was centrifuged at 100 000 g. The supernatant was used as the enzyme source.

TH protein content in the enzyme sample was measured by EIA using a monospecific polyclonal antibody against bovine adrenal medulla TH (Mogi *et al.*, 1984, 1986) and a standard TH protein purified from human adrenal medulla (Kojima *et al.*, 1984). Preparation of the immunoreagents and the immunoassay system, were essentially similar to those described previously (Mogi *et al.*, 1984).

RESULTS

Table 5 shows TH content (ng/mg of total protein), TH activity (V_{max}, pmol of DOPA formed/min/mg of total protein), and TH "homospecific activity" (nmol of DOPA formed/min/mg of enzyme protein) of the nigro-striatal region (caudate nucleus, putamen, and substantia nigra) from control and Parkinsonian patients. TH content and TH activity of the nigro-striatal regions from Parkinsonian patients were significantly lower than those from the control patients. TH content in the caudate nucleus alone was also significantly reduced in Parkinsonian brains. TH activity in the caudate nucleus alone also greatly reduced in the Parkinsonian patients, but the reduction was not significant due to great variations .

The "homospecific activity" (activity per enzyme protein) was significantly increased approximately 3–4 fold in the nigro-striatal region and in the caudate nucleus alone. Each one sample of the substantia nigra from control and Parkinsonian patients was available, both TH content and TH activity were reduced in parallel, but the homospecific activity of the Parkinsonian patient was similar to that of the control patient.

Table 5.
TH contents, TH activity, and "homospecific activity" of TH of the nigro-striatal regions from control and Parkinsonian patients*

Brains (n)	TH content (ng/mg total protein)	TH activity (pmol/min/mg total protein)	TH homospecific activity (nmol/min/mg of TH protein)
Controls (9)	661 ± 180	22.5 ± 9.1	36.4 ± 7.0
Caudate nucleus (6)	840 ± 246	26.5 ± 13.6	30.5 ± 8.1
Putamen (2)	321	18.1	59.3
Substantia nigra (1)	268	6.8	25.5
Parkinson's disease (7)	24.2 ± 7.4[b]	1.8 ± 0.6[a]	118.2 ± 43.4[a]
Caudate nucleus (4)	23.0 ± 12.0[a]	2.0 ± 0.7	120.7 ± 37.0[a]
Putamen(2)	17.1	1.6	159.5
Substantia nigra (1)	43.0	1.1	25.8

*Mean ±SEM. Difference from control group indicated: [a]$p < 0.05$; [b]$p < 0.01$ (Student's t-test).

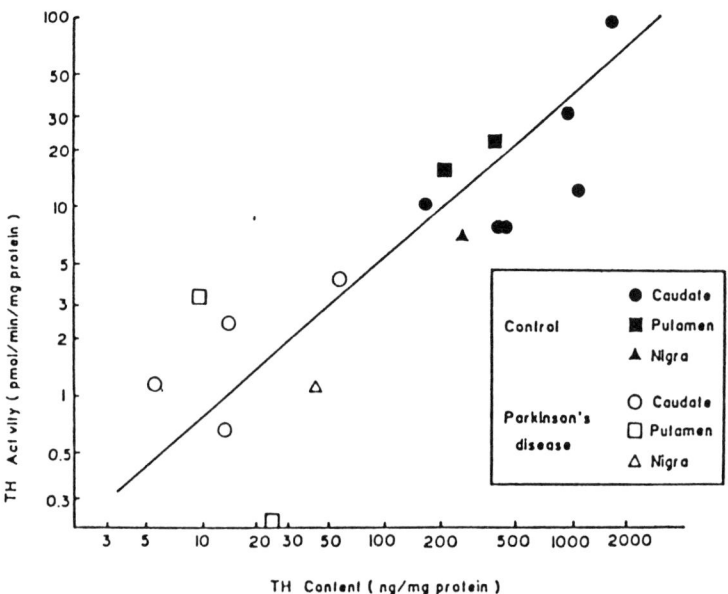

Figure 4. Comparison between TH content determined by EIA and TH activity in the nigro-striatal region of control and Parkinsonian patients. $Y = 0.05X - 5.95$, $r = 0.74$.

As shown in Fig. 4 a significant positive correlation ($r = 0.74$) was observed between the TH activity and the TH protein in the tissues from control and Parkinsonian patients, and significant reductions in TH protein content and TH activity were observed in Parkinsonian brains, as compared with control brains.

DISCUSSION

The reduction of TH protein content in the nigro-striatal region of Parkinsonian brains was proved immunochemically for the first time. This fact is consistent with the previously reported results on reduction in TH activity, and supports the concept that TH may disappear due to the cell loss of the nigro-striatal dopaminergic neurons.

An interesting finding in the present study is that the "homospecific activity" (unit of activity per unit of enzyme protein) was significantly increased in the Parkinsonian brain. This suggests a compensatory TH activation following the reduction of TH protein. We showed that TH activity and TH protein decreased in parallel at the 9th day following repeated daily MPTP administration for 8 days to mice, but could not show any significant change in the "homospecific activity" of TH in MPTP-treated mice (Mogi *et al.*, 1987).

On the other hand, two months after MPTP treatment in monkeys, the ratio of HVA/DA was reported to be elevated in the putamen and caudate nucleus, suggesting a compensatory increase in transmitter turnover in the surviving neurons (Elsworth *et al.*, 1987). It is conceivable, therefore, that the compensatory increase in the "homospecific activity" of TH in the striatum in Parkinson's disease as well as in MPTP-Parkinsonism may occur in some later stages during the progress of disease.

The molecular mechanism for the increase in the "homospecific activity" in TH molecular in Parkinson's disease remains to be further elucidated. One possibility is increased phosphorylation of TH molecules to activate TH. Another probable molecular change would be expression of different forms of mRNA for TH by alternative mRNA splicing. Production of multiple forms of human TH mRNA from a single gene by alternative splicing has been found recently (Grima *et al.*, 1987; Kaneda *et al.*, 1987). Different forms of TH would also change the regulation of the activity by phosphorylation and dephosphorylation.

REFERENCES

Aldeson P. S. R. and Hedrick B. (1983). Two forms of striatal tyrosine hydroxylase from DEAE - cellulose chromatography. *Brain Res.* **268**, 129–137.

Bradford M. M. (1976). Rapid and sensitive method for quantitation of microgram quantities of protein utilizing principle of protein-dye binding. *Anal. Biochem.* **72**, 248–254.

Edelman A. M., Raese J. D., Lazar M. A. and Barchas J. D. (1981). Tyrosine hydroxylase: Studies on the phosphorylation of a purified preparation of the brain enzyme by the cyclic AMP-dependent protein kinase. *J. Pharmacol. Exp. Therap.* **216**, 647–653.

Elsoworth J. D., Deutch A, Y., Redmond D. E., Jr, Sladek J. R. Jr and Roth R. H. (1987). Effects of 1-methyl-4-phenyl-1,2,3,6-tetrahydropyridine (MPTP) on catecholamines and metabolites in primate brain and CSF. *Brain Res.* **415**, 293–299..

Engvall, E. and Perlmann, P. (1971). Enzyme-linked immunosorbent assay (ELISA). Quantitative assay of immunoglobulin. G. *Immunochemistry* **8**, 871–877.

Grima, B., Lamouroux, A., Bori C., Julien J-F., Lavoy-Agid F. and Mallet J. (1987). A single human-gene encoding multiple tyrosine hydroxylases with different predicted functional characteristics. *Nature* **326**, 707–711.

Hornykiewicz O. (1966). Dopamine (3-hydroxytyramine) and brain function. *Physiol. Rev.* **18**, 925–964.

Ishikawa, E. and Kato, K. (1978). Sandwich enzyme immunoassay. *Scand. J. Immunol.* **8**, 43–55.

Ishikawa K., Narita O., Saito H. and Kato K. (1982). Determination of ferritin in urine and in serum of normal adults with a sensitive enzyme immunoassay. *Clin. Chim. Acta* **123**, 73–81.

Joh T. H., Park D. H. and Reis D. J. (1978). Direct phosphorylation of brain tyrosine hydroxylase by cyclic AMP-dependent protein kinase: Mechanism of enzyme activation. *Proc. Natl. Acad. Sci. USA* **75**, 4744–4748.

Kaneda N., Kobayashi K., Ichinose H., Kishi F., Nakazawa A., Kurosawa Y., Fujita K. and Nagatsu T. (1987). Isolation of a novel cDNA clone for human tyrosine hydroxylase alternative RNA splicing products four kinds of mRNA from a single gene. *Biochem. Biophys. Res. Commun.* **146**, 971–975.

Kojima K., Mogi M., Oka. K. and Nagatsu T. (1984). Purification and immunochemical characterization of human adrenal tyrosine hydroxylase. *Neurochem. Int.* **6**, 475–480.

Lloyd K. G., Davison L., Hornykiewicz O. (1975). The neurochemistry of Parkinson's disease: effect of L-DOPA therapy. *J. Pharm. Exp. Therap.* **195**, 453–464.

Markey K. A., Kondo S., Shenkman L. and Goldstein M. (1980). Purification and characterization of tyrosine hydroxylase from a clonal pheochromocytoma cell line. *Mol. Pharmacol.* **17**, 79–85.

McGeer P. L., McGeer E. G. (1976). Enzymes associated with the metabolism of catecholamines, acetylcholine and GABA in human controls and patients with Parkinson's disease and Huntington's chorea. *J. Neurochem.* **26**, 65–76.

Mogi M., Kojima K. and Nagatsu T. (1984). Detection of inactive or less active forms of tyrosine hydroxylase in human adrenals by a sandwich enzyme immunoassay. *Anal. Biochem.* **138**, 125–132.

Mogi M., Kojima K., Harada M. and Nagatsu T. (1986). Purification and immunochemical properties of tyrosine hydroxylase in human brain. *Neurochem. Int.* **8**, 423–428.

Mogi M., Harada M., Kojima K., Kiuchi K., Nagatsu I. and Nagatsu T. (1987). Effects of repeated systemic administration of 1-methyl-4-phenyl-1,2,3,6-tetrahydropyridine (MPTP) on striatal tyrosine hydroxylase activity *in vitro* and tyrosine hydroxylase content. *Neurosci. Lett.* **80**, 213–218.

Mogi M., Harada M., Kiuchi K., Kondo T., Narabayashi H., Rausch D., Riederer P., Jellinger K. and Nagatsu T. (1988). Homospecific activity (activity per enzyme protein) of tyrosine hydroxylase increases in Parkinsonian brain. *J. Neural Transm.* **72**, 77–81.

Nagatsu T., Kato T., Numata (Sudo) Y., Ikuta K., Sano M., Nagatsu I., Kondo Y., Inagaki S., Iizuka R., Hori A. and Narabayashi H. (1977). Phenylethanolamine N-methyltransferase and other enzymes of catecholamine metabolism in human brain. *Clin. Chim. Acta* **75**, 221–232.

Nagatsu T., Oka K. and Kato T. (1979). Highly sensitive assay for tyrosine hydroxylase activity by high performance liquid chromatography. *J. Chromatogr.* **163**, 247–252.

Oakley B. R., Kirsch D. R. and Morris N. R. (1980). A simplified ultra-sensitive silver stain for detecting proteins on polyacrylamide gels. *Anal. Biochem.* **105**, 361–363.

O'Farrell, P. H. (1975). High resolution two-dimensional electrophoresis of proteins. *J. Biol. Chem.* **250**, 4007–4021.

Oka K., Ashiba G., Sugimoto T., Matsuura S. and Nagatsu T. (1982). Kinetic properties of tyrosine hydroxylase purified from bovine adrenal medulla and bovine caudate nucleus. *Biochim. Biophys. Acta* **706**, 188–196.

Okuno S. and Fujisawa H. (1982). Purification and some properties of tyrosine 3-monooxygenase from rat adrenal. *Eur. J. Biochem.* **122**, 49–55.

Okuno S. and Fujisawa H. (1984). The activity of tyrosine 3-monooxygenase is usually suppressed by an endogenous factor in brain. *Biochem. Biophys. Res. Commun.* **124**, 223-228.

Park D. H. and Goldstein M. (1975). Purification of tyrosine hydroxylase from pheochromocytoma tumors. *Life Sci.* **18**, 55–60.

Riederer P., Rausch W.-D., Birkmayer W., Jellinger K. and Seemann D. (1978). CNS modulation of adrenal tyrosine hydroxylase in Parkinson's disease and metabolic encephalopathies. *J. Neural Transm.* **14** (Suppl.), 121–131.

Rush R. A., Kindler, S. H. and Udenfriend, S. (1974). Homospecific activity, an immunologic index of enzyme homogeneity; changes during the purification of dopamine-beta-hydroxylase. *Biochem. Biophys. Res. Commun.* **61**, 38–44.

Smith S. C., Kemp B. E., Mcadem W. J., Mercer J. F. B. and Cotton R. G. H. (1984). Two apparent molecular weight forms of human and monkey phenylalanine hydroxylase are due to phosphorylation. *J. Biol. Chem.* **259**, 11284–11289.

Towbin H., Staehelin T. and Gordon J. (1979). Electrophoretic transfer of proteins from polyacrylamide gels to nitrocellulose sheets: Procedure and some applications. *Proc. Natl. Acad. Sci. USA* **76**, 4350–4354.

Vulliet P. R., Langman T. A. and Weiner, N. (1980). Tyrosine hydroxylase: A substrate of cyclic AMP-dependent protein kinase. *Proc. Natl. Acad. Sci. USA* **77**, 92–96.

Yamauchi T. and Fujisawa H. (1979). *In vitro* phosphorylation of bovine adrenal tyrosine hydroxylase by adenosine 3':5'-monophosphate dependent protein kinase. *J. Biol. Chem.* **254**, 503–507.

Yoshitake S., Hamaguchi Y. and Ishikawa E. (1979). Efficient conjugation of rabbit Fab' with beta-D-galactosidase from *E-coli*. *Scand. J. Immunol.* **10**, 81–86.

Tyrosine Hydroxylase, pp. 125–133
M. Naoi *et al.* (Eds)
© VSP 1993

MONITORING OF TYROSINE HYDROXYLASE ACTIVITY IN CONSCIOUS RAT STRIATUM WITH A MICRODIALYSIS

TAKESHI KATO, SHOJI HISAMATSU and MAN XU

Laboratory of Molecular Recognition, Graduate School of Integrated Science, Yokohama City University, Kanazawa-ku, Yokohama 236, Japan

Abstract—Using a microdialysis technique, it is possible to monitor *in vivo* tyrosine hydroxylase activity in rat brain regions. Although DOPAC/dopamine ratio in the brain dialysate might be correlated with *in vivo* tyrosine hydroxylase activity, a more accurate method for it have been developed. After the perfusion of 10 μM NSD1015, a DOPA decarboxylase inhibitor, with Ringer solution, L-DOPA level was accumulated about 0.5 μM. Perfusion of NSD1015 is very useful to test whether or not some drugs activate tyrosine hydroxylase activity *in vivo*.

INTRODUCTION

Development of the assay method for *in vitro* tyrosine hydroxylase activity has been reported by many investigators (Nagatsu *et al.*, 1964; Nissbrandt *et al.*, 1989). Until 1979, *in vitro* assay method for tyrosine hydroxylase activity had been measured by radioisotope method (Nagatsu *et al.*, 1964; Coyle and Axelrod, 1972). Owing to a development of a sensitive electrochemical detection (Kissinger *et al.*, 1973), we are able to determine a small amount of catechol compounds in tissues and in plasma. Using this system with HPLC, our groups had established highly sensitive assay methods for catechol synthesizing enzymes including tyrosine hydroxylase (Nagatsu *et al.*, 1979; Matsui *et al.*, 1981).

Since a very high-sensitive assay method to determine catechols (L-DOPA, dopamine, DOPAC, and HVA) has been developed, we applied this method to monitor *in vivo* tyrosine hydroxylase activity using a microdialysis technique.

MATERIALS AND METHODS

Materials. Male Wistar rats weighing 250–350 g were purchased and housed with ad libitum access to food and water. Oxotremorine, NSD1015 (*m*-hydroxybenzylhydrazine HCl), lobeline, and forskolin were purchased from Sigma.

Microdialysis procedure. Rats were stereotaxically implanted with U-shaped microdialysis cannulae while the animals were anesthetized with pentobarbital

(50 mg/kg, *i.p.*). Hollow fibers (IP-U cellulose, Asahi Medical supply) with an outer diameter of 300 μm and 90% cut off at 50 000–70 000 dalton were used to prepare a 3 mm length of dialysis tube as reported previously (Xu *et al.*, 1989, 1990). The skull was exposed and the cannula was implanted into the right side of the striatum A -0.2, L $+2.7$, and V -6.0 mm from the bregma according to the atlas of Paxinos and Watson (1986). On day 3 after the implantation, the perfusion was carried out. The cannula was continuously perfused with Ringer solution (147 mM Na^+, 2.3 mM Ca^{2+}, 4 mM K^+, and 155.6 mM Cl^-) at a speed of 4 μl/min. To obtain a stable level of dopamine and its metabolites in the perfusate, the cannula was perfused for about 2–3 h. The following perfusate (dialysate) was collected at 15 min intervals as one fraction (60 μl) and injected 50 μl to HPLC. At least 4 fractions were collected for a basal level before drug treatment. Drugs treated were dissolved in Ringer solution. L-DOPA, dopamine, 3,4-dihydroxy-phenylacetic acid (DOPAC), and homovanillic acid (HVA) in the dialysates were analyzed by HPLC with electrochemical detections. The concentrations of dopamine and its metabolites in the dialysates were calculated as percentages of mean values of four stable fractions obtained before drug treatment (basal level) or without any drug treatment (control). Data are shown as mean $\pm$ S.E.M.

RESULTS

Changes in the levels of dopamine and DOPAC after treatment with drugs. In the previous reports, turnover of dopaminergic neurons which is the ratio of DOPAC/dopamine in tissue had been believed to be an indicator for tyrosine hydroxylase activity. In the present study we examined changes in the level of dopamine in the dialysates after treatments with acetylcholine receptor agonists (oxotremorine and lobeline) and forskolin, an adenylate cyclase activator.

As shown in Fig. 1B, treatment with 1 mM oxotremorine, a muscarinic receptor agonist, immediately increased dopamine release about 100%, but the metabolite DOPAC did not increase simultaneously. During the perfusion of the drug, DOPAC level gradually increased and reached 150% at 1.5 h.

Treatment with a nicotinic receptor agonist lobeline showed a different pattern to that of oxotremorine. As shown in Fig. 1C, treatment with 0.1 mM lobeline rapidly increased dopamine release and then after gradually decreased it. And after then dopamine level was suppressed 30% for 2 h after the treatment. DOPAC level in the dialysate slightly increased for 1 h after the treatment and returned to the basal level.

Intracellular activation of the adenylate cyclase treated with 0.1 μM forskolin induced a high amount of dopamine release (210%), while the metabolite level slightly increased after the treatment (25%).

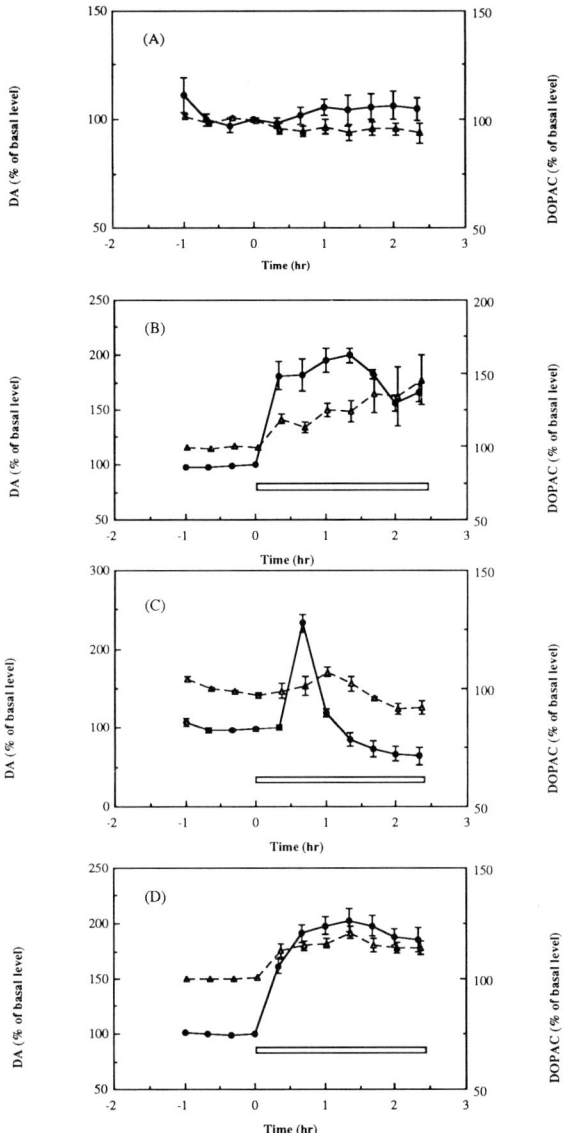

Figure 1. Effects of oxotremorine, lobeline, and forskolin on dopamine and DOPAC levels in the dialysate of rat striatum. (A) control, (B) oxotremorine (1 mM), (C) lobeline (0.1 mM), (D) forskolin (0.1 μM). Each animal was perfused for 1.5–2 h, and further perfused for 1 h and 20 min to obtain a stable basal dopamine and DOPAC levels. The percentages in figures were calculated from the mean values of the peak heights in those four fractions of the dialysate. ●—●: dopamine, △- -△: DOPAC.

Effect of NSD1015 on dopaminergic nerve terminals in rat striatum. Treatment with 10 μM NSD1015, an aromatic amino acid decarboxylase inhibitor, gradually increased L-DOPA level and reached about 500 nM (Fig. 2A), indicating that concentration of L-DOPA 2 h after the treatment with 10 μM NSD1015 is 100 times higher than that of dopamine level (6 nM). As shown in Fig. 2B, treatment with 10 μM NSD1015 did not affect dopamine level in the dialysate. However, the concentrations of the metabolites, DOPAC (Fig. 2C) and HVA (Fig. 2D), decreased about 30 and 20%, respectively, after the NSD1015 treatment.

Effects of oxotremorine and forskolin in the presence of NSD1015. In the presence of 10 μM NSD1015, administration of oxotremorine into the striatum through the perfusate appeared an unknown peak on the same retention time to that of L-DOPA. Therefore, we could not find any change in the level of L-DOPA after the treatment. Treatment with 0.3 mM oxotremorine increased about 50% dopamine release (Fig. 3A), but did not increase DOPAC (Fig. 3B) or HVA (data not shown) level under the preperfusion of 10 μM NSD1015.

In the presence of NSD1015, forskolin (0.1 μM) treatment slightly affected on the dopaminergic nerve terminals in rat striatum. As shown in Fig. 4A, L-DOPA level did not increase but slightly decreased (Fig. 4A). Treatment with the forskolin slightly increased dopamine level (Fig. 4A) and decreased DOPAC level (Fig. 4C). But these changes were not statistically significant.

DISCUSSION

In the present study we showed that forskolin simultaneously increases both dopamine and DOPAC levels in the dialysate (Fig. 1D), while oxotremorine increases dopamine release but does not increase DOPAC level simultaneously (Fig. 1C). These differences might be due to the different activating mechanism of tyrosine hydroxylase. It is well known that forskolin induces increase of dopamine release via the activation of adenylate cyclase. Forskolin also activates tyrosine hydroxylase by the stimulation of protein kinase A. The data measured in the current experiments confirm previous reports that forskolin increases dopamine release and metabolism through the activation of protein kinase A.

Westerink *et al.* (1990) have reported that preperfusion of 100 μM NSD1015 monitors the L-DOPA formation in the striatum as an index of tyrosine hydroxylase activity. Since treatment with a high concentration (100 μM) of NSD1015 completely inhibits aromatic amino acid decarboxylase activity, concentration of dopamine in the nerve terminals may be very low, indicating that reduction of dopamine level in the nerve terminals greatly affects the regulation of tyrosine hydroxylase activity because tyrosine hydroxylase is inhibited by dopamine (Nagatsu *et al.*, 1964). Westerink and de Vries (1991) have recently reported that

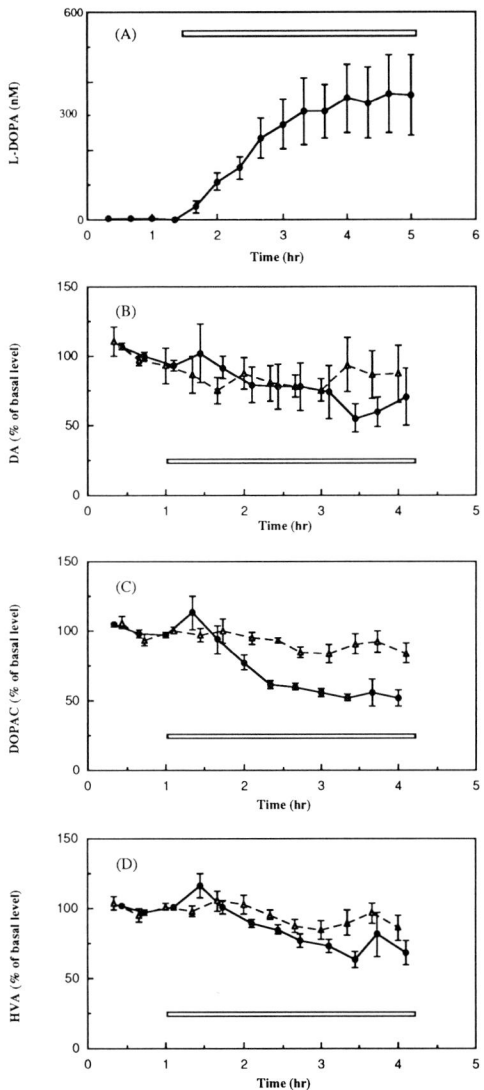

Figure 2. Effect of aromatic amino acid decarboxylase inhibitor, NSD1015, on L-DOPA, dopamine, DOPAC, and HVA levels in the dialysate. Administration of NSD1015 (10 μM) into rat striatum changed the levels of (A) L-DOPA, (B) dopamine, (C) DOPAC, (D) HVA. Other conditions are the same as those for Fig. 1. L-DOPA concentration was expressed as nmol/L. $\triangle$- -$\triangle$: without NSD, •—•: 10 μM NSD1015 treatment.

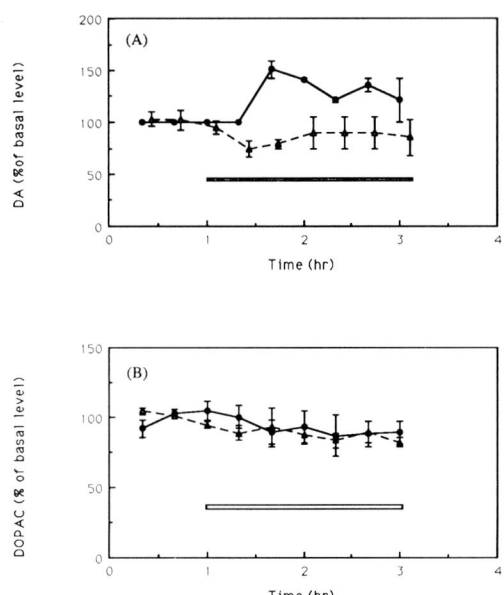

Figure 3. Effect of oxotremorine on dopamine and DOPAC levels in the dialysate of rat striatum in the presence of NSD1015. Under the perfusion of 10 μM NSD1015, oxotremorine (0.3 mM) was coperfused. (A) dopamine, (B) DOPAC. L-DOPA peak was not determined in the present experiment because an unknown peak appeared on DOPA peak after the oxotremorine treatment. Other conditions are the same as those for Fig. 2. △– –△: NSD alone, ●—●: NSD + oxotremorine.

treatment with a low concentration (10 μM) of NSD1015 is useful to study not only tyrosine hydroxylase but also tryptophan hydroxylase activities.

In the present report, we coperfused 10 μM NSD1015 with Ringer solution to determine tyrosine hydroxylase activity in rat striatum. Although perfusion of 10 μM NSD1015 accumulated L-DOPA formation about 500 nM (Fig. 2A), concentrations of dopamine metabolites, DOPAC and HVA, slightly decreased (Fig. 2C, D). From these data, we assume that even in a low concentration (10 μM) of NSD1015, *in vivo* tyrosine hydroxylase activity is suppressed.

There are a few interesting reports that peripheral and intracerebral administrations of tyrosine (substrate) and tetrahydrobiopterin (a cofactor) stimulate *in vivo* tyrosine hydroxylase activity (During *et al.*, 1988; Koshimura *et al.*, 1990). In our recent experiment (Xu *et al.*, 1989), we have also reported that some muscarinic receptors might activate *in vivo* tyrosine hydroxylase. To investigate the above possibility, we coadministered 10 μM NSD1015 in the perfusate. In the presence of 10 μM NSD1015, we could not measure change in the L-DOPA level after the treatment with 0.3 mM oxotremorine, because an unknown peak derived

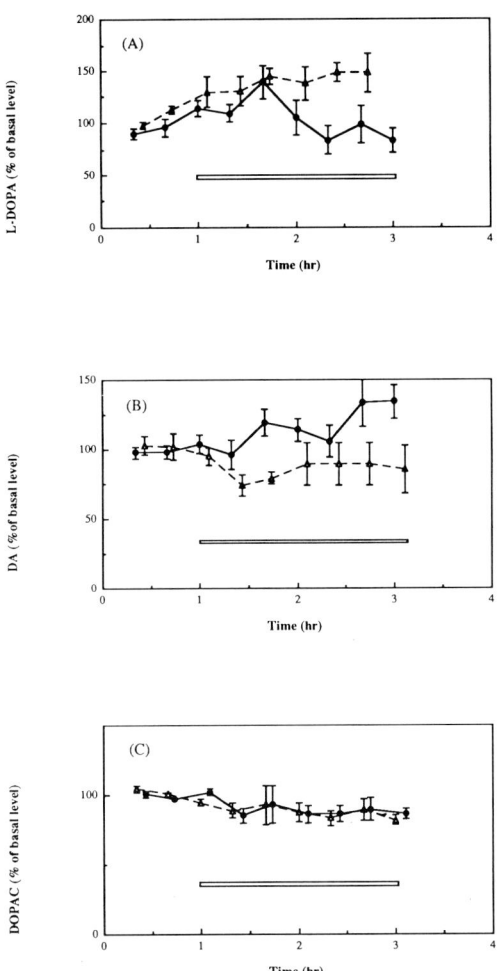

Figure 4. Effect of forskolin on L-DOPA, dopamine, and DOPAC levels in the dialysate of rat striatum in the presence of NSD1015. Under the perfusion of 10 μM NSD1015, forskolin (0.1 μM) was coperfused. (A) L-DOPA, (B) dopamine, (C) DOPAC. Other conditions are the same as those for Fig. 2. $\triangle$- -$\triangle$: NSD alone, ●—●: NSD + forskolin.

from the latter drug interfered the measurement of L-DOPA. The oxotremorine treatment increases 50% dopamine release but does not increase DOPAC level (Fig. 3A, B). In the absence of NSD1015, oxotremorine increases both dopamine release and DOPAC level (Fig. 2B) (Xu *et al.*, 1989).

Since it is well known that forskolin activates tyrosine hydroxylase, we applied 0.1 μM forskolin into rat striatum in the presence of 10 μM NSD1015 (Fig. 4). However, the treatment with 0.1 μM forskolin does not increase L-DOPA level (Fig. 4A). Furthermore, in the absence of NSD1015, 0.1 μM forskolin increases dopamine release and DOPAC level, but in the presence of 10 μM NSD1015 the forskolin slightly increases dopamine release (Fig. 4B).

From these results, we assume the following meaning(s) for NSD1015 treatment. (1) A high concentration of L-DOPA produced by 10 μM NSD1015 treatment inhibits tyrosine hydroxylase *in vivo*. (2) Decrease of dopamine concentration changes regulation mechanism(s) of *in vivo* tyrosine hydroxylase. (3) NSD1015 affects some intracellular signal transduction systems.

In conclusion, the present data indicate that *in vivo* microdialysis technique is useful to study the inter- and intra-cellular regulation mechanisms of neurotransmitters (Xu *et al.*, 1989, 1990; Ajima *et al.*, 1990; Westerink and de Vries, 1991; Kato *et al.*, 1992), and the interaction between neuronetworks and behaviors.

REFERENCES

Ajima, A., Yamaguchi, T. and Kato, T. (1990). Modulation of acetylcholine release by D1, D2 dopamine receptors in rat striatum under freely moving conditions. *Brain Res.* **518**, 193–198.

Coyle, J. L. and Axelrod, J. (1972). Tyrosine hydroxylase in rat brain: developmental characteristics. *J. Neurochem.* **19**, 1117–1123.

Kato, T., Otsu, Y., Furune, Y. and Yamamoto, T. (1992). Different effects of L-, N- and T-type calcium channel blockers on striatal dopamine release measured by microdialysis in freely moving rats. *Neurochem. Internationals* **21**, 99–107.

Kissinger, P. T., Refshange, C., Dreiling, R. and Adams, R. N. (1973). An electrochemical detector for liquid chromatography with picogram sensitivity. *Anal. Lett.* **6**, 465–477.

Koshimura, K., Miwa, S., Lee, K., Fujiwara, M. and Watanabe, Y. (1990). Enhancement of dopamine release *in vivo* from the rat striatum by dialytic perfusion of 6R-L-erythro-5,6,7,8- tetrahydrobiopterin. *J. Neurochem.* **54**, 1391–1397.

Matsui, H., Kato, T., Yamamoto, C., Fujita, K. and Nagatsu, T. (1981). Highly sensitive assay for dopamine β-hydroxylase activity in human cerebrospinal fluid by high performance liquid chromatography-electrochemical detection: properties of the enzyme. *J. Neurochem.* **37**, 289–296.

Nagatsu, T., Levitt, M. and Udenfriend, S. (1964). Tyrosine hydroxylase. The initial step in norepinephrine biosynthesis. *J. Biol. Chem.* **239**, 2910–2917.

Nagatsu, T., Oka, K. and Kato, T. (1979). Highly sensitive assay for tyrosine hydroxylase activity by high-performance liquid chromatography. *J. Chromatogr.* **163**, 247–252.

During, M. J., Acworth, I. N. and Wurtman, R. J. (1988). Effects of systemic L-tyrosine on dopamine release from rat corpus striatum and nucleus accumbens. *Brain Res.* **452**, 378–380.

Nissbrandt, H., Sundstrom, E., Jonsson, G., Hjorth, S. and Carlsson, A. (1989). Synthesis and release of dopamine in rat brain: comparison between substantia nigra pars compacta, pars reticulata, and striatum. *J. Neurochem.* **52**, 1170–1182.

Paxinos, G. and Watson, C. (1986). *The Rat Brain in Stereotaxic Coordinates.* Academic Press, Sydney.

Watanabe, S., Toru, M., Ichiyama, A. and Kataoka, T. (1981). The activity of rat pineal and brain tyrosine hydroxylase during the daily cycle of light and darkness as determined by the modified 14-CO_2 assay method. *J. Neurochem.* **36**, 266–275.

Westerink, B. H. C., de Vries, J. B. and Duran, R. (1990). Use of microdialysis for monitoring tyrosine hydroxylase activity in the brain of conscious rats. *J. Neurochem.* **54**, 381–387.

Westerink, B. H. C. and de Vries, J. B. (1991). Effect of precursor loading on the synthesis rate and release of dopamine and serotonin in the striatum: a microdialysis study in conscious rats. *J. Neurochem.* **56**, 228–233.

Xu, M., Mizobe, F., Yamamoto T. and Kato, T. (1989). Differential effects of M1- and M2- muscarinic drugs on striatal dopamine release and metabolism in freely moving rats. *Brain Res.* **495**, 232–242.

Xu, M., Yamamoto T. and Kato, T. (1990). *In vivo* striatal dopamine release by M1 muscarinic receptors is induced by activation of protein kinase C. *J. Neurochem.* **54**, 1917–1919.

Tyrosine Hydroxylase, pp. 135–151
M. Naoi *et al.* (Eds)
© VSP 1993

PURIFICATION OF TYROSINE HYDROXYLASE

KOHICHI KOJIMA

Hatano Research Institute, Food and Drug Safety Center, 729-5 Ochiai, Hadano Kanagawa 257, Japan

INTRODUCTION

Tyrosine hydroxylase (tyrosine 3-monooxygenase, TH) [L-tyrosine, tetrahydropteridine: oxygen oxidoreductase (3-hydroxylation), EC 1,14,16,2] catalyzes the first and rate-limiting step in the biosynthesis of catecholamines, dopamine, noradrenaline and adrenaline.

In the middle of the 1960s, some important researches on TH were initiated by several research groups. These researches began with the discovery of TH by Nagatsu *et al.* in 1964. After this discovery, there have been many reports on the purification of TH from various animal species.

Various investigators have studied TH from different tissues and species. For more than a decade a number of attempts to purify and characterize the enzyme were not successful because of its lability and its tendency to aggregate during purification, until Markey *et al.* (1980) succeeded in obtaining a pure form from cultured rat pheochromocytoma cells. However, it is still difficult to purify the native form of TH from enzyme sources of large animals such as bovine adrenal medulla and human tissues due to its tendency to aggregate during purification and owing to the presence of multiple forms with different molecular weights.

History of TH purification (Table 1)

During the early years, TH purification was limited to partial purification or concentration of TH, because the techniques available during those years had limited application to purification. Thus, the specific activity of TH purified was lower than 100 nmol/min/mg protein. However, the kinetic properties and other characteristics were investigated extensively for partially purified TH.

Petrack *et al.* (1968) have reported that approximately 90% of the total TH activity of bovine adrenal medulla homogenates is localized in the particulate fraction of the homogenate. The particulate enzyme has been solubilized by incubation with trypsin, and the solubilized enzyme has been partially purified.

Poillon (1971) has reported the absolute requirement for a reduced tetrahydropteridine as a cofactor and the effect of Fe^{2+} and suggested that intermolecular

Table 1.
History of TH purification

Origin	Steps	Comments	Reference
Sheep brain caudate nuclei	0.2 % Triton X-100	apparent K_m : 0.1 mM (L-Tyr)	Poillon
	Ultrafiltration	0.33 mM (DMPH$_4$)	(1971)
	Iodotyrosine modified Sepharose 4B		
Bovine adrenal medulla	Lysed granule	* Mr : ~ 40 000	Shiman et al.
	α-chymotrypsin digestion	** sp. act. : 65 (40% pure)	(1971)
	46–62% (NH$_4$)$_2$SO$_4$		
	and		
	Sephadex G-150		
	0–53% and 53–60% (NH$_4$)$_2$SO$_4$		
	or		
	3-Iodo-L-tyrosine modified Sepharose 4B		
Human pheochromocytoma	0–80% (NH$_4$)$_2$SO$_4$	Yield : 9%	Park and
	Sepharose 4B	sp. act. : 129.1	Goldstein
	Ultrafiltration	apparent K_m : (1976)	
	Centrifugation on a linear	9.1×10^{-5} M (Tyr, DMPH$_4$)	
	sucrose density gradient	1.9×10^{-5} M (Tyr, BH$_4$)	
		10.1×10^{-5} M (DMPH$_4$)	

Table 1. (continued)

Origin	Steps	Comments	Reference
Bovine adrenal medulla	Lysed granule	85–90 % pure	Hoeldtke and Kaufman (1977)
	α-Chymotrypsin digestion	Mr : 34 000	
	46–62% (NH$_4$)$_2$ SO$_4$	0.50 to 0.75 mol of iron / mol of enzyme	
	Sephadex G-150		
	DEAE-Cellulose	sp. act. : 360	
	Concentrated with pressure dialysis and collodion bag	Yield : 5.7%	
	Sucrose gradient centrifugation		
	Second sucrose gradient centrifugation		
Rat caudate nuclei	0–80% (NH$_4$)$_2$ SO$_4$	Mr : 62 000	Joh *et al.* (1978)
	Dialysis	sp. act. : 252	
	25–50% (NH$_4$)$_2$ SO$_4$	Yield : 4.4%	
	Dialysis		
	Phenyl-Sepharose		
	DEAE-Cellulose		
	DEAE-Cellulose		
	Polyacrylamide gel electrophoresis		

Table 1. (continued)

Origin	Steps	Comments	Reference
Rat pheochromocytoma	30–42% $(NH_4)_2SO_4$	sp. act. : 368	Vulliet et al.
	Dialysis	K_m : 74 μM (Tyr)	(1980)
	DEAE-Cellulose		
	0–44% $(NH_4)_2SO_4$	M_r : 60 000	
	Hydroxyapatite		
	0–50% $(NH_4)_2SO_4$		
	Sucrose density gradient		
Pheochromocytoma	0–80% $(NH_4)_2SO_4$	M_r : 62 000	Markey et al.
PC12 cloned cells	Dialysis	(210 000–220 000)	(1980)
	25–35% $(NH_4)_2SO_4$	sp. act. : 373	
	Sepharose 4B	Yield : 9 %	
	Sucrose density gradient	isoelectric point : 5.3 ± 0.1 (4°C)	
Bovine corpus	DEAE-Cellulose	sp. act. : 7.41	Edelman et al.
striatum	Hydroxyapatite	Yield : 0.7%	(1981)
	CM Sephadex C-50	M_r : 60 000	
	Glycerol density gradient		

Table 1. (continued)

Origin	Steps	Comments	Reference
Rabbit adrenal	Heparin-Sepharose 4B Phenyl-Sepharose DEAE-Cellulose 5–40% $(NH_4)_2SO_4$ Sepharose CL-4B I-125 Protein column (HPLC)	sp. act. : 40.0 Yield : 24 %	Lloyd and Walega (1981)
Bovin corpus striatum	Edelman *et al.* (1981) modified DEAE-Cellulose Dialysis CM Sephadex C-50 Glycerol density gradient	90 % or more activity lost within 72 h at 4°C Mr : 60 000 & 62 000 pH optimum of phosphorylated TH : 6.6	Lazar *et al.* (1982)
Rat adrenal	30–40% $(NH_4)_2SO_4$ Ultrogel AcA 22 pH fractionation (pH : 5.8–5.2) Heparin-substituted Sepharose 4B Phosphocellulose Ultrogel AcA 22	sp. act. : 1 600 Mr : 59 000 isoelectric point : 6.7 (with 8 M urea) 6.6 (without 8 M urea)	Okuno and Fujisawa (1982)

Table 1. (continued)

Origin	Steps	Comments	Reference
Bovine adrenal medulla	DEAE-Sephacel 0–40% $(NH_4)_2$ SO_4 Bio-Gel A-1.5m Heparin-Sepharose CL-6B Bio-Gel A-1.5m	Mr : 60 000 sp. at. : 1 880 Yield : 9.3 % isoelectric point : 6.0 N-terminal : Glu	Oka et al. (1983)
Rat adrenal gland Rat caudate nucleus	DEAE-Sephacel 0–40% $(NH_4)_2SO_4$ Bio-Gel A-1.5m Heparin -Sepharose	sp. act. : 812 Yield : 21 % Mr : 60 000	Togari et al. (1983)
Human adrenal	DEAE-Sephacel 0–40% $(NH_4)_2$ SO_4 Bio-Gel A-1.5m Heparin-Sepharose Cl-6B or less active forms	Mr : 60 000 Yield : 5 % sp. act. : 315 Active and inactive	Mogi et al. (1984)
Human adrenal	DEAE-Sephacel 0–40% $(NH_4)_2$ SO_4 Bio-Gel A-1.5m Heparin-Sepharose	sp. act. : 310 Yield : 4.4 % Mr : 60 000 (280 000)	Kojima et al. (1984)

Table 1. (continued)

Origin	Steps	Comments	Reference
Rat striatal	Phenyl-Sepharose Heparin-Sepharose NPG-Sepharose	Mr : 61 300 st. act. : 406 Yield : 8.58	Richtand *et al.* (1985)
Rat adrenal (treat with or without reserpine)	DEAE-Sephacel 0–40% $(NH_4)_2 SO_4$ Bio-Gel A-1.5m Heparin-Sepharose	Mr : 60 000 3-fold increase with reserpine treatment	Togari *et al.* (1985)
Human brain (caudate nucleus and putamen)	DEAE-Sephacel 0–40% $(NH_4)_2SO_4$ Bio-Gel A-1.5m Heparin-Sepharose CL-6B sp. act. : 12.8 Yeild : 1.3%	Mr : 60 000 (280 000) isoelectric point : 6.0 (active) and 5.3 to 5.8 (inactive) active : inactive = 1 : 8	Mogi *et al.* (1986)
Bovine adrenal medulla	Phenyl-Sepharose 4B DNA-cellulose Mr : 62 400 and 61 100	sp. act. : 180 Yield : 58 %	Nelson and Kaufman (1987)

Table 1. (continued)

Origin	Steps	Comments	Reference
TRat Brain	Phenyl-Sepharose 4B	sp. act. : 91.6	Nelson and
	DNA-cellulose	Yield : 18%	Kaufman
	Mr : 63 000		(1987)
	0.07 mol of phosphate/63 000		
PC12 cells	30–42% (NH$_4$)$_2$ SO$_4$	Mr : 60 000	Kuhan and
	Sepharose CL-6B	isoelectric point : 5.4, 5.8 and 5.9	Billingsley
	DE-52	sp. act. : 258	(1987)
	Heparin-Sepharose		
Bovine adrenal	25–35% (NH$_4$)$_2$ SO$_4$	sp. act. : 425	Haavik et al.
medulla	DEAE-Sephacel	Yield : 4.5 %	(1988)
	Heparin-Sepharose	Mr : 60 000 (210 000)	
	L-Tyrosine-agarose		

* Mr : molecular weight of monomer (molecular weight of tetramer).

** sp. act. : specific activity (nmol/min/mg protein).

aggregation of the TH protein occurred as a concomitant of the purification procedure.

Shiman *et al.* (1971) have found that the stoichiometry of the reaction catalyzed by TH varies with different pterin cofactors. They have concluded that TH which was stimulated by Fe^{2+} can be explained by the known ability of Fe^{2+} to decompose H_2O_2.

Park and Goldstein (1976) have described the purification of TH, the production of TH antibody, and the kinetic properties of human pheochromocytoma TH.

Hoeldtke and Kaufman (1977) have characterized a number of physical and kinetic properties of the enzyme from the bovine adrenal.

Joh *et al.* (1978) have demonstrated that the enzyme could be phosphorylated by a cyclic AMP-dependent protein kinase, that such phosphorylation increases the catalytic activity of the enzyme, and that the kinetics of activation suggest that the effect is due to conversion of inactive enzyme to active enzyme.

Vulliet *et al.* (1980) have reported that TH purified from rat pheochromocytoma was phosphorylated in the presence of ATP, Mg^{2+}, and the catalytic subunit of cyclic AMP-dependent protein kinase. The phosphorylation of the enzyme was closely correlated with enhancement of the enzyme activity in the presence of subsaturating concentration of the reduced pterin cofactor.

Markey *et al.* (1980) have described the purification and properties of rat pheochromocytoma TH as well as the immunoreactivity of rabbit anti-rat TH with TH from various species.

Edelman *et al.* (1981) have shown that the purified preparation of striatal TH can be phosphorylated *in vitro* by cyclic AMP-dependent protein kinase which also activated the enzyme. This was the first report on the stoichiometry of phosphorylation of TH from the brain.

Lloyd and Walega (1981) have reported the application of high pressure liquid chromatography to the purification of native TH.

Lazar *et al.* (1982) have made a detailed presentation of the kinetic properties of TH from the brain which were determined using a highly purified preparation from bovine corpus striatum.

Okuno and Fujisawa (1982) have purified TH to a much higher specific activity of 1600 nmol/min/mg protein from rat adrenal medullae than the former reports of 373 nmol/min/mg protein from rat pheochromocytoma cells. Purification of the enzyme caused considerable changes in its kinetic properties, yielding an active state distinct from that of the native enzyme or the phosphorylated enzyme by the action of cyclic AMP-dependent protein kinase.

Richtand *et al.* (1985) have demonstrated that the lability of the purified enzyme was primarily due to proteolysis as it approached homogeneity. Use of a combination of protease inhibitors has made possible a purification protocol of high yield which results in a stable enzyme preparation. This has allowed them to characterize TH from the rat brain and to compare the effects of effectors, including cyclic AMP-dependent phosphorylation, heparin, sodium chloride, and phosphatidylinositol with the objective of specifying whether their effects represent a single common mechanism of action.

Nelson and Kaufman (1987) have found that with the use of nucleic acid-substituted cellulose, TH can be purified in a highly reproducible manner from the whole brain in a rapid, two-column procedure.

Kuhn and Billingsley (1987) have purified and characterized TH from cultured PC12 cells. Furthermore, antibodies to TH have been produced in rabbits and the properties of the antibody have also been studied in some details.

Haavik *et al.* (1988) have reported a new procedure that permits large-scale purification of TH from the cytosolic fraction of bovine adrenal medulla. Their physiochemical studies have suggested that the enzyme dose not undergo such major substrate- or cofactor-induced conformational changes as had been reported for the related enzyme, phenylalanine hydroxylase.

TH PURIFICATION BY NAGATSU'S GROUP

In 1982, TH from bovine adrenal medulla was purified to apparent homogeneity (Oka *et al.*, 1982). Finally in 1983 (Oka *et al.*, 1983), TH was purified to apparent homogeneity from the soluble fraction of bovine adrenal medulla. The specific activity of this preparation was 1880 nmol/min/mg protein (Table 2), which is almost 5-fold higher than in the previous report (Oka *et al.*, 1982), and was the highest among the reported values. This TH had an apparent molecular weight (Mr) of about 280 000 by Bio-Gel A-1.5 m chromatography, and gave a single band with a Mr of 60 000 by sodium dodecyl sulfate (SDS) polyacrylamide gel electrophoresis. The enzyme was considered to be composed of four identical subunits. Isoelectric point of purified enzyme was pH 6.0. The amino acid composition of the enzyme was characterized by fairly high contents of glutamic acid and alanine residues. The N-terminal amino acid was determined to be glutamic acid. In this study, we have used Bio-Gel A-1.5 m column chromatography twice and washed the heparin-Sepharose CL-6B column in the buffer of pH 8.3, and we finally have obtained a highly purified TH. Although we could not completely rule out the presence of inactive forms of the enzyme in this enzyme preparation of the highest specific activity, a single component in SDS polyacrylamide gel electrophoresis and a single N-terminal amino acid support both the homogeneity and the highest purity of the native form of the enzyme by the present purification. After this excellent purification method was established, his group has conducted further studies using purified TH. In order to study the molecular properties of the enzyme modified in physiologically or pharmacologically *in vivo*, a micro-purification method is required in yielding a highly purified enzyme preparation from the tissue samples of the animals under various physiological or pharmacological conditions. According to the established method, a rapid and simple simultaneous micro-purification procedure of TH and dihydropteridine reductase (DPR) has been developed from soluble supernatants of 1 to 2 g of rat adrenal gland or caudate nucleus (Togari *et al.*, 1983). All purification procedures for the two enzymes were complete within 3 days. The recovery of TH and DPR was reproducible, being approximately 20 and 40%, respectively. Purification procedure for TH involved chromatographies with DEAE-Sephacel, Bio-Gel A-1.5 m and heparin-Sepharose CL-6B. As judged by gel filtration and SDS polyacrylamide gel electrophoresis, the enzyme purified from each tissue appeared to be homogeneous and was composed of an identical subunit, each possessing a Mr of 60 000. With DEAE-Sephacel column chromatography, TH was separated completely from DPR. DPR was purified by subsequent chromatographies with Sephadex G-50 and blue-Sepharose to a purity of 50%. DPR in adrenals and brain was found to be a NADH-dependent type. This micro-purification procedure is applicable to assess the molecular properties of TH modified physiologically or pharmacologically *in vivo*, and to obtain a small amount of pure enzyme as antigen for producing its antibody.

Table 2.
Purification of tyrosine hydroxylase from bovine adrenal medulla*

Purification step	Total protein (mg)	Total activity (units)**	Specific activity (units/mg protein)	Yield (%)
supernatant (100 000 × g)	1738	22 675	13.0	100
DEAE-Sephacel	264	13 192	50.0	58.2
0–40% $(NH_4)_2SO_4$	83.9	12 845	153	56.6
Bio-Gel A-1.5 m	32.1	9 048	282	39.9
heparin-Sepharose CL-6B	2.94	5 408	1839	23.9
Bio-Gel A-1.5 m	1.12	2 106	1880	9.3

*: Starting material; bovine adrenal medulla, 69.3 g.
**: Units are expressed in nmol of 3,4-dihydroxyphenylalanine (dopa) formed/min at 30 °C. Tyrosine, 0.1 mM; 6-methyltetrahydropterin, 1 mM.

A sensitive sandwich enzyme immunoassay (EIA) for TH from bovine and human adrenals has been developed (Mogi *et al.*, 1984). Anti-TH antibody was prepared from bovine adrenal TH. The assay system consisted of an antibody F(ad')2 immobilized on polystyrene beads as a solid phase and of β-D-galactosidase-conjugated antibody. This method is highly sensitive and specific for assay of TH. Human adrenal TH level was determined by similar sensitivity as bovine adrenal TH, suggesting the presence of common antigenic sites between human and bovine adrenal enzyme. The presence of inactive or less active forms of TH in human adrenals was ascertained by purification of the enzyme and monitoring with this enzyme immunoassay as well as with enzyme activity assay.

In 1984, Kojima et al. have reported a procedure for purification of human adrenal TH as an extension of the previously reported bovine enzyme purification (Oka et al., 1982). TH was purified from the soluble fraction of human adrenal glands. The enzyme in human adrenal glands that was purified to apparent homogeneity had an apparent Mr of about 280 000. SDS polyacrylamide gel electrophoresis gave a single band with a Mr of 60 000 similar to the Mr of bovine adrenal enzyme. The enzyme was considered to be composed of four identical subunits. The specific activity of the final preparation was approximately 310 nmol 3,4-dihydroxyphenylalanine (DOPA) formed/min/mg protein. The use of the "Western Blot" method showed that human adrenal TH did not aggregate as rapidly as bovine adrenal TH. This was the first report on purification of homogeneous human TH.

To answer the questions "was TH in adrenal medulla that was newly formed in response to reserpine the same as the preexisting enzyme?" and "to what extent did TH induced by reserpine contribute to the increase in in situ, TH activity?" Togari et al. (1985) have described the molecular and physiological properties of newly synthesized TH induced by reserpine in rat adrenals, using both the developed micro-purification method of this enzyme and the tissue slice method for measuring tyrosine hydroxylation in situ under physiological conditions by HPLC with electrochemical detection. When repeated doses of reserpine were given at daily intervals for three days, the enzyme activity measured in homogenates of the adrenal glands was increased 3-fold. Furthermore, when TH in the adrenal glands from both control and reserpine-treated rats was purified, both the total activity of the enzyme and the enzyme protein content purified from reserpine-treated rats were also about 3-fold higher than those of the control rats (Table 3). The two purified enzymes revealed similar properties: a single subunit with a Mr of 60 000 was observed by SDS polyacrylamide gel electrophoresis, and the K_m value for pterin cofactor, 6-methyltetrahydropterin was about 300 μM. In contrast, in situ TH activity measured under physiological conditions at pH 7.2 in adrenal tissue slices was elevated 6-fold by reserpine pretreatment for 3 days, and was stimulated by carbachol (0.1 mM) and elevated K^+ (52 mM) in a roughly proportional rather than additive way relative to slices from untreated rats. These results indicate that newly synthesized TH induced by reserpine in rat adrenal gland has properties similar to the enzyme in control rats and that reserpine increases not only the amount of TH molecular but also the in situ activity of TH. Since reserpine also increases the biosynthesis of tetrahydrobiopterin as demonstrated by Viveros and co-workers, this 6-fold increase in situ TH activity may depend both upon the 3-fold increase in the amount of enzyme molecules and upon the increase of the physiologically available tetrahydrobiopterin in the adrenal gland.

Table 3.
Purification of tyrosine hydroxylase in the soluble fraction of adrenal glands from rats treated with or without reserpine

	Steps	Total protein (mg)	Total activity (nmol/min)	Yield (%)
Control	supernatant (100 000 × g)	38.8	40.8	100
	DEAE-Sephacel	6.0	23.2	56
	0–40% (NH$_4$)$_2$SO$_4$	1.4	19.7	49
	Bio-Gel A-1.5 m	0.4	10.4	26
	heparin-Sepharose CL-LB	0.01	8.5	21
Reserpine	supernatant (100 000 × g)	50.6	110.9	100
	DEAE-Sephacel	6.3	66.8	60
	0–40% (NH$_4$)$_2$SO$_4$	2.1	54.8	49
	Bio-Gel A-1.5 m	0.5	28.6	26
	heparin-Sepharose CL-BL	0.03	24.9	22

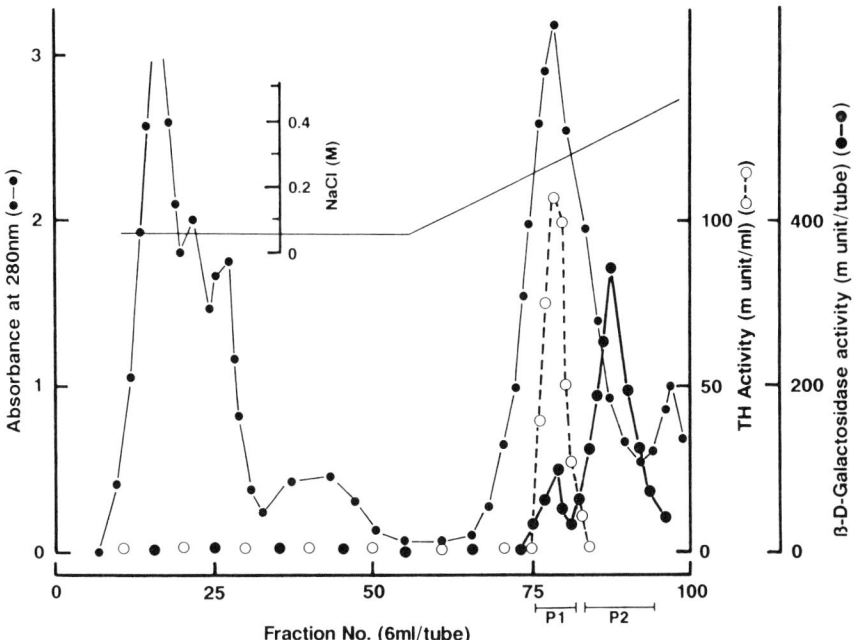

Figure 1. Chromatography of human brain TH on a DEAE-Sephacel column, o—o TH activity; •—•, absorbance at 280 nm; •—•, human brain TH protein as detected with a sandwich EIA and expressed as β-D-galactosidase activity bound to the solid phase.

As an extension of the purification of the human enzymes previously reported (Kojima *et al.*, 1984), Mogi *et al.* (1986) succeeded in homogeneously purifying TH from the human brain (caudate nucleus + putamen) for the first time. The major form of the active enzyme in the cytoplasmic fraction was purified to apparent homogeneity (Fig. 1). The molecular weight of the purified enzyme was estimated to be 280 000 by gel filtration. SDS polyacrylamide gel electrophoresis of the purified enzyme gave a single subunit with molecular weight of 60 000, which was similar to the subunit of human adrenal TH. Using a sandwich EIA, the presence of inactive form(s) of TH in the human brain was demonstrated, and the total content of this immunoinactive form(s) was approximately 8 times higher than that of the active form. By Western blot technique after two-dimensional electrophoresis, TH in the crude fraction of the human brain was found to consist of multiple forms with different pI-values and with the same molecular weight. The pI of the major spots ranged from 5.3 to 5.8 and that of the minor spot was 6.0. Because the pI of the purified enzyme preparation was 6.0, this protein with pI at 6.0 may be the active form of TH.

High performance liquid chromatography (HPLC) has been extensively developed during last 15 years, and great advances have also been made in the study of enzymes and proteins by HPLC. One application of HPLC to enzymes is the assay of enzyme activity, and the other is the purification of enzyme proteins. The application of HPLC to the assay of the enzyme activity has become one of the most important and valuable techniques in enzymology. The assay for catecholamine-metabolizing enzymes has been summarized (Nagatsu *et al.*, 1987) and the assay of TH activity by HPLC is shown in the appendix of this chapter. The isolation and partial purification of catecholamine-synthesizing enzymes by HPLC have been described (Kojima *et al.*, 1985), and have suggested that HPLC gives sharp and rapid separation effectively.

CONCLUSION

Enzymes of catecholamine biosynthesis have been purified to homogeneity and extensively studied in biochemistry, pharmacology, physiology, and neurobiology until 1985. However, large scale purification is still difficult and the determination of primary structure of these enzymes has been impossible by means of protein chemistry. In the research of catecholamine-metabolizing enzymes, the molecular biological method has been introduced, and the DNA sequences of many catecholamine-metabolizing enzymes have been analyzed in the last several years. The molecular biological method has become quite a useful and powerful technique, while the usefulness of the protein purification method is shrinking slightly. However, both methods are interdependent for further research of biochemistry. The research for catecholamines is becoming very important for clinical problems. The research of this area will be extensively developed by future fundamental and clinical studies.

REFERENCES

Edelman, A. M., Raese, J. D., Lazar, M. A. and Barchas, J. D. (1981). Tyrosine hydroxylase: Studies on the phosphorylation of a purified preparation of the brain enzyme by the cyclic AMP-dependent protein kinase. *J. Pharmacol. Exp. Ther.* **216**, 647–653.

Haavik, J., Andersson, K. K., Petersson, L. and Flatmark, T. (1988). Soluble tyrosine hydroxylase (tyrosine 3-monooxygenase) from bovine adrenal medulla: Large-scale purification and physicochemical properties. *Biochim. Biophys. Acta* **953**, 142–156.

Hoeldtke, R. and Kaufman, S. (1977). Bovine adrenal tyrosine hydroxylase Purification and properties. *J. Biol. Chem.* **252**, 3160–3169.

Joh, T. H., Park, D. H. and Reis, D. J. (1978). Direct phosphorylation of brain tyrosine hydroxylase by cyclic AMP-dependent protein kinase: Mechanism of enzyme activation. *Proc. Natl. Acad. Sci. USA* **75**, 4744–4748.

Kojima, K., Mogi, M., Oka, K. and Nagatsu, T. (1984). Purification and immunochemical characterization of human adrenal tyrosine hydroxylase. *Neurochem. Int.* **6**, 475–480.

Kojima, K., Parvez, S., Parvez, H., Kato, Y. and Nagatsu, T. (1985). Application of HPLC for analysis and purification of catecholamine-synthesizing enzymes. In: *Progress in HPLC*. **Vol. 1.** H. Parvez, Y. Kato and S. Parvez (Eds). VNU Science Press, Utrecht, pp. 149–155.

Kuhn, D. M. and Billingsley, M. L. (1987). Tyrosine hydroxylase: Purification from PC12 cells, characterization and production of antibodies. *Neurochem. Int.* **11**, 463–475.

Lazar, M. A., Lockfeld, A. J., Truscott, R. J. W. and Brachas, J. D. (1982). Tyrosine hydroxylase from bovine striatum: Catalytic properties of the phosphorylated and nonphosphorylated forms of the purified enzyme. *J. Neurochem.* **39**, 409–422.

Lloyd, T. and Walega, M. A. (1981). Purification of tyrosine hydroxylase by high-pressure liquid chromatography. *Anal. Biochem.* **116**, 559–563.

Markey, K. A., Kondo, S., Shenkman, L. and Goldstein, M. (1980). Purification and characterization of tyrosine hydroxylase from a clonal pheochromocytoma cell line. *Mol. Pharmacol.* **17**, 79–85.

Mogi, M., Kojima, K. and Nagatsu, T. (1984). Detection of inactive or less active forms of tyrosine hydroxylase in human adrenals by a sandwich enzyme immunoassay. *Anal. Biochem.* **138**, 125–132.

Mogi, M., Kojima, K., Harada, M. and Nagatsu, T. (1986). Purification and immunochemical properties of tyrosine hydroxylase in human brain. *Neurochem. Int.* **8**, 423–428.

Nagatsu, T., Kojima, K., Parvez, S. and Parvez, H. (1987). Analysis of enzyme activities by electrochemical detection coupled with HPLC. In: *Progress in HPLC*, **Vol. 2.** H. Parvez, M. Bastrat-Malsot, S. Parvez, T. Nagatsu and G. Carpentier (Eds). VNU Science Press, Utrecht, pp. 293–306.

Nagatsu, T., Levitt, M. and Udenfriend, S. (1964). Tyrosine hydroxylase. The initial step in norepinephrine biosynthesis. *J. Biol. Chem.* **239**, 2910–2917.

Nelson, T. J. and Kaufman, S. (1987). Interaction of tyrosine hydroxylase with ribonucleic acid and purification with DNA-cellulose of poly(A)-Sepharose affinity chromatography. *Arch. Biochem. Biophys.* **257**, 69–84.

Oka, K., Ashiba, G., Sugimoto, T., Matsuura, S. and Nagatsu, T. (1982). Kinetic properties of tyrosine hydroxylase purified from bovine adrenal medulla and bovine caudate nucleus. *Biochim. Biophys. Acta.* **706**, 188–196.

Oka, K., Kojima, K. and Nagatsu. T. (1983). Characterization of tyrosine hydroxylase from bovine adrenal medulla. *Biochem.* **7**, 387–393.

Okuno, S. and Fujisawa, H. (1982). Purification and some properties of tyrosine 3-monooxygenase from rat adrenal. *Eur. J. Biochem.* **122**, 49–55.

Park, D. H. and Goldstein, M. (1976). Purification of tyrosine hydroxylase from pheochromocytoma tumors. *Life Sci.* **18**, 55–60.

Petrack, B., Sheppy, F. and Fetzer, V. (1968). Studies on tyrosine hydroxylase from bovine adrenal medulla. *J. Biol. Chem.* **243**, 743–748.

Poillon, W. N. (1971). Kinetic properties of brain tyrosine hydroxylase and its partial purification by affinity chromatography. *Biochem. Biophys. Res. Commun.* **44**, 64–70.

Richtand, N. M., Inagami, T., Misono, K. and Kuczenski, R. (1985). Purification and characterization of rat striatal tyrosine hydroxylase — Comparison of the activiation by cyclic AMP-dependent phosphorylation and by other effectors. *J. Biol. Chem.* **260**, 8465–8473.

Shiman, R., Akino, M. and Kaufman, S. (1971). Solubilization and partial purification of tyrosine hydroxylase from bovine adrenal medulla. *J. Biol. Chem.* **246**, 1330–1340.

Togari, A., Kano, H., Oka, K. and Nagatsu, T. (1983). Simultaneous simple purification of tyrosine hydroxylase and dihydropteridine reductase. *Anal. Biochem.* **132**, 183–189.

Togari, A., Kojima, K. and Nagatsu, T. (1985). Molecular and physiological properties of tyrosine hydroxylase induced by reserpine in adrenal gland. *Life Sci.* **37**, 1605–1611.

Vulliet, P. R., Langan, T. A. and Weiner, N. (1980). Tyrosine hydroxylase: A substrate of cyclic AMP-dependent protein kinase. *Proc. Natl. Acad. Sci. USA* **77**, 92–96.

APPENDIX

Analysis of TH activity — HPLC method using L-tyrosine as a substrate

Since the activity of TH is very low, TH has been measured only by radioisotopic methods. In the late 1970s, a highly sensitive method based on HPLC-electrochemical detection (ECD) was developed.

Principle

HPLC-ECD method uses L-tyrosine as the natural substrate and D-tyrosine for the blank, since TH is specific for L-tyrosine.

This method combines the simple and specific isolation of enzymatically formed L-DOPA by our double-column procedure (the top column of Amberlite CG-50 and the bottom column of aluminum oxide) with the highly sensitive assay of DOPA by HPLC-ECD. α-methyl DOPA as an internal standard is added to each sample after TH incubation. Use of α-methyl DOPA as an internal standard makes the assay highly accurate. Both the double columns and HPLC permit nearly complete isolation of DOPA, and thus the blank values are very low. The only interfering substance is endogenous DOPA in crude tissues and nonenzymatically formed DOPA from both L- and D-tyrosine, and this blank value can be completely cancelled by the control incubation with D-tyrosine plus 3-iodo-L-tyrosine (an enzyme inhibitor). TH activity in less than 1 mg of a brain nucleus can be assayed by this method.

Procedures

The standard incubation mixture consists of the following components in a total volume of 100 μl (final concentrations in parentheses); 20 μl of 1 M acetate buffer pH 6.0 (2.0 M), 20 μl of 1 mM L-tyrosine in 0.01 M HCl (0.2 mM), 10 μl (6*RS*)-methyl-5,6,7,8-tetrahydropterin (1 mM) in 1 M 2-mercaptoethnol (100 mM), 30 μl of 0.25 M sucrose (75 mM) containing the enzyme, 10 μl of 1 mg/ml catalase (10 μg/100 μl) or 10 μl of 10 mM ferrous ammonium sulfate (1 mM), water. For the blank incubation, D-tyrosine is used as substrate instead of L-tyrosine and 50 pmol of DOPA are added to another blank incubation as an internal standard for DOPA.

Incubation is done at 37 °C for 10 min, and the reaction is stopped with 600 μl 0.5 M perchloric acid containing 50 pmol α-methyl DOPA as an internal standard in an ice-bath. After 10 min, 20 μl 0.2 M EDTA and 300 μl 1 M potassium carbonate are added to adjust the pH to 8.0–8.5, and the mixture is centrifuged at 1 600 × g for 10 min at 4 °C. The clear supernatant is passed through the double columns with the upper column containing 200 μl Amberlite CG-50 (12.5 × 0.5 cm ID) and the bottom column containing 100 mg aluminum oxide (12.5 × 0.4 cm ID), fitted together sequentially. The effluent through both columns is discarded. Both columns are washed once with 1.5 ml water, and the washings are discarded DOPA and α-methyl DOPA pass through the first Amberlite column and are absorbed on the second aluminum oxide column, which is separated and washed with 1.5 ml of water twice and with 100 μl of 0.5 M HCl once. DOPA and α-methyl DOPA are eluted with 200 μl of 0.5 M HCl.

A 100 μl aliquot of the eluate is injected into the high-performance liquid chromatograph with an electrochemical detector and a column (25 × 0.4 cm ID) packed with ODS (particle size 5 μm). The mobile phase is 0.1 M potassium phosphate buffer (pH 3.5) with a flow-rate of 0.6 ml/min; the detector potential is set at 0.8 V vs the Ag/AgCl electrode. Under these conditions the retention times are: solvent front, 1.8 min; DOPA, 3.8 min: and α-methyl DOPA, 5.5 min. Limit of sensitivity for DOPA is 0.1 pmol. The DOPA formed enzymatically by TH is calculated by the equation

$$\frac{R(L) - R(D)}{R(D + S) - R(D)} \times 50 \text{ pmol,}$$

where R is the ratio to peak heights (peak height of DOPA/peak height of α-methyl DOPA), $R(L)$ being that from L-tyrosine incubation, $R(D)$ from D-tyrosine incubation, and $R(D+S)$ that of D-tyrosine plus DOPA (internal standard, 50 pmol).

The highest sensitivity can be obtained by the double column procedure to reduce the interfering peaks in HPLC, but the first Amberlite CG-50 column can be omitted in regular TH assay, and only the aluminum oxide column can be used.

Tyrosine Hydroxylase:
Its molecular studies

Tyrosine Hydroxylase, pp. 155–162
M. Naoi *et al.* (Eds)
© VSP 1993

PROGRESS IN TYROSINE HYDROXYLASE GENE RESEARCH

TONG H. JOH

*Laboratory of Molecular Neurobiology, Cornell University Medical College,
The W. M. Burke Medical Research Institute, White Plains,
New York 10605, USA*

INTRODUCTION

It all started in 1953. Professor Sydney Udenfriend observed that there is an enzyme which catalyzes the reaction of L-tyrosine to L-dihydroxyphenylalanine (L-dopa) (Udenfriend *et al.*, 1953), and named the enzyme tyrosine hydroxylase (TH). It was the beginning of the study on catecholamine biosynthesis. In 1964, Professor Toshiharu Nagatsu along with Professor Udenfriend, identified and characterized this enzyme (Nagatsu *et al.*, 1964). It initiated a keen interest in TH which is one of the most widely studied enzymes in pharmacology, biochemistry and neurobiology. Fortunately, Professor Sydney Udenfriend, discusses this fact in his chapter in this book as a tribute to T. Nagatsu. Professor Sydney Udenfriend, my undisputed idol in science, is the right person to talk about the importance of this enzyme in these various fields. As we all know, the biochemical identification of TH (Udenfriend *et al.*, 1953; Nagatsu *et al.*, 1964), and the concept of its being as the rate limiting enzyme in the catecholamine biosynthesis (Levitt *et al.*, 1965) have been the foundations for the enormous amounts of research on this enzyme which have taken place in the past 30 years. Professor Toshiharu Nagatsu and Professor Udenfriend opened the field to many scientists including this author. In the past 25 years, I have been either a primary or secondary author for many scientific articles, and the majority of these publications are related to TH or have included TH in them. Undoubtedly, this is not unique since there are many scientists who have similar publication records. The studies on the characterization and regulation of this enzyme have produced thousands of publications, and the results of these publications have been utilized not only for the basic sciences but also for human health research. Professor Nagatsu, a major contributor and leader in this field, is a person that I highly regard. I am most honored to contribute a chapter to this book.

REGULATION OF TYROSINE HYDROXYLASE: A HISTORICAL REVIEW

It has been known for some time that the steady state level of TH in the catecholaminergic neurons and cells is not fixed but in a constant state of fluctuation.

It would be impossible to try to summarize all the documented evidence for the changes in the TH activity in this chapter. However, some historically important observations should not be overlooked in this brief review on the subject. In the past 20 years, many drugs, chemicals and physiological manipulations have been used to induce or activate TH in catecholaminergic neurons and cells *in vivo*, and in cultured tissues and cells. The main observation leading to the majority of these studies was that the TH activity increased several fold when the discharge of sympathetic neurons is directly or reflexly augmented for several days (Mueller *et al.*, 1969; Thoenen *et al.*, 1969; Kvetňanský *et al.*, 1977). Because the increase in the TH activity depends on the integrity of preganglionic innervation, the increase is referred to as trans-synaptic induction (Thoenen *et al.*, 1969). The term, induction, was used in the field of pharmacology, at this time, as the induced increase of enzyme activity. In 1973, Joh *et al.*first demonstrated that the increase of the TH activity is due to the elevated level of the enzyme protein. The investigation of the mechanisms underlying the accumulation of the TH protein by these stimuli required a basic molecular biological technology. It took more than 10 years to prove that the increased level of the TH protein is due to the elevated steady state level of the TH mRNA (Black *et al.*, 1985; Stachowiak *et al.*, 1985, 1988, 1990; Faucon-Biguet *et al.*, 1986; Zigmond, 1989).

Since 1970, molecular biology and genetic engineering technology have created a completely new dimension in the bioscientific fields. Biochemical pharmacology and neurobiology have also adopted this new discipline and have rapidly expanded into the fields of molecular and cellular biology. In 1982, the first cloning of the rat TH cDNA was announced (Lamouroux *et al.*, 1982). Shortly after the announcement, a partial length of the rat cDNA clone was obtained (Lewis *et al.*, 1983). Grima *et al.* (1985) published a complete nucleotide sequence of the rat TH cDNA, and the information became an important contribution to both catecholamine biochemists and molecular biologists. It was especially valuable to those scientists in the field of the TH research. However, it was found that the TH mRNA is not a single species and that there are multiple human mRNA forms (Grima *et al.*, 1985). Professor Nagatsu's laboratory was the first to announce that there are four human TH mRNA forms which are the products of alternative splicing (Kaneda *et al.*, 1987; O'Malley *et al.*, 1987; Kobayashi *et al.*, 1988). Once again, Professor Nagatsu's leadership in the TH research was displayed. In 1988, D'Mello *et al.* published a complete sequence of the bovine TH cDNA. Thus, the structure of TH in various species has been completely characterized, and their amino acid sequences deduced from the nucleotide sequences were identified. Because of these developments, much new data regarding the structure and function of the TH protein became available. This included the possible amino acid sites for the essential TH phosphorylation. Using these cDNAs as probes for *in situ* hybridization and/or Northern blotting, quantitative measurements of TH mRNAs *in situ* and *in vitro* are widely used in the neurobiological, pharmacological and biochemical analyses of the TH induction.

Table 1.
Tyrosine hydroxylase cDNAs

Species	Length	References	
rat	2.1 kb	Lamouroux *et al.*	1982
rat		Lewis *et al.*	1983
rat		Grima *et al.*	1985
human	3 diff.	Grima *et al.*	1987
human	4 diff.	Kaneda *et al.*	1987
human	4 diff.	Kobayashi *et al.*	1988
bovine	2.0 kb	D'Mello *et al.*	1988

STRUCTURE/FUNCTION ANALYSIS OF THE TH GENE

In this section, I do not intend to summarize all the publications written on this subject. I understand that several contributors to this book with this same subject will summarize the past and the most recent development in the field. However, this section will focus on the important theme which has been developing in this area of study.

Among many interesting molecular biological subjects in TH gene research, three areas of research will be highlighted in this chapter. In my opinion, these three areas are both basic and important. They are the investigations for (a) cis-acting elements in the TH gene for tissue/cell specific expression; (b) cis-acting elements for TH gene transcription; and (c) expression and regulation of various TH mRNA forms, especially in the human brain. Even though this book contains more specific functional analyses of these elements, a short summary is included in this chapter. There are several recent reviews describing some of these subjects which are available (Joh, 1990; Costa and Joh, 1991).

(a) cis-acting elements in the TH gene for the tissue/cell specific expression

As shown in Fig. 1, there are potential cis-acting motifs located in the 5' proximal region of both rat and human genes. They are AP-1, AP-2, POU/OCT, SP-1 and cyclic AMP-response element (CRE) (Kobayashi *et al.*, 1988; Cambi *et al.*, 1989; Carroll *et al.*, 1991a). The relative distance from the transcription initiation sites and the orders of the location of these motifs in both human and rat TH genes are highly conserved. The more upstream sequence has not been well characterized and it is premature to suggest that they are highly conserved. However, our recent characterization of these regions indicated the high conservation of the sequence (unpublished observation). Thus, it is possible to predict that the transcriptional regulation and tissue/cell specific expression are similar in both species. Whether all of these cis-acting elements are necessary for tissue/cell specific expression is unclear. There are several reports which indicate the importance of the AP-1 site (Harrington *et al.*, 1987; Cambi *et al.*, 1989; Gandelman *et al.*, 1991) or the 3' region (Gandelman *et al.*, 1991) of the gene for tissue/cell specific expression. The conclusive answer has to come from the expression in transgenic mice. The expression of the human TH gene in transgenic mice was first performed by Professor Nagatsu's group (Kaneda *et al.*, 1991). They used an 11 kb DNA fragment consisting of 2.5 kb of a 5' upstream region, the entire exon-intron structure and 0.5 kb of the 3' flanking region. The transgenes were expressed specifically in the brain and adrenal gland (Kaneda *et al.*, 1991). Since this group used an entire TH gene, the exact sequences required for the specificity are still not defined. Thus, there is no consensus as to the position of the sequences in the rat or human gene that direct TH expression in a strictly tissue/cell type specific manner. Further studies are necessary.

(b) cis-acting elements in the TH gene for the transcriptional activity

Two cis-acting motifs in the TH gene have been implicated as the transcriptional regulatory elements. They are CRE, which is located at −45 bp upstream of the transcription initiation site of both human and rat TH genes; and AP-1, which is located at −205 bp and −207 bp upstream of the human and rat TH gene, respectively.

The elevation of intracellular cyclic AMP leads to an increase in the transcriptional activity of several eukaryotic genes via CRE, a consensus octamer, 5'TCACGTAC3'. The TH gene is one of these eukaryotic genes. The TH activity (Acheson *et al.*, 1984) and the mRNA in bovine chromaffin cells (Carroll *et al.*, 1991b) and PC12 cells (Lewis *et al.*, 1983; Tank *et al.*, 1986) were increased by the elevation of cAMP. The rate of transcription of the TH gene is also increased in PC12 cells followed by an increase in the TH protein (Lewis *et al.*, 1987). The phorbol 12-myristate 13-acetate (TPA) also leads to an increase in TH mRNA in chromaffin cultures (Hong *et al.*, 1989; Stachowiak *et al.*, 1990; Carroll *et al.*, 1991). Other groups have suggested the AP-1 sequences for enhanced TH expression in PC12 cells (Harrington *et al.*, 1987; Cambi *et al.*, 1989; Gandelman *et al.*, 1991). However, Carroll *et al.* (1991a, 1991b) in recent publications have shown the involvement of CRE in the induction of the TH gene by the elevation of intracellular cAMP. These results are consistent with other neuronal genes which are regulated by cAMP (Comb *et al.*, 1986; Montminy *et al.*, 1986; Hyman *et al.*, 1988; Roesler *et al.*, 1988). The importance of CRE participation in the TH gene transcription is further demonstrated by Kim *et al.* (1992) which showed that both the basal and inducible transcription of the TH gene are dependent upon CRE.

A

CTTAGGAAATCAGCATGGTTCTCCCTGTGTGCCCTGGTTTGGTTA

GAGAGCTCTAGCGGTCTCCTGTCCCACATAATACCAGCCAGCCCC

TGCCCTACGTCGT **GCCTCGGG** CTGAGGG **TGATTCA** GAGGCAG
             AP-2   AP-1

GTGCCTGTGACAGTGG **ATGCAATT** AGATCTAATGGGACGGAGGC
                   POU/OCT

CTTTCTCGTCGCCCTCGCTCCATGCCCACCCCCGCCTCCCTCAGG

CACAGCAGGCGTGGAGAGGATGCGCAGGAGGTAGGAGGTGGGG

GACCCAGAGGGGCTT **TGACGTCAG** CCTGGCC **TTTAAA** GAGGGC
                      CRE
                       -45

GCCTGCCTGGCGAGGGC TGTGGAGACAGAACTCGGGACCACC
                          +1

B

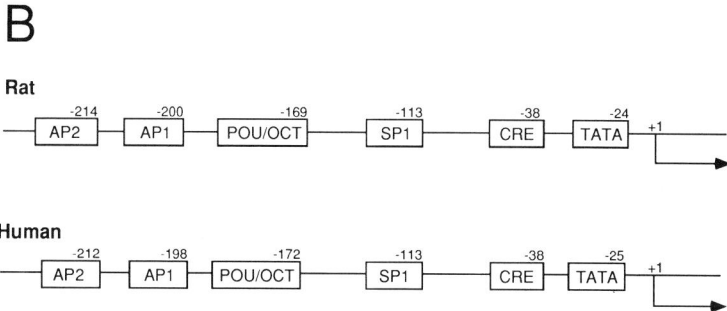

Figure 1. (A) The nucleotide sequences of 5′ upstream of the rat tyrosine hydroxylase gene (Cambi *et al.*, 1988), and (B) comparison of the location of various cis-acting elements in 5′ upstream of the rat and human tyrosine hydroxylase genes (Kobayashi *et al.*, 1988).

It has become apparent that the protein products of the proto-oncogenes c-fos and c-jun are components of the mammalian transcription factor activator protein-1 (AP-1) (Curran and Franza, 1988). There is much physiological and molecular biological evidence indicating the AP-1 in the TH gene is involved in the induction of TH gene transcription (Harrington *et al.*, 1987; Cambi *et al.*, 1989; Gandelman *et al.*, 1991). As previously mentioned, however, this still requires further studies. This author is inclined to believe the possibility that both CRE and AP-1 play a synergistic role in the induction of the TH gene, since the induction of the c-fos gene is directly related to the CRE-motif (Dash *et al.*, 1991). This means that CRE motifs in both the TH and c-fos genes are active in response to the elevation of cellular cAMP level, leading to the induction of the TH gene through the participation of both CRE and AP-1 motifs in the gene.

CONCLUSION

For the conclusion of this short summary of the progress in TH gene research, I would like to extend my heartfelt congratulations to Professor Nagatsu for his long-term participation and the extensive publication record in the field of TH research. We all anticipate Professor Nagatsu's future contributions to this exciting field in the years to come.

REFERENCES

Acheson, A., Naujoks, K. and Thoenen, H. (1984). NGF mediated enzyme induction in primary cultures of bovine adrenal chromaffin cells: specificity and level of regulation. *J. Neurosci.* **4**, 1771–1780.

Black, I. B., Chikaraishi, D. M. and Lewis, E. J. (1985). Trans-synaptic increase in RNA coding for tyrosine hydroxylase in a rat sympathetic ganglion. *Brain Res.* **339**, 151–153.

Cambi, F., Fung, B. and Chikaraishi, D. M. (1989). 5′ Flanking sequences direct cell-specific expression of rat tyrosine hydroxylase. *J. Neurochem.* **53**, 1656–1659.

Carroll, J. M., Kim, K. S., Kim, K. T., Goodman, H. M. and Joh, T. H. (1991a). Effects of second messenger system activation on functional expression of tyrosine hydroxylase fusion gene constructs in neuronal and nonneuronal cells. *J. Mol. Neurosci.* **3**, 65–74.

Carroll, J. M., Evinger, M. J., Goodman, H. M. and Joh, T. H. (1991b). Differential and coordinate regulation of TH and PNMT mRNAs in chromaffin cell cultures by second messenger system activation and steroid treatment. *J. Mol. Neurosci.* **3**, 75–83.

Comb, M., Birnberg, N., Seasholtz, A., Herbert, E. and Goodman, H. M. (1986). A cyclic AMP and phorbol ester-inducible DNA element. *Nature* **323**, 353–356.

Costa, E. and Joh, T. H. (1991). *Neurotransmitter Regulation of Gene Transcription., Fidia Research Foundation Symposium Series,* Vol. 7., Thieme Medical Publication Inc., New York.

Curran, T. and Franja, B. R., Jr (1988). Fos and Jun: the AP1 connection. *Cell* **55**, 395–397.

Dash, P. K., Karl, K. A., Colicos, M. A., Orywes, R. and Kandel, E. R. (1991). cAMP response element-binding protein is activated by Ca^{2+}/calmodulin- as well as cAMP-dependent protein kinase. *Proc. Natl. Acad. Sci. USA* **88**, 5061–5065.

D'Mello, S. R., Weisberg, E. P., Stachowiak, M. K., Turzai, L. M., Gioro, A. E. and Kaplan, B. B. (1988). Isolation and nucleotide sequence of a cDNA clone encoding bovine adrenal tyrosine hydroxylase: comparative analysis of tyrosine hydroxylase gene products. *J. Neurosci. Res.* **19**, 440–449.

Faucon-Biguet, N., Buda, M., Lamouroux, A., Samolyk, D. and Mallet, J. (1986). Time course of changes in tyrosine hydroxylase mRNA in rat brain and adrenal medulla after a single injection of reserpine. *EMBO J.* **5**, 281–287.

Gandelaman, K-Y., Coker, G. T. and O'Malley, K. (1991). Species and regional differences in expression of cell type specific elements in the rat and human tyrosine hydroxylase gene loci. *J. Neurochem.* **55**, 2149–2152.

Grima, B., Lamouroux, A., Blanot, F., Faucon-Biguet, N. and Mallet, J. (1985). Complete coding sequence of rat tyrosine hydroxylase mRNA. *Proc. Natl. Acad. Sci. USA* **82**, 617–621.

Harrington, C., Lewis, E. and Chikaraishi, D. (1987). Identification of cell-type specificity of the tyrosine hydroxylase gene promotor. *Nucleic Acid Res.* **15**, 2363–2384.

Hong, S., Tuominen, R., Kaplan, B., Poisner, A. and Stachowiak, M. (1989). Stimulation of protein kinase C increases TH and PNMT expression in cultured adrenal chromaffin cells. *Soc. Neurosci. Abst.* **15**, 987.

Hyman, S., Comb, M., Lim, Y. S., Pearlberg, J., Green, M. R. and Goodman, H. M. (1988). A common trans-acting factor is involved in transcriptional regulation of neurotransmitter genes by cyclic AMP. *Mol. Cell. Biol.* **8**, 4225–4233.

Joh, T. H., Gagman, C. and Reis, D. J. (1973). Immunochemical demonstration in increased accumulation of tyrosine hydroxylase protein in sympathetic ganglia and adrenal medulla elicited by reserpine. *Proc. Natl. Acad. Sci. USA* **70**, 2767–2771.

Joh, T. (Ed.) (1990). *Catecholamine genes, Neurology and Neurobiology,* Vol. 57, Wiley-Liss, New York.

Kaneda, N., Kobayashi, K., Ichinose, H., Kishi, F., Nakazawa, A., Kurosawa, Y., Fujita, K. and Nagatsu, T. (1987). Isolation of a novel cDNA clone for human tyrosine hydroxylase: Alternative RNA splicing produces four kinds of mRNA from a single gene. *Biochem. Biophys. Res. Commun.* **146**, 971–975.

Kaneda, N., Sasaoka, T., Kobayashi, K., Kiuichi, K., Nagatsu, I., Kurosawa, Y., Fujita, K., Yokoyama, M., Nomura, T., Katsuki, M. and Nagatsu, T. (1991). Tissue-specific and high-level expression of the human tyrosine hydroxylase gene in transgenic mice. *Neuron* **6**, 583–594.

Kim, K. S., Kim, M. K., Carroll, J. and Joh, T. H. (1992). The basal transcription of the tyrosine hydroxylase gene are dependent upon a cAMP-response element. Submitted for publication.

Kobayashi, K., Kaneda, N., Ichinose, H., Kishi, F., Nakazawa, A., Kurosawa, Y., Fujita, K. and Nagatsu, T. (1988). Structure of the human tyrosine hydroxylase gene: Alternative splicing from a single gene accounts for generation of four mRNA types. *J. Biochem.* **103**, 907–912.

Kvetňanský, R. (1973). Trans-synaptic and humoral regulation of adrenal catecholamine synthesis in stress. In: *Frontiers in Catecholamine Research.* E. Usdin and S. Snyder (Eds), Pergamon Press, New York, pp. 223–229.

Lamouroux, A., Vigny, A., Faucon-Biguet, N., Samolyk, D., Pryvat, A., Salomon, J. C., Pujol, J. F. and Mallet, J. (1982). Identification of cDNA clones coding for rat tyrosine hydroxylase antigen. *Proc. Natl. Acad. Sci. USA* **79**, 3881–3885.

Levitt, M., Spector, S., Sjoerdsma, A. and Udenfriend, S. (1965). Elucidation of the rate limiting step in norepinephrine biosynthesis in the perfused guinea-pig heart. *J. Pharmacol. Exp. Therap.* **148**, 1–8.

Lewis, E. J., Tank, A. W., Weiner, N. and Chikaraishi, D. M. (1983). Regulation of tyrosine hydroxylase mRNA by glucocorticoid and cyclic AMP in a rat pheochromocytoma cell line: Isolation of a cDNA clone tyrosine hydroxylase mRNA. *J. Biol. Chem.* **258**, 14632–14637.

Lewis, E. J., Harringto, C. A. and Chikaraishi, D. M. (1987). Transcriptional regulation of the tyrosine hydroxylase gene by glucocorticoids and cyclic AMP. *Proc. Natl. Acad. Sci. USA* **84**, 3553–3557.

Montminy, M., Servarino, J., Wagner, G. and Goodman, H. (1986). Identification of cyclic AMP responsive element of the somatostatin gene. *Proc. Natl. Acad. Sci. USA* **83**, 6682–6686.

Mueller, R. A., Thoenen, H. and Axelrod, J. (1969). Increase in tyrosine hydroxylase activity after reserpine administration. *J. Pharmacol. Exp. Therap.* **169**, 74–79.

Nagatsu, T., Levitt, M. and Udenfriend, S. (1964). Tyrosine hydroxylase: The initial step in norepinephrine biosynthesis. *J. Biol. Chem.* **239**, 2910–2917.

O'Malley, K. L., Anhalt, M. J., Martin, B. M., Kelsoe, J. R., Winfield, S. L. and Ginns, E. I. (1987). Isolation and characterization of the human tyrosine hydroxylase gene: identification of alternative splice sites responsible for multiple mRNAs. *Biochem.* **26**, 6910–6914.

Roesler, W. J., Vandenbark, G. R. and Hanson, R. W. (1988). Cyclic AMP and the induction of eukaryotic transcription. *J. Biol. Chem.* **263**, 9063–9066.

Stachowiak, M., Sebbane, R., Stricker, E., Zigmond, M. and Kaplan, B. B. (1985). Effect of chromic cold exposure on TH mRNA in rat adrenal gland. *Brain Res.* **359**, 356–359.

Stachowiak, M. K., Rigaul, R. J., Lee, P. H. K., Viveros, O. H. and Hong, J. S. (1988a). Regulation of tyrosine hydroxylase and phenyethanolamine N-methyltransferase mRNA levels in sympathoadrenal system by the pituitary-adrenocortical axis. *Mol. Brain Res.* **3**, 275–286.

Stachowiak, M., Stricker, E., Zigmond, M. and Kaplan, B. (1988b). A cholinergic antagonist blocks cold stress induced alterations in rat adrenal TH mRNA. *Mol. Brain Res.* **3**, 93–106.

Stachowiak, M. K., Hong, J. S. and Viveros, O. H. (1990). Coordinate and differential regulation of phenylethanolamine N-methyltransferase, tyrosine hydroxylase and proenkephalin mRNAs by neuronal and hormonal mechanisms in cultured adrenal medullary cells. *Brain Res.* **510**, 277–288.

Tank, A. W., Curella, P. and Ham, L. (1986). Induction of tyrosine hydroxylase mRNA by cyclic AMP and glucocorticoid in rat pheochromocytoma cell line: evidence for regulation of TH by multiple mechanisms. *Mol. Pharmacol.* **30**, 497–503.

Thoenen, H., Mueller, R. A. and Axelrod, J. (1969). Trans-synaptic induction of tyrosine hydroxylase. *J. Pharmacol. Exp. Therap.* **160**, 249–254.

Udenfriend, S., Cooper, J. R., Clark, C. T. and Baer, J. E. (1953). *Science* **117**, 663–666.

Zigmond, R. E. (1989). A comparison of long term and short term regulation of tyrosine hydroxylase activity. *J. Physiol. (Paris)* **83**, 267–271.

Tyrosine Hydroxylase, pp. 163–175
M. Naoi *et al.* (Eds)
© VSP 1993

MULTIPLICITY OF HUMAN cDNAs OF TYROSINE HYDROXYLASE

KAREN L. O'MALLEY

Department of Anatomy and Neurobiology, Washington University School of Medicine, 660 S. Euclid Avenue, St. Louis, MO 63110, USA

INTRODUCTION

Since the discovery of tyrosine hydroxylase (TH) by Drs Nagatsu and Udenfriend in 1964 its kinetic properties, distribution, structure and regulation have been intensely investigated. Understanding the biochemical and physiological aspects involved in the regulation of TH have been further prompted by their possible benefit to studies of human diseases and their treatment. For example, atypical catecholamine function has been implicated in a number of neuropsychiatric and neurological disorders, in particular schizophrenia, affective disorders and Parkinson's disease. The development of gene probes for human TH has allowed the direct testing of these hypotheses. Additionally, full length human TH cDNAs and/or genomic clones have facilitated the development of vector systems by which TH may become a therapeutic tool in diseases such as Parkinson's. This chapter reviews some of the key features of TH gene expression and their significance in human disorders.

TH AND THE AROMATIC AMINO ACID HYDROXYLASE GENE FAMILY

TH cDNAs and genomic structures. The isolation and characterization of TH cDNAs from a variety of species has allowed detailed structural comparisons to be made. After optimal alignment, rat and human TH share 92% amino acid sequence identity (Grima *et al.*, 1985; Brown *et al.*, 1987; Grima *et al.*, 1987; O'Malley *et al.*, 1987; Kobayashi *et al.*, 1988); quail (Fauquet *et al.*, 1988) and mammalian TH are 75–77% identical; *Drosophila* TH (Neckameyer and Quinn, 1989) has only 48–49% identity with the vertebrate hydroxylases.

The deduced amino acid sequence of phenylalanine hydroxylase (Ledley *et al.*, 1985) exhibited an overall 49% identity with human TH confirming earlier suggestions that these enzymes are structurally related. Sequence analysis of rabbit tryptophan hydroxylase (Grennett *et al.*, 1987) added another member to the aromatic amino acid hydroxylase gene family. The latter enzyme is 53% and 48%

identical to PH and TH, respectively. Alignment of all three aromatic amino acid hydroxylases reveals that the mid to C-terminal portion of the proteins is highly conserved, thus, representing the catalytic domain of the protein. The less conserved N-terminal region codes for regulatory sites that are different for each protein.

The conservation of protein sequences among the aromatic amino acid hydroxylases extends to their genomic structures. We have determined the organization of the rat (Brown *et al.*, 1987) and human TH genes (O'Malley *et al.*, 1987); D'Mello *et al.* (1989) determined the bovine TH structure. The TH gene from all species isolated so far is single copy containing 13 primary exons spanning approximately 8 kb (Fig. 1). PH and tryptophan hydroxylase (TrpH) are also single copy with 9-10 identical splice sites within these three genes (Fig. 2). Intron sizes, however, are much larger so that the PH and TrpH span *c.* 90 kb (DiLella *et al.*, 1986) and 21 kb (Stoll and Goldman *et al.*, 1991), respectively. The observation that the majority of introns in PH, TH and TrpH match exactly as to position within the gene and to the phase that each intron interrupts the codon further delineates the evolutionary relationship of these genes.

TH is part of a gene cluster on human chromosome 11p 15.5. Moss *et al.* (1986) mapped TH to the short arm of human chromosome 11, close to the insulin (INS) and H-*ras*-1 genes (Kittur *et al.*, 1985). In order to more precisely determine the

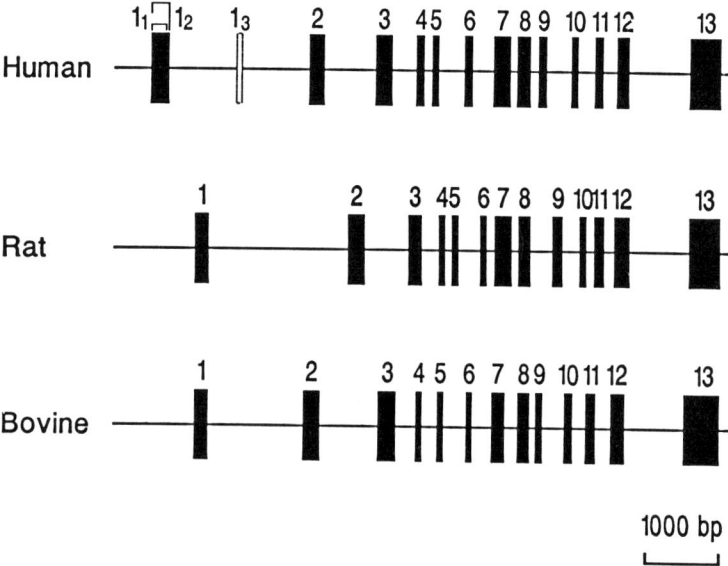

Figure 1. Structures of the human, rat and bovine genes. Exons are depicted by solid boxes, introns by straight lines.

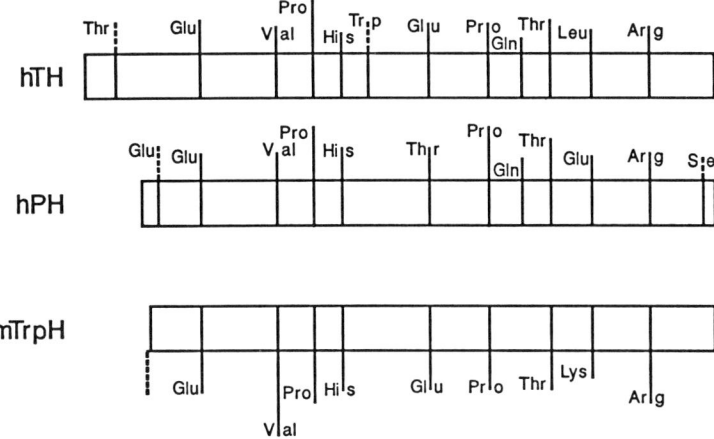

Figure 2. Conservation of splice sites between human TH and PH and murine TrpH genes. Horizontal bars represent protein structures aligned to maximize amino acid homology. The solid vertical lines indicate the shared intron positions. Dashed lines represent non-conserved splice sites.

location of the TH gene we mapped it relative to other loci on chromosome 11p. Surprisingly, we found that the TH gene is 5′ to INS and is separated by only 2.7 kb of flanking DNA (O'Malley and Rotwein, 1988). Both genes have the same transcriptional polarity and form a head-to-tail linkage group with insulin-like growth factor 2 (IGF-2; Fig. 3). The close physical proximity of the three genes is intriguing. INS is part of a gene family which includes relaxin (Blundell *et al.*, 1980) as well as IGF-1 and -2. PH has been mapped to chromosome 12q (Lidsky *et al.*, 1984) adjacent to IGF-1 (Ledley *et al.*, 1988). Based on a similarity in banding patterns and the conservation of linkage groups, previous reports have speculated that human chromosomes 11 and 12 are ancestral homologues created by tetraploidy. As the linkage group TH-INS-IGF-2 is on the short arm of 11 and PH-IGF-1 are on the long arm of 12, a pericentric inversion may have created this arrangement. TrpH has also been mapped to the middle of chromosome 11p (Ledley *et al.*, 1987) suggesting additional rearrangements have occurred.

TH polymorphisms and their possible association with bipolar affective disorder.
Because of the tight linkage of the TH-INS-IGF-2 genes, previously described polymorphisms for INS are identical to those observed with TH. Most of these are due to the hypervariable region of DNA within the 5′ flanking region of the human INS gene (for example, see Rotwein *et al.*, 1981). These include Bgl II, Eco RI, Hinc II and Sac I. Additional restriction site polymorphisms include Rsa I and Taq I, all of which would vary in frequency according to the racial

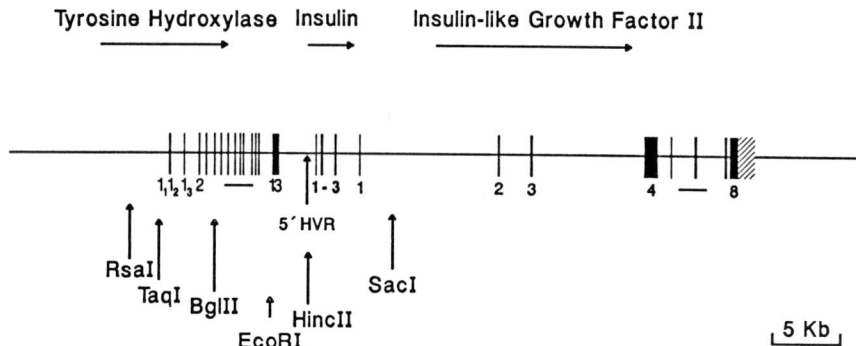

Figure 3. Organization of TH-INS-IGF-2 locus. Head to tail arrangement of the 3 genes spanning approximately 42 kb of chromosomal DNA. The positions of RFLPs are indicated by arrows. The thickened arrow depicts the hypervariable region (5′ HVR) 5′ to the INS gene transcription initiation site.

group studied (Chakravarti *et al.*, 1986). The Rsa I and Taq I polymorphisms are located at the 5′ end of the TH gene within the promoter region (O'Malley *et al.*, 1987; O'Malley and Rotwein, 1988). Since mutations associated with the control of promoter regions may effect levels of gene expression we tested the 5′ RFLPs for their involvement in bipolar affective disorder in the general population.

RFLPs are usually used in large pedigree studies to demonstrate cosegregation of a particular marker with a particular phenotype. The rationale behind this approach is that if a given genetic disorder has only one or a few causes in the general population, then an RFLP of the disease gene itself will be associated with the disease in a population study. In a study group of 51 controls and 30 bipolar affective disorder patients we found no significant associations between affected status and allele frequencies (Todd and O'Malley, 1989). In contrast, using an increased population size (50 bipolar patients), Leboyer *et al.* (1990) found a positive association between the Taq I RFLP and affective disorder. Recently we repeated these experiments with an increased sample size (55 bipolars). We still see no association between TH and the bipolar disorder (O'Malley and Todd, unpublished results). These results suggest either a difference in diagnostic criteria or a population difference in the French subjects used for the Leboyer *et al.* study.

ALTERNATIVE SPLICE SITES GENERATE MULTIPLE HUMAN mRNAs

In contrast to rat and bovine TH, the human gene is alternatively spliced creating at least four different messages (Ginns *et al.*, 1987; Grima *et al.*, 1987; O'Malley *et al.*, 1987; Kaneda *et al.*, 1987). Analysis of the sequence at the end of exon 1 reveals two donor dinucleotides (GT), one 12 nucleotides downstream of the other generating a 3′ extension (Fig. 4). The analogous sequences in the rat (O'Malley *et al.*, 1987) and bovine (D'Mello *et al.*, 1989) are not present. An additional human transcript is created by alternative splicing of 81 nucleotides from within intron 1. Therefore this messenger initiates at the normal ATG codon in exon 1 and includes an additional 27 amino acids before joining exons 2 (Fig. 4). A fourth human TH cDNA would be predicted from the splicing pattern of HTH-2 and HTH-3 (Fig. 4). This was subsequently confirmed by Kaneda *et al.* (1987) by isolation of the cDNA.

We used the unparalleled sensitivity of the polymerase chain reaction (PCR) to confirm and extend these observations in a variety of tissues (see below; Coker *et al.*, 1990). Because none of these alternative exons are present in rat or bovine genes and since all represent variations and modifications of exon 1, we have elected to use terminology reflecting these observations. Therefore, the alternatively spliced exons are referred to as 1_1, 1_2 and 1_3. Exon 1_1 corresponds to the human cDNA sequences analogous to rat and bovine which comprise amino acids 1–30 (Grima *et al.*, 1985; Brown *et al.*, 1987). Exon 1_2 signifies the additional 4 amino acids at the end of exon 1_1; exon 1_3 is used to designate the additional 27 amino acids expressed in certain human tissues (Fig. 4, and see below). We believe this numerology, allows for the most coherent reference between species.

The functional significance of the multiple mRNAs for the human TH gene remains to be determined. A variety of expression systems have been utilized to assess the kinetic properties of each isotype. Horellou *et al.* (1988) used *in vitro* transcription to generate full length TH transcripts which were then injected into *Xenopus* oocytes. While a detailed kinetic analyses is difficult to perform with these extracts, it appears that HTH-1 had the highest activity. In contrast, using both a baculovirus expression paradigm as well as stable transfection in psi-2 cells and NIH 3T3 fibroblasts, in our hands both HTH-1 and HTH-2 had similar activity profiles (17 versus 20 pmol/min/mg and 0.1 mM vs 0.2 mM Km_{Tyr}, respectively; Ginns *et al.*, 1988; 1990). Kobayashi *et al.* (1988) have reported Km_{Tyr} values of 0.1 and 0.16 mm for HTH-1 and HTH-2 in transfected COS cells.

Many studies have demonstrated that multiple phosphorylation sites at the amino-terminal portion of rat TH serve to regulate the activity of this enzyme. All of these sites have been conserved in the human TH isotypes (Grima *et al.*, 1987; O'Malley *et al.*, 1987). Previously, we and others speculated that the additional four amino acids present in HTH-2 would create a new calcium-calmodulin type II protein kinase site at the adjacent serine (Arg-Gly-Gln-Ser) which might have functional significance. However, the recent observation that this same serine (Ser31) is phosphorylated *in vivo* via nerve growth factor and/or phorbol dibutyrate mediator mechanisms (Haycock, 1990) suggests other factors

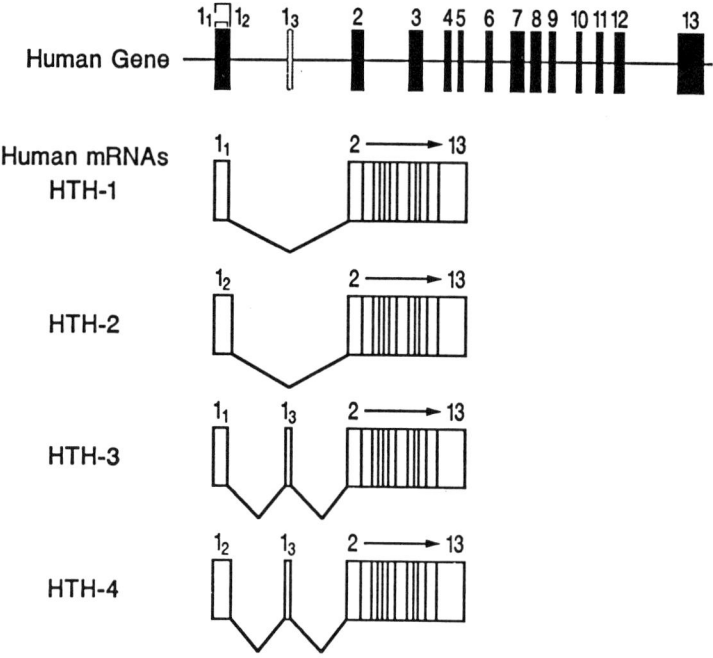

Figure 4. Transcriptional pattern of the human TH gene. Single initiation site but variable splicing generates multiple human TH mRNAs. The various forms of human TH mRNA are schematically depicted below the genomic arrangement from which they are derived.

may be involved. Indeed, it may be that the additional amino acids inserted in HTH-2, -3, and -4 block the phosphorylation of Ser31 instead of leading to its activation. Further, studies are necessary to clarify which mechanism operates *in vivo*.

Tissue distribution of multiple human TH transcripts. In order to determine the tissue-specific expression of the multiple human transcripts we performed a detailed study of the distribution and expression of the different forms in various neuroendocrine tissues. Because HTH-2 is only 12 bp larger than HTH-1, standard methodologies for detecting these 2 messages would not have been successful. Instead, we adapted the technique of amplification of RNA transcripts via the polymerase chain reaction. The RNA-PCR method is a rapid, specific and extremely sensitive methodology. By using primers complementary to exons flanking the alternative transcripts these mRNAs will be simultaneously co-amplified if present in the tissue. The location and predicted size of the resulting PCR fragments are shown in Fig. 5.

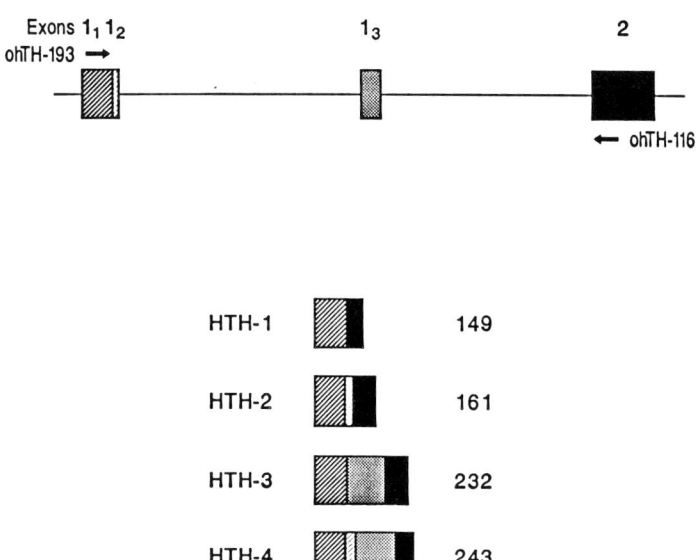

Figure 5. PCR strategy for detecting multiple human transcripts. Predicted size (bp) of PCR fragments are shown on the right using indicated primers.

Using PCR analysis we determined that many human neuronal tissues express all 4 forms of human TH. In addition to the tissues normally recognized as having substantial levels of TH such as the superior cervical ganglia, the adrenal gland, the adrenergic C1 region, locus coeruleus, substantia nigra and hypothalamus, the caudate and putamen also have appreciable levels of TH (Fig. 6). In every tissue tested, the major forms of TH mRNA expressed were HTH-1 and -2 in approximately a 50:50 ratio. HTH-3 and -4 accounted for only minor amounts of the total TH mRNA. Forms 3 and 4 were highest in pheochromocytoma tissue (3.5 and 5.1%, respectively) followed by adrenal medullary tissue (2.3–2.8%) and on average were less than 1% of all other tissues surveyed. These results are based on a single individual and thus may not be completely representative of the population as a whole. Further studies are in progress to address this point.

The reason for the absence of alternative exons in the rat and bovine TH genes and their presence in human is not clear. The donor site at exon 1_2 would be predicted to be the most stable suggesting that splicing should occur predominately at that position. These 12 bp share 50% identity with the rat splice site junction. However, the human donor bases (100–108) share only 22% identity with the analogous rat sequences indicating this site probably arose by multiple mutations. Therefore, alternative splicing at the TH locus appears to be a relatively recent evolutionary trend.

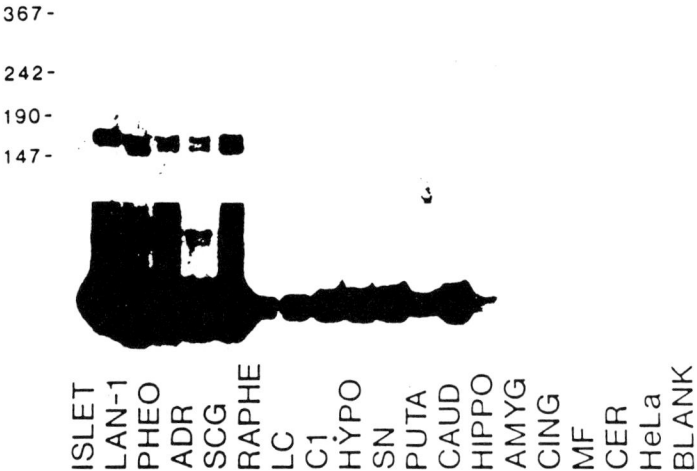

367-

242-

190-

147-

ISLET LAN-1 PHEO ADR SCG RAPHE LC C1 HYPO SN PUTA CAUD HIPPO AMYG CING MF CER HeLa BLANK

Figure 6. Tissue distribution of human TH alternative transcripts. Total RNA was analyzed (Coker *et al.*, 1990) using the polymerase chain reaction to detect the four human transcripts. A representative experiment is shown above where the same membrane was exposed for 3 h (top) or 2 days (bottom). The top panel demonstrates that the major forms of the TH mRNA are HTH-1 and HTH-2 which are expressed in approximately a 1 to 1 ratio. Markers indicate location of standards (bp). Abbreviations: ADR, adrenal; AMYG, amygdala; CAUD, caudate; CER, cerebellum; CING, cingulate cortex; HIPPO, hippocampus; HYPO, hypothalamus; LC, locus coeruleus; PHEO, pheochromocytoma; PUTA, putamen; SCG, superior cervical ganglia; SN, substantia nigra.

TH AS A THERAPEUTIC TOOL

As the first step towards the development of methods combining gene transfer and brain graft therapy, we have begun to develop techniques and cell lines amenable to implantation that express (or over-express) TH. Such cell lines (or methods of introducing TH into primary cultures) may provide a purified population of cells expressing this pivotal enzyme in the production of dopamine and other catecholamines. These cells could then be used for reimplantation trials in animal models of Parkinson's disease or other disorders. Eventually these cells might become an anatomically specific analogue of systemic L-DOPA therapy in humans.

L-DOPA production in cells genetically modified by TH/retroviral vectors. In collaboration with Dr Fred Gage (UCSD) we have utilized a retroviral vector system for the stable and high efficiency transfer of TH into possible candidate cells. The rat TH cDNA was inserted into a Moloney murine leukemia virus derived vector (Wolff *et al.*, 1989) and transmissible virus was generated by

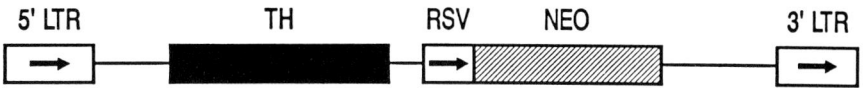

Figure 7. Structure of TH/retroviral vector. This vector expresses TH from the 5' LTR and the selectable bacterial *neo* gene from an internal RSV promoter.

transfection of appropriate helper cells (Fig. 7). Supernatant from these cells was then used to infect various cell lines.

Initially, fibroblasts were tested for their ability to take up and express TH. A fibroblast cell line, 208F, was infected with the retroviral TH construction and screened for production of L-DOPA. Cell extracts of 208F cells infected with the TH vector had significant levels of TH activity (10.8 pmoles L-DOPA/min/mg protein) and produced L-DOPA (1.3 ng/mg protein). These 208F cells expressing TH *in vitro* were implanted into the caudate of animals with unilateral 6-OHDA lesions in the nigrostriatal bundle. Prior to implantation, these animals demonstrated the expected asymmetry in rotational behavior. Following implantation, we found significant decreases in rotational asymmetry for the TH expressing 208F cells but not for the untransfected parent cell line. These experiments represent the first successful alleviation of behavioral impairment using genetically modified cells expressing the missing neurotransmitters.

TH-expressing Schwann cells. Another possible host cell is the Schwann cell, the glial cell responsible for myelinating the peripheral nervous system. In addition to normally being found in association with neurons, it is possible to obtain Schwann cells from peripheral nerve biopsies that could serve as a source of autologous cells for transplantation. Therefore, using the recombinant TH retrovirus previously developed (Wolff *et al.*, 1989). We established an immortalized Schwann cell line that stably expressed high levels of TH and secreted L-DOPA (Owens *et al.*, 1991).

We subsequently established cultures of primary Schwann cells that expressed high levels of TH immunoreactivity and also produced L-DOPA. When co-cultured with dorsal root ganglion neurons in myelin promoting medium, the TH-expressing Schwann cells retained the capacity to differentiate (Owens *et al.*, 1991). These studies demonstrate the feasibility of developing peripheral cells as an autologous source that can be genetically modified and reimplanted in a CNS lesion site.

L-DOPA production from HSV/TH vectors. Despite these encouraging results there are problems associated with the retroviral vector systems. These include: low titers, the requirement for replicating cells and the inherent cumbersomeness of the system. New vector systems and methodologies are being developed that may prove superior. Herpes simplex virus, in particular, looks promising since it 1) has a wide host range, 2) can infect postmitotic cells such as neurons, 3) can be maintained indefinitely in a latent state, and 4) much of its large genome can be deleted suggesting a higher capacity for foreign gene insertion. Geller and Breakefield (1988) have developed defective herpes simplex virus, type 1 (HSV-1) vectors to introduce genes into neurons. Using the Lac Z reporter gene, Geller and Freese (1990) have demonstrated stable β-galactosidase activity in primary cultures of spinal cord, cerebellum, thalamus, basal ganglia, hippocampus as well as various cortical regions. Furthermore, Boothman *et al.* (1990) have shown that HSV-1 vectors can express β-galactosidase in a wide range of human cells, including several of neural origin. Since other genes could replace Lac Z, this type of vector may prove useful for delivering genes into the CNS directly.

In collaboration with Dr A. Geller (Dana-Farber Institute) we have inserted the human TH cDNA into the pHSV derived vector (Fig. 8). The pHSV/TH plasmid was then used to generate virus for the introduction of TH into candidate cells.

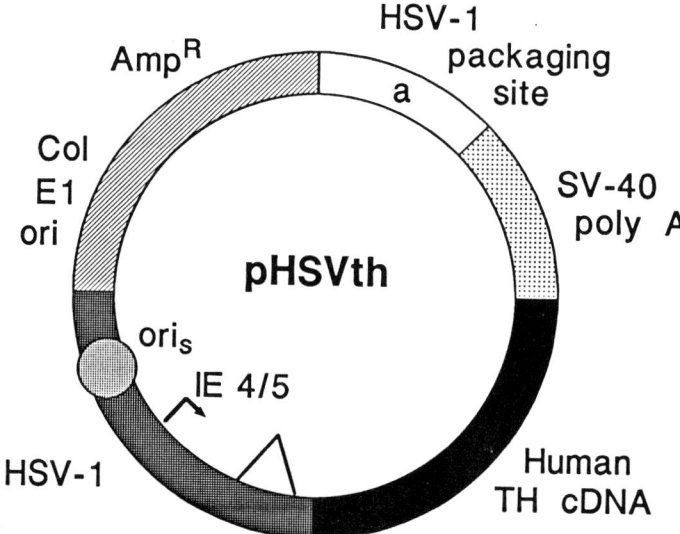

Figure 8. The structure of the pHSVth vector. The transcription unit in pHSVth is composed of the HSV-1 IE 4/5 promoter (arrow), the intervening sequence following that promoter (triangle), the human TH cDNA (black segment), and the SV-40 early region polyadenylation site (dotted segment). Two sequences are required for propagation of pHSVlac in a HSV-1 virus stock: The HSV-1 origin of DNA replication, ori$_s$ (circle filled with dots), supports replication of pHSVth DNA. The HSV-1 a sequence (clear segment) contains the packaging site which is responsible for subsequently packaging pHSVth DNA into HSV-1 virus particles. 3. Sequences from pBR322 (diagonal line segment) support the growth of pHSVth DNA in *E. coli.*

Initially, we have used fibroblasts to test the effectiveness of the HSV/TH virus and have preliminary evidence that infected cells express TH. Specifically, we have shown that HSV/TH DNA is properly packaged into HSV-1 virus particles, that pHSV/TH DNA is transcribed to yield RNA containing the TH gene, that TH protein is synthesized as assayed by immunoreactivity, and that functional TH enzyme is synthesized as demonstrated by a TH enzyme assay.

Since HSV-1 vectors can stably express a gene in cultured striatal neurons (Geller and Freese, 1990) and in neurons in the adult rat brain following stereotaxic injection of virus particles (Geller *et al.*, 1989), it may be possible to synthesize functional TH enzyme in striatal neurons, both in culture and in the adult brain. Therefore, we infected dissociated rat striatal neurons with the pHSVth virus. Immunocytochemistry with anti-TH antibodies revealed approximately 70% of the cells expressed TH as compared with mock infected controls. Even more promisingly, it appears that dopamine levels 6-fold over background can be released from these cultures in a regulated fashion. By extrapolation of these values with the number of infected neurons per culture, it would appear that the amount of L-DOPA produced per infected neuron would be in the same range as proved effective in the rat model of Parkinsonism described above (Wolff *et al.*, 1989). Therefore, pHSV/TH should be an effective therapeutic tool if it can be delivered into a sufficient number of cells *in vivo*. These studies are now in progress.

SUMMARY

TH continues to be an enzyme of sustaining interest for many researchers. Since Dr Nagatsu's pioneering work in the 1960's, TH has been studied not only for its possible role in disorders such as schizophrenia and manic-depression but also as a therapeutic tool in its own right for illnesses such as Parkinson's disease. Without a doubt the next 30 years of TH research will be as exciting and provocative as the last.

Acknowledgements

This work has been supported in part by the American Parkinson's Disease Association, The Monsanto Corporation, The Theodore and Vada Stanley Foundation and MH 50081.

REFERENCES

Blundell, T. L. and Humbel, R. E. (1980). Hormone families: pancreatic hormones and homologous growth factors. *Nature (London)* **287**, 781–787.

Boothman, D. A., Geller, A. I. and Pardee, A. B. (1989). Expression of the *E. coli* Lac Z gene from a defective HSV-1 vector in various human normal, cancer-prone and tumor cells. *FEBS Lett.* **258**, 159–162.

Brown, E. R., Coker, G. T. III and O'Malley, K. L. (1987). Organization and evolution of the rat tyrosine hydroxylase gene. *Biochemistry* **26**, 5208–5212.

Chakravarti, A., Elbein, S. C. and Permutt, M. A. (1986). Evidence for increased recombination near the human insulin gene: implication for disease association studies. *Proc. Natl. Acad. Sci. USA* **83**, 1045–1049.

Coker, G. T., III, Studelska, D., Harmon, S., Burke, W. and O'Malley, K. L. (1990). Analysis of tyrosine hydroxylase and insulin transcripts in human neuroendocrine tissues. *Mol. Brain Res.* **8**, 93–98.

D'Mello, S. R., Turzai, L. M., Gioio, A. E. and Kaplan, B. B. (1989). Isolation and structural characterization of the bovine tyrosine hydroxylase gene. *J. Neurosci. Res.* **23**, 31–40.

DiLella, A., Kwok, S. C. M., Ledley, F. D., Marvit, J., and Woo, S. L. C. (1986). Molecular structure and polymorphic map of the human phenylalanine hydroxylase gene. *Biochemistry* **25**, 743–749.

Fauquet, M., Grima, B., Lamourous, Z. and Mallet, J. (1988). Cloning of quail tyrosine hydroxylase: amino acid homology with other hydroxylases discloses functional domains. *J. Neurochem.* **50**, 142–148.

Geller, A. I. and Freese, A. (1990). Infection of cultured central nervous system neurons with a defective herpes simplex virus 1 vector results in stable expression of *Escherichia coli* β-galactosidase. *Proc. Natl. Acad. Sci. USA* **87**, 1149–1153.

Geller, A. I. and Breakefield, X. O. (1988). A defective HSV-1 vector expresses *Escherichia coli* β-galactosidase in cultured peripheral neurons. *Science* **241**, 1667–1669.

Ginns, E. I., Rehavi, M., Martin, B. M., Weller, M., O'Malley, K. L., LeMarca, M. E., McAllister, C. G. and Paul, S. M. (1988). Expression of human tyrosine hydroxylase cDNA in invertebrate cells using a baculovirus vector. *J. Biol. Chem.* **263**, 7406–7410.

Ginns, E. I., Kelsoe, J. R., Martin, B. M., Winfield, S. L., LaMarca, M. E., Luu, M. D., Paul, S. M. and O'Malley, K. L. (1987). Cloning and expression of multiple cDNA's for human tyrosine hydroxylase. *Soc. Neurosci. Abs.* **13**, 859.

Ginns, E. I., Rehavi, M., Martin, B. M., O'Malley, K. L., LeMarca, M. E., Stubblefield, B. K., Winfield, S. and Paul, S. M. (1990). Isolation and expression of cDNA clones for human tyrosine hydroxylase. In: *The Catecholamine Genes.* T. Joh (Ed.). Wiley-Liss, Inc., New York, pp. 23–42.

Grenett, H. E., Ledley, F. D., Reed, L. I. and Woo, S. L. C. (1987). Full-length cDNA for rabbit tryptophan hydroxylase: functional domains and evolution of aromatic amino acid hydroxylases. *Proc. Natl. Acad. Sci. USA* **84**, 5530–5534.

Grima, B., Lamouroux, A., Blanot, F., Biguet, N. F. and Mallet, J. (1985). Complete coding sequence of rat tyrosine hydroxylase mRNA. *Proc. Natl. Acad. Sci. USA* **82**, 617–621.

Grima, B., Lamouroux, A., Boni, C., Julien, J.-F., Javoy-Agid, F. and Mallet, J. (1987). A single human gene encoding multiple tyrosine hydroxylases with different predicted functional characteristics. *Nature (London)* **326**, 707–711.

Haycock, J. W. (1990). Phosphorylation of tyrosine hydroxylase *in situ* at serine 8, 19, 31, and 40. *J. Biol. Chem.* **265**, 11682–11691.

Horellou, P., Le Bourdelles, B., Clot-Humbert, J., Guibert, B., Leviel, V. and Mallet, J. (1988). Multiple human tyrosine hydroxylase enzymes, generated through alternative splicing, have different specific activities in *Xenopus* oocytes. *J. Neurochem.* **51**, 652–655.

Kaneda, N., Kobayashi, K., Ichinose, H., Kishi, F., Nakazawa, A., Kurosawa, Y., Fujita, K. and Nagatsu, T. (1987). Isolation of a novel cDNA clone for human tyrosine hydroxylase: alternative RNA splicing produces four kinds of mRNA from a single gene. *Biochem. Biophys. Res. Commun.* **146**, 971–975.

Kittur, S. D., Hoppener, J. W. M., Antonarakis, S. E., Daniels, J. D. J., Meyers, D. A., Maestri, N. E., Jansen, M., Korneluk, R. G., Nelkin, B. D. and Kazazian, H. H., Jr (1985). Linkage map of the short arm of human chromosome 11: location of the genes for catalase, calcitonin, and insulin-like growth factor II. *Proc. Natl. Acad. Sci. USA* **82**, 5064–5067.

Kobayashi, K., Kaneda, N., Ichinose, H., Kishi, F., Nakazawa, A., Kurosawa, Y., Fujita, K. and Nagatsu, T. (1988). Structure of the human tyrosine hydroxylase gene: alternative splicing from a single gene accounts for generation of four mRNA types. *J. Biochem.* **103**, 907–912.

Leboyer, M., Malafosse, A., Boularand, S., Campion, D., Gheysen, F., Samolyk, D., Henriksson, B., Denise, E., des Lauriers, A., Lepine, J. P., Zarifian, E., Clerget-Darpoux, F. and Mallet, J. (1990). Tyrosine hydroxylase polymorphisms associated with manic depressive illness. *Lancet* **335**, 1219.

Ledley, F. D., DiLella, A. G., Kwok, S. C. M. and Woo, S. L. C. (1985). Homology between phenylalanine and tyrosine hydroxylases reveals common structural and functional domains. *Biochemistry* **24**, 3389–3394.

Ledley, F. D., Grenett, H. E., Bartos, D. P., van Tuinen, P., Ledbetter, D. H. and Woo, S. L. (1987). Assignment of human tryptophan hydroxylase locus to chromosome 11: gene duplication and translocation in evolution of aromatic amino acid hydroxylases. *Somat. Cell Mol. Genet.* **13**, 575–580.

Lidsky, A. S., Robson, K. J. H., Thirumalachary, C. H., Barker, P. E., Ruddle, F. H. and Woo, S. L. C. (1984). The PKU locus in man is on chromosome 12. *Am. J. Hum. Genet.* **36**, 527–533.

Moss, P. A. H., Davies, K. E., Boni, C., Mallet, J. and Reeders, S. T. (1986). Linkage of tyrosine hydroxylase to four other markers on the short arm of chromosome 11. *Nucl. Acids Res.* **14**, 9926–9932.

Neckameyer, W. S. and Quinn, W. G. (1989). Isolation and characterization of the gene for *Drosophila* tyrosine hydroxylase. *Neuron* **2**, 1167–1175.

O'Malley, K. L. and Rotwein, P. (1988). Human tyrosine hydroxylase and insulin gene are contiguous on chromosome 11. *Nucl. Acids Res.* **16**, 4437–4446.

O'Malley, K. L., Anhalt, M. J., Martin, B. M., Kelsoe, J. R., Winfield, S. L. and Ginns, E. I. (1987). Isolation and characterization of the human tyrosine hydroxylase gene: identification of 5' alternative splice sites responsible for multiple mRNAs. *Biochemistry* **26**, 6910–6914.

Owens, G. C., Johnson, R., Bunge, R. P. and O'Malley, K. L. (1991). L-3,4-Dihydroxyphenylalanine synthesis by genetically modified Schwann cells. *J. Neurochem.* **56**, 1030–1036.

Rotwein, P., Chyn, R., Chirgwin, J., Cordell, B., Goodman, H. M. and Permutt, M. A. (1981). Polymorphism in the 5'-flanking region of the human insulin gene and its possible relation to Type 2 Diabetes. *Science* **213**, 1117–1120.

Stoll, J. and Goldman, D. (1991). Isolation and structural characterization of the murine tryptophan-hydroxylase gene. *J. Neurosci.* **28**, 457–465.

Todd, R. D. and O'Malley, K. L. (1989). Population frequencies of tyrosine hydroxylase restriction fragment length polymorphisms in bipolar affective disorder. *Biol. Psychiatry* **25**, 626–630.

Tyrosine Hydroxylase, pp. 177–191
M. Naoi *et al.* (Eds)
© VSP 1993

THE HUMAN TYROSINE HYDROXYLASE GENE

TOSHIHARU NAGATSU,[1] NORIO KANEDA,[2] KAZUTO KOBAYASHI,[1]
HIROSHI ICHINOSE,[1] TOSHIKUNI SASAOKA,[2] KAZUTOSHI KIUCHI,[3]
KEISUKE FUJITA[1] and YOSHIKAZU KURÓSAWA[1]

[1] *Institute for Comprehensive Medical Science, School of Medicine,*
Fujita Health University, Toyoake, Aichi 470-11, Japan
[2] *Department of Biochemistry, Nagoya University School of Medicine,*
Nagoya 466, Japan
[3] *Radioisotope Center Medical Branch, Nagoya University, Nagoya 466, Japan*

Abstract—Tyrosine hydroxylase (TH) is the first, rate-limiting enzyme for catecholamine biosynthesis, and it is a tetrahydropterin-requiring, iron-containing monooxygenase. Both cDNAs and genomic DNA of human TH have been cloned and the nucleotide sequences have been determined. As a result, four similar but distinct mRNAs (types 1–4) are found to be produced through alternative mRNA splicing from a single gene. The human TH gene is split into 14 exons interrupted by 13 introns, spanning approximately 8.5 kilobase pairs (kbp). The 12-bp insertion sequence in type 2 and type 4 mRNA is encoded by the 3′-terminal portion of the first exon. The 81-bp sequence in type 3 and type 4 mRNA corresponds to the second exon. Two kinds of alternative splicing are involved: the alternative use of the two donor sites in the first exon and the inclusion/exclusion of the second exon. The four types were expressed in COS cells, and all had enzyme activities. The type 1 enzyme had the highest homospecific activity (activity per enzyme protein), the values for the other enzymes ranging from 30 to 40%. In human brains and arenal medulla, the distributions of the four types are: type 2 ≥ type 1 ≫ type 4 ≫ type 3. Type 3 is almost undetectable. We have produced transgenic mice carrying human TH gene and characterized the features. The expression level of human TH mRNA in brain and adrenals was about 50-fold and 5-fold higher than that of endogenous mouse TH mRNA. A large difference was observed in brain between the amount of human TH mRNA detected by *in situ* hybridization (about 50-fold) and the amount of human TH protein detected by immunohistochemistry (about 5-fold). Furthermore, catecholamine content did not change significantly. These results in transgenic mice suggest that the enzymatic activity of TH may be regulated in brain so as to minimize the effect of transgene and to maintain neural homeostasis even though the aberrant high level expression of TH mRNA has been brought about.

Key words: human tyrosine hydroxylase; cDNAs; genomic DNA; alternative splicing; multiple mRNAs; gene expression; transgenic mice.

PROPERTIES AND PHYSIOLOGICAL ROLES OF TH

Tyrosine hydroxylase (TH) catalyzes the first step of catecholamine biosynthesis (Nagatsu et al., 1964). TH requires both L-tyrosine and molecular oxygen as substrates and a tetrahydropterin (2-amino-4-hydroxy-5,6,7,8-tetrahydropteridine) as cofactor, producing L-dopa, H_2O, and a quinonoid dihydropterin as the products. The natural pterin cofactor of pterin-dependent monooxygenases is tetrahydrobiopterin, which is also most effective as the cofactor of TH reaction (Brenneman and Kaufman, 1964).

TH is purified from bovine adrenal medulla (Nagatsu and Oka, 1987), rat adrenals (Fujisawa and Okuno, 1987) and rat pheochromocytoma (Tank and Weiner, 1987). We purified human TH from adrenals (Mogi et al., 1984) and brain (Mogi et al., 1986). Rat, bovine or human TH is a tetrameric protein of about 240 kDa, consisting of a subunit of about 60 kDa. Each subunit is assumed to have a catalytic center that binds the substrates, tyrosine and molecular oxygen, and the pterin cofactor, tetrahydrobiopterin. The catalytic domain is localized in the C-terminal region, while a regulatory domain is localized in the N-terminal region (Ledley et al., 1985). Phosphorylation at the N-terminal regulatory domain by various protein kinases activated TH and dephosphorylation by phosphatase inactivates TH. TH in human adrenals and brain is composed of both active and less active forms (Mogi et al., 1984, 1986; Ishii et al., 1990). The less active forms can be detected by enzyme immunoassay and Western blot analysis. In human brain, besides the active form three less active forms were detected by two-dimensional electrophoresis and Western blot analysis. Both active and less active forms had similar Mr but different pI values. The active form had the pI at pH 6.0, whereas the pI of the less active forms ranged from 5.3 to 5.8 (Mogi et al., 1986). The multiple forms could not be observed in bovine TH. As described below, human TH has 4 types of mRNA encoding different proteins. The question is whether or not the active and less active forms correspond to the different types of human TH.

TH is acutely activated following the release of the catecholamine neurotransmitter: (1) by the removal of feedback inhibition following the release of catecholamines which inhibit TH in competition with the pterin cofactor (Nagatsu et al., 1964; Udenfriend et al., 1965); and (2) by phosphorylation mainly catalyzed by cyclic AMP-dependent protein kinase and Ca^{2+}/calmodulin-dependent protein kinase II (Review, Zigmond et al., 1989). Under chronic activation of catecholamine-containing cells, the enzyme protein is induced (Mueller et al., 1969). The activation of the gene expression is regulated by cyclic AMP, glucocorticoids or nerve growth factor (NGF) (Review, Dahlström et al., 1988).

TH may play an important role in the etiology of some diseases attributed to the presumed impairment of central catecholaminergic neurons such as Parkinson's disease, manic depressive illness, and schizophrenia. TH as well as the biopterin cofactor are decreased in the nigrostriatal dopamine neurons of Parkinsonian patients (Nagatsu *et al.*, 1981; Nagatsu *et al.*, 1984). Linkage analysis of an autosomal dominant type of manic-depressive illness indicated that a mutant gene is closely linked to the TH gene locus on chromosome 11, suggesting that some defect in the TH gene may cause the manic-depressive illness (Egeland *et al.*, 1987).

MULTIPLE mRNAs OF HUMAN TH

Because it was difficult to obtain sufficient amounts of TH protein to elucidate the complete amino acid sequence, the primary structure of TH should have been deduced from the nucleotide sequence of TH cDNA. TH cDNA has been isolated and characterized from rat (Grima *et al.*, 1985), mouse (Ichikawa *et al.*, 1991), cow (D'Mello *et al.*, 1988), quail (Fauquet *et al.*, 1988), and *Drosophila melanogaster* (Neckameyer *et al.*, 1989). The TH mRNA from non-primate animals is single and has high homology among various species. The rat TH mRNA has the open reading frame, including the initiation code, of 1,494 base pairs (bp) that code 498 amino acids (Grima *et al.*, 1985). Grima *et al.* (1987) and we (Kaneda *et al.*, 1987; Kobayashi *et al.*, 1987) have found 4 types of human TH (types 1–4) by cDNA cloning (Fig. 1). Nucleotide sequence analysis of full-length cDNA of types 1 and 2 (Grima *et al.*, 1987), type 3 (Kobayashi *et al.*, 1987) and type 4 (Kaneda *et al.*, 1987) was completed to deduce the amino acid sequences. These mRNAs are constant for the major part but are distinguishable from one another, as is the insertion/deletion of 12- and 81-bp sequences, respectively between the 90th and 91st nucleotides of type 1. Type 1 is the shortest and is similar to TH mRNA of various animal species. Type 4 TH mRNA codes the longest human TH molecules and has 93-bp sequence composed of the 12- and 81-bp sequence inserted into type 1. This insertion does not alter the reading frame of the protein-coding region. Type 2 and type 3 have the 12- and 81-bp insertion sequences, respectively.

Southern blot analysis of human genomic DNA suggested that human TH gene exists as a single gene per haploid DNA, indicating that these different mRNAs are produced through alternative mRNA splicing from a single primary transcript (Kaneda *et al.*, 1987). We (Ichikawa *et al.*, 1990) detected two types of TH mRNA corresponding to type 1 and type 2 in adrenals and brain of two species of monkeys (*Macaca irus* and *Macaca fuscata*), but only a single form of TH mRNA in *Sunkus murinus*, one species of insectivore, and rat, one species of rodent. Thus we suggest that the multiplicity of TH mRNA is primate-specific. We believe that the TH multiplicity offers an excellent model for studying the evolution of an alternative mRNA splicing.

We (Kobayashi *et al.*, 1988) as well as other workers (O'Malley *et al.*, 1987; Le Bourdelles *et al.*, 1988) isolated genomic clones encoding the human TH gene

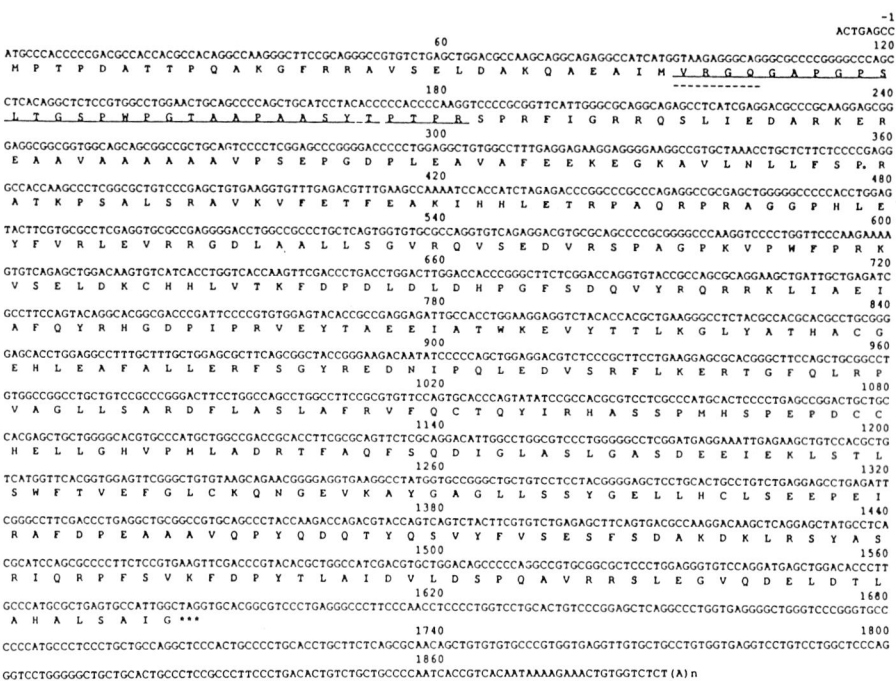

Figure 1. Nucleotide sequence and the deduced amino acid sequence of type 4 human TH cDNA. Type 2 and type 3 human TH cDNA contain the 12-bp (4 amino acids) sequence on the dotted line alone and the 81-bp (27 amino acids) sequence on the solid line alone, respectively. Type 1 human TH cDNA lacks the 93-bp (31 amino acids) sequence on the both solid and dotted lines. (Kaneda et al., 1987).

and determined the nucleotide sequence. The nucleotide sequence of the human TH gene indicated that it is composed of 14 exons interrupted by 13 introns, spanning approximately 8.5 kb (Fig. 2). We determined the nucleotide sequences of all exons and their surrounding regions. The nucleotide sequence of the coding region was the same as that of type 4 cDNA. We predicted that the inserted 81-bp fragment is encoded by an independent exon, but that the inserted 12-bp fragment is encoded by part of an exon, and the exon/intron organization of the human TH gene confirmed these predictions. The 12-bp insertion sequence is derived from the 3'-terminal portion of exon 1 and the 81-bp insertion sequence is encoded by exon 2. The N-terminal region is encoded by the 5'-portion of exon 1, and the remaining region from exon 3 to exon 14 is common to all four kinds of mRNA. Since the transcription of this gene seems to start at a single position, we concluded that the four types of human TH mRNA are produced through alternative splicing from the same primary transcript. There are two modes of alternative splicing. One is the alternative use of two donor sites in exon 1 (the sequence from 117 to 125 and the sequence from 129 to 135). The selection of these potential donor sites determines the insertion/deletion of the 12-bp sequence. The other mode is the inclusion/exclusion of an entire exon 2. Expression of type 1/2 or type 3/4 human TH mRNA is determined by exclusion or inclusion of exon 2 in the spliced products. All the other 12 exons downstream from exon 2 are spliced and incorporated into mature mRNA. We have proposed the potential secondary structure of the primary transcript, which may be involved in the inclusion/exclusion of exon 2 (Fig. 3). Computer-assisted analysis of the secondary structure of the primary transcript led to prediction of 4 stable hairpin loops in introns 1 and 2. In intron 1, the sequence from 299 to 325 complements that from 1,005 to 1,029 (structure A). In intron 2, there are two hairpin loops: the sequence from 1197 to 1213 complements that from 1536 to 1553 (structure B), and the sequence from 1426 to 1443 is complementary to that from 1842 to 1859 (structure C). Furthermore the sequence from 1004 to 1017 in intron 1 pairs with that from 1500 to 1512 in intron 2. These potential secondary structures may account for the inclusion/exclusion of exon 2. Hairpin loops of A, B, and C may facilitate the inclusion of exon 2 into mRNA by juxtaposing exons 1 and 2 as well as exons 2 and 3. When the hairpin loop of structure D is formed, the hairpin loop of structure A should be destroyed, because the sequence from 1005 to 1016 is common to structure A and D. The hairpin loop of structure D may be involved in the joining of exon 1 directly to exon 3, preventing the inclusion of exon 2 into mRNA by physically separating exon 2 from exon 1 and 3. Concerning the mechanism of tissue-specific differentiation of alternative mRNA splicing, we assume the presence of trans-acting factors that stabilize the hairpin structure and discriminate between structures A and D. Transcription of the TH gene is regulated in a tissue-specific manner and by inducible factors. We (Kobayashi *et al.*, 1988) compared the 5'-flanking region of the human TH gene with that of the rat TH gene (Lewis *et al.*, 1987). They show approximately 70% homology and there are many conserved sequences. One conserved sequence is homologous to the consensus sequence of the binding site of the transcription factor, Sp1, which is known to activate the transcription initiation of many viral and mammalian genes (Dynan *et al.*, 1985). Another sequence is highly homologous to the conserved sequence required for the cyclic AMP response element (CRE).

EXPRESSION OF HUMAN TH IN CELLS

We (Kobayashi *et al.*, 1988) expressed the type 1–4 human TH in COS cells. Expression vectors plasmid pAS-TH 1–4, having human TH type 1–4 cDNA, respectively, were constructed and COS cells were transfected with plasmid DNA. COS cells transfected with each of pAS-TH 1–4 gave major immunoreactive

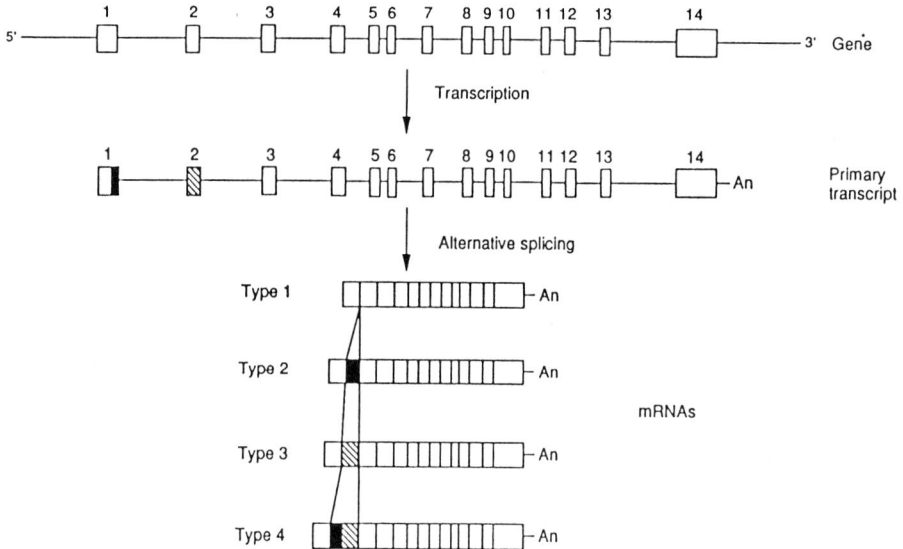

Figure 2. Schematic diagram of regulation of the alternative splicing pathway producing the 4 types of human TH mRNA from a single gene. The 3'-terminal portion of exon 1, which corresponds to the 12-bp insertion sequence, is indicated by a filled box. The hatched box shows exon 2 that encodes the 81-bp insertion sequence. (Kobayashi *et al.*, 1988).

bands at 61, 61, 65 and 65 kDa, respectively, by Western blotting, and exhibited TH activity. The K_m values for tyrosine ranged from 100 to 250 μM, and those for (6RS)-methyltetrahydropterin from 180 to 250 μM, and they are similar among the 4 types. The relative homospecific activity (TH activity (V_{max}) per TH protein) are different among the 4 types of human TH. Type 1 human TH had the highest homospecific activity, the values for the other enzymes ranging from about 0.3 to 0.4. Thus the insertion sequences appear to inhibit the TH activity.

In collaboration with us, Uchida, Kohsaka and co-workers (Uchida *et al.*, 1988; Uchida *et al.*, 1989; Uchida *et al.*, 1990; Ishii *et al.*, 1990) have been attempting to use genetically engineered non-neuronal cells as donor cells for intracerebral grafting that produce catecholamines or the precursor amino acid, L-DOPA and recover the motor function in Parkinsonian model animals having deficiency in the nigrostriatal dopamine neurons. C6 cells transfected with type 2 human TH cDNA (Uchida *et al.*, 1988) and type 1 human TH cDNA (Uchida *et al.*, 1989) were able to express active TH molecules and to release L-DOPA from the cells into the culture medium. Since C6 cells lack aromatic L-amino acid decarboxylase, DOPA formed from tyrosine is not decarboxylated to dopamine. NRK fibroblast cell lines transfected with type 2 human TH cDNA expressed active TH molecules. Fibroblasts contain neither aromatic L-amino

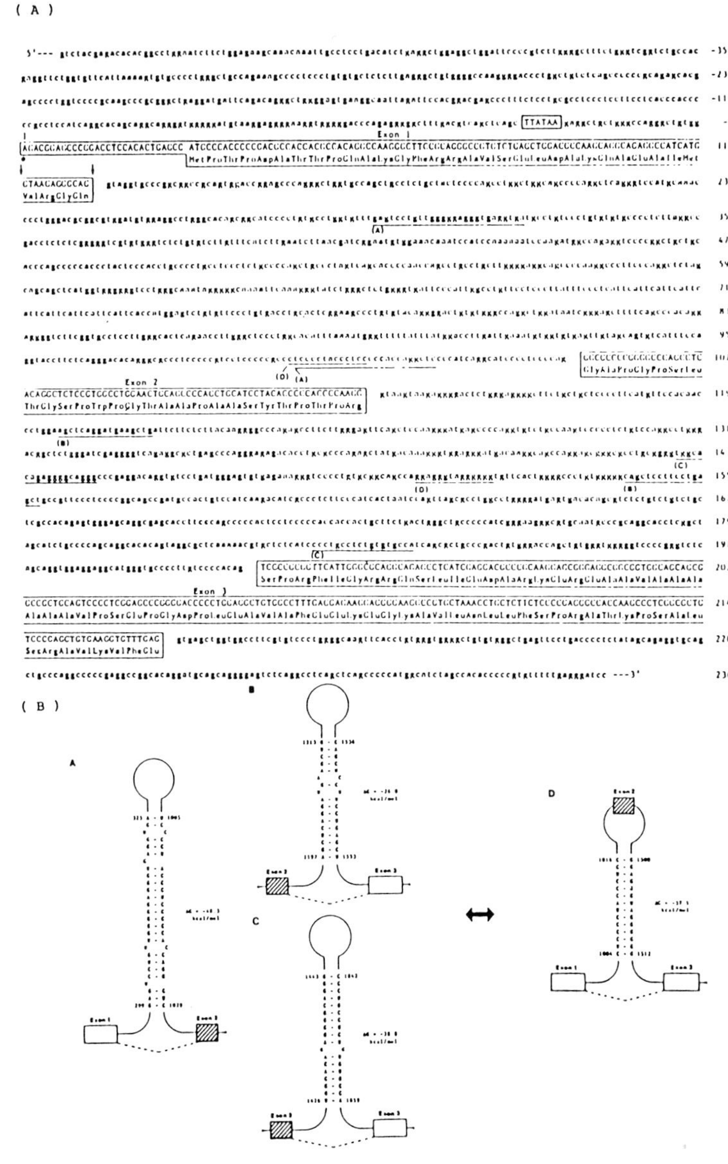

(A)

(B)

Figure 3. Nucleotide sequence from the upstream region of exon 1 to the downstream region of exon 3. Exons 1, 2 and 3 are boxed with thin lines. The typical TATA box is boxed with a thick line. Arrows show the alternative donor sites in exon 1. The sequence which complement each other to form the hairpin loops (structures A, B, C and D shown below) are underlined. The 4 (structures A, B, C and D) potential secondary structures in introns 1 and 2 of the primary transcript are analyzed with the computer program with minimum matching length = 14-bp and matching percentage > 85%. (Kobayashi *et al.*, 1988).

acid decarboxylase nor tetrahydrobiopterin, the cofactor of TH. Thus, addition of tetrahydrobiopterin is necessary to produce L-DOPA by the expressed TH. The natural (6R) form of tetrahydrobiopterin is more active as cofactor than the unnatural (6S) form of tetrahydrobiopterin (Ishii *et al.*, 1990). Horellou *et al.* (1988) inserted the full coding sequences of type 1–4 human TH cDNA into SP6 vector and used it for the synthesis of mRNA. These workers injected pure type 1–4 human TH mRNA into frog *Xenopus* oocytes and succeeded in yielding active TH. Homospecific activity (TH activity per TH protein) was highest in type 1 human TH and lower in type 2–4 human TH, respectively. Their results on homospecific activity of type 1–4 human TH expressed in frog oocytes are similar to our data in COS cells. The type 2 human TH was also expressed in cells using a baculovirus vector by Ginnes *et al.* (1988), and the results of their kinetic analysis are similar to our data on type 2 human TH expressed in COS cells.

EXPRESSION OF HUMAN TH IN TRANSGENIC MICE

We (Kaneda *et al.*, 1991) have succeeded in producing transgenic mice carrying human TH gene, possessing 2.5 kb upstream to the transcription initiation site and 0.5 kb downstream from the polyadenylation site without any plasmid sequences considered to interfere with transgene expression in mice. We found that this DNA fragment covers a sequence(s) sufficient to permit tissue-specific and high-level expression in brain and adrenal gland of this transgenic line. Analysis of the transcripts in brain and adrenal gland revealed that the integrated human gene is transcribed from the correct initiation site in transgenic mice. Primer extension analysis showed two significant extension products that corresponded exactly to type 1 and type 2 human TH mRNAs. We (Ichikawa *et al.*, 1991) have cloned a single species of mouse TH cDNA, which revealed 97% homology with rat TH cDNA. These results would show that the fundamental cellular machinery necessary for the alternative splicing of human TH mRNA is present and functioning in mouse brain and adrenal gland and produces multiple forms of the enzyme from the human mRNA sequences. The expression level of human TH mRNA in brain was about 50-fold higher than that of endogenous mouse TH mRNA. However, in adrenal gland, expression of mRNA from the transgene was about 5-fold higher than that of endogenous mouse TH mRNA. Because both tissues have the same copy number of the transgene, this observation suggests the existence of different regulatory mechanisms for transcription of the TH gene in brain and adrenal gland of the mouse. Furthermore, the over-expression of mRNA in brain would suggest that nuclear factors necessary for TH gene transcription are sufficiently present in an appropriate cell type of the brain. The localization of TH mRNA in brain was examined by *in situ* hybridization. TH mRNA of the non-transgenic mouse was detected specifically in the substantia nigra and ventral tegmental area, which are the predominant dopamine neurons in the midbrain. *In situ* hybridization in slices from transgenic brain clearly demonstrated an enormous expression of the human TH mRNA in these nuclei, which indicates that the transgenes were expressed in a region-specific manner in the brain of transgenic mice. Most interestingly, a large difference was observed between the amount of human TH mRNA detected by *in situ* hybridization and the amount of TH protein detected by immunocytochemistry. When TH protein was localized by immunochemistry, most of the increment of TH (immunoreactivity) was found in the substantia nigra and ventral tegmental area. However, the increase in TH immunoreactivity was less than that of the mRNA itself in these regions of transgenic brain and did not reflect the enormous increment of transcripts from the transgene detected by *in situ* hybridization. These findings were generally supported by Western blot analysis and by the measurement of TH activity in various brain regions. Since the integrity of transcribed human TH mRNA has been proven by Northern blot and primer extension analyses, this finding suggests the presence of an unknown mechanism(s) at the translational level of TH mRNA and/or the presence of other post-translational mechanisms, such as rapid degradation of human TH protein in mouse brain. The mechanisms underlying this discrepancy remain to be investigated in more detail. As regards the relation between TH activity and catecholamine level, we also observed a discrepancy in transgenic mice. Striatum from transgenic mice exhibited 3,4-fold higher *in vitro* TH activity than that of non-transgenic mice, but the dopamine content was only 1.05-fold higher in transgenic striatum. Similar observations were made with mesencephalon-diencephalon. TH activity in this region was increased 3.6-fold in transgenic mice, but dopamine and noradrenaline levels were only 1.09 and 1.02-fold higher, respectively, than those of non-transgenic mice. Since *in vitro* TH activity was measured under optimal conditions, the activity observed represents the maximum activity (V_{max}) and reflects the amount of the enzyme protein. In fact, the increase in TH activity was almost consistent with the increase in the amount of total TH protein estimated by Western blot analysis. In any case, the enzymatic activity of TH seemed to be regulated in brain so as to minimize the effect of the transgene and to maintain neural homeostasis even though the aberrant high-level expression of TH mRNA has been brought about. In contrast to brain, changes in TH activity and noradrenaline level were generally paralleled in adrenal gland (about 1.2-fold increase), though some decrease in the adrenaline level was induced by an unknown mechanism(s) in transgenic mice. Therefore, these results suggest that the regulation mechanism of TH activity in brain is distinct from that in adrenal gland in the *in vivo* situation.

We (Nagatsu *et al.*, 1991) have also observed the transgene expression of TH mRNA not only in the catecholamine neurons but also in non-catecholamine neurons in the brain of transgenic mice by *in situ* hybridization. In adult transgenic mice, human TH was atypically expressed in the olfactory (the anterior olfactory nucleus and pyriform cortex) and visual (nucleus suprachiasmaticus and nucleus parabigeminalis) system, in addition to typical catecholamine neurons. These results suggest the possibility that TH plays some novel roles in sensory system. It should be noted that these sensory areas expressed TH transiently during normal development.

We have performed preliminary studies to examine possible changes in blood pressure, circadian rhythms and behavioral activity in these transgenic mice. However, under normal conditions, transgenic mice exhibited no significant phenotype abnormalities. We are further examining possible changes in the phenotypes under various conditions.

HUMAN TH mRNAs IN TISSUES

We have been analyzing TH mRNA extracted from adrenal (medulla) and brain (substantia nigra and locus coeruleus) by the primer extension method (Kaneda *et al.*, 1990). Figure 4 shows schematically the experimental design and the results on young and old subjects. The expected lengths of the extension products are 154, 166, 235, and 247 bp for types 1, 2, 3 and 4, respectively, and the relative amounts of each extension product was estimated by autoradiography and by densitometry. In adult adrenal medulla all types of TH mRNA was detected. The relative ratio of type 1:2:3:4 was about 0.9:1.0:0.1:0.3. On the other hand, in brain (substantia nigra), type 3 was undetectable and the relative ratio of each type was about 0.5:1.0:0.0:0.1 in old brains. In both tissues, types 1 and 2 were major species and types 3 and 4 were minor species, and type 3 was almost undetectable. In contrast to brain and adrenal, type 3 was much more expressed in human pheochromocytoma, in which the ratio was about 0.9:1.0:0.3:0.1. We analyzed the pattern of TH isozyme in neonate (2 days) or fetus (24 weeks after gestation), and in the brain stem of these very young tissues, the relative ratio of type 1 TH mRNA appears to be larger than in aged brain.

Changes in the isozymes of TH in neuropsychiatric diseases are of interest. Reduction in TH in the nigrostriatal dopamine region is a characteristic change in Parkinson's disease (Lloyd *et al.*, 1975; McGeer and McGeer, 1976; Nagatsu *et al.*, 1977). It has been generally assumed that the decrease in TH activity in Parkinsonian brain is due to a reduction in TH protein as a result of cell death of the nigrostratal dopamine neurons by unidentified mechanisms. In fact, we found TH activity and TH protein to be decreased in parallel in Parkinsonian brain. However, we found in Parkinsonian brain at later stages a more marked decrease in TH protein than in TH activity. Thus the homospecific activity (TH activity per TH protein) of residual TH in Parkinsonian brains appears to be increased (Mogi *et al.*, 1988). This result is in contrast to constant homospecific activity of TH in the striatum of Parkinsonian mouse produced by 1-methyl-4-phenyl-1,2,3,6-tetrahydropyridine (MPTP), at least in an early stage (Mogi

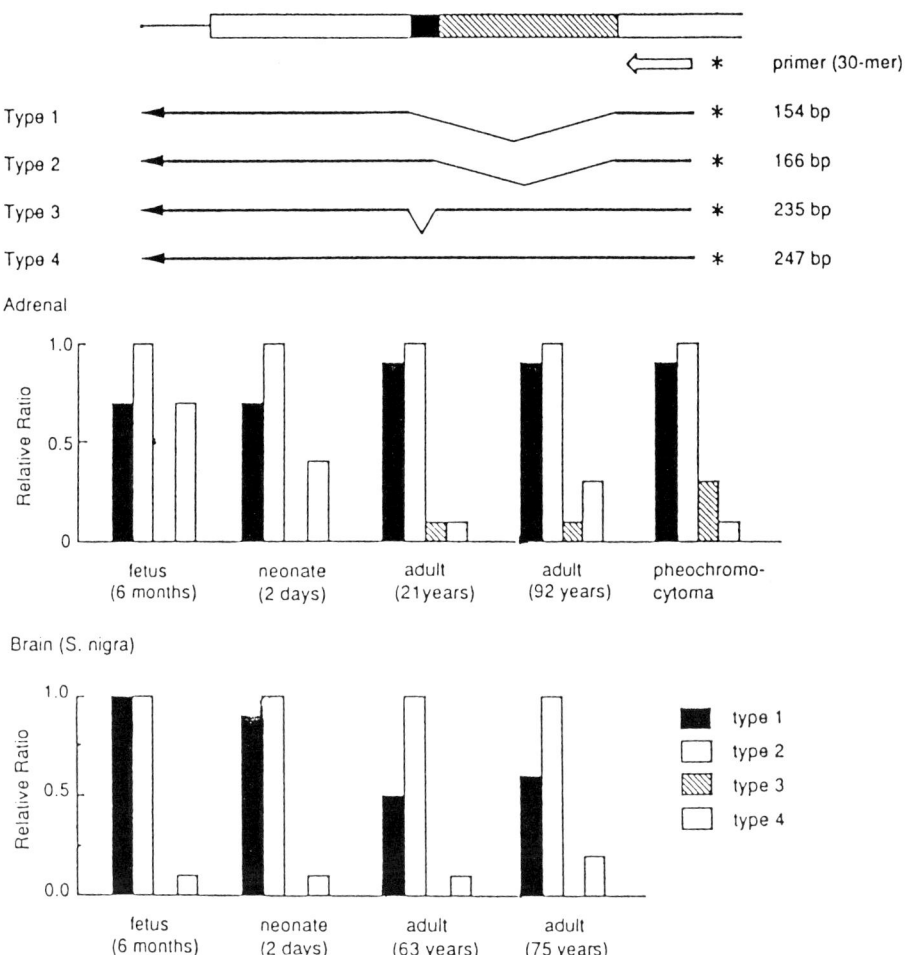

Figure 4. Primer extension analysis of TH mRNA types in human adrenal and brain. (A) The expected lengths of the extension products in each type of mRNA. A synthetic oligonucleotide (30-mer) complementary to nucleotide 189–218 in type 4 mRNA was used as the primer. (B) Expression pattern of four types of TH mRNA in human adrenal and brain (substantia nigra). Total RNA (~ 50 μg) extracted from each tissue was subjected to the primer extension analysis, and the relative amount of each mRNA type was estimated by densitometry. The amount of type 2 mRNA was normalized to 1.0. (Kaneda *et al.*, 1990).

et al., 1987). These results suggest some molecular changes in TH protein in Parkinson's disease, which may be related to the multiple forms of human TH. Since we have recently shown that multiple TH mRNAs may be primate-specific, the molecular changes of TH in Parkinson's disease would be better estimated in MPTP-Parkinsonism of monkeys.

CONCLUSION

Multiple mRNAs with different coding regions have been found in humans and monkeys, and they are produced by alternative splicing from a single gene. We believe that the TH multiplicity offers an excellent model for studying the evolution of an alternative splicing. The multiple forms of human TH may introduce another new regulatory mechanism to TH regulation. The molecular biological approach to TH should be useful to examine probable genetic alternation of TH in various neuropsychiatric disorders, such as Parkinson's disease, affective disorders, and schizophrenia.

Acknowledgements

This work is supported by Grants-in-Aid for Scientific Research from the Japanese Ministry of Education, Science and Culture.

REFERENCES

Brenneman, A. R. and Kaufman, S. (1964). The role of tetrahydropteridines in the enzymatic conversion of tyrosine to 3,4-dihydroxyphenylalanine. *Biochem. Biophys. Res. Commun.* **17**, 177–181.

Dahlström, A., Belmaker, R. H. and Sandler, M. (Eds) (1988). Progress in Catecholamine Research, Part A: Basic Aspects and Peripheral Mechanisms. Alan R. Liss, New York.

D'Mello, S. R., Weisberg, E. P., Stachowiak, M. K., Turzai,L .M., Gioio, A. E. and Kaplan, B. B. (1988). Isolation and nucleotide sequence of a cDNA clone encoding bovine adrenal tyrosine hydroxylase: comparative analysis of tyrosine hydroxylase gene products. *J. Neurosci. Res.* **19**, 440–449.

Dynan, W. S., Saffer, J. D., Lee, W. S. and Tjian, R. (1985). Transcription factor Sp1 recognizes promoter sequences from the monkey genome that are similar to the simian virus 40 promoter. *Proc. Natl. Acad. Sci. USA* **82**, 4915–4919.

Egeland, J. A., Gerhard, D. S., Pauls, D. L., Sussex, J. N., Kid, K. K., Allen, C. R., Hostetter, A. M. and Housman, D. E. (1987). Bipolar affective disorders linked to DNA markers on chromosome 11. *Nature* **325**, 783–787.

Fauquet, M., Grima, B., Lamouroux, A. and Mallet, J. (1988). Cloning of quail tyrosine hydroxylase: amino acid homology with other hydroxylases discloses functional domains. *J. Neurochem.* **50**, 142–148.

Fujisawa, H. and Okuno, S. (1987). Tyrosine 3-monooxygenase from rat adrenals. In: *Methods in Enzymology*. S. Kaufman (Ed). Academic Press, New York, p. 63–71.

Ginns, E. I., Rehari, M., Martin, B. M., Weller, M., O'Malley, K. L., La Marca, M. E., McAllister, C. G. and Paul, S. M. (1988). Expression of human tyrosine hydroxylase cDNA in invertebrate cells using a baculovirus vector. *J. Biol. Chem.* **263**, 7406–7410.

Grima, B., Lamouroux, A., Blanot, F., Biguet, N. F. and Mallet, J. (1985). Complete coding sequence of rat tyrosine hydroxylase mRNA. *Proc. Natl. Acad. Sci. USA* **82**, 613–621.

Grima, B., Lamouroux, A., Boni, C., Jullien, J.-F., Javoy-Agid, F. and Mallet, J. (1987). A single human gene encoding multiple tyrosine hydroxylases with different predicted functional characteristics. *Nature* **326**, 707–711.

Horellou, P., Le Bourdelles, B., Clot-Humbert, J., Guibert, B., Level, V. and Mallet, J. (1988). Multiple human tyrosine hydroxylase enzymes, generated through alternative splicing have different specific activities in *Xenopus* oocytes. *J. Neurochem.* **51**, 652–655.

Ichikawa, S., Ichinose, H. and Nagatsu, T. (1990). Multiple mRNAs of monkey tyrosine hydroxylase. *Biochem. Biophys. Res. Commun.* **173**, 1331–1336.

Ichikawa, S., Sasaoka, T. and Nagatsu, T. (1991). Primary structure of mouse tyrosine hydroxylase deduced from its cDNA. *Biochem. Biophys. Res. Commun.* **176**, 1610–1616.

Ishii, A., Hagihara, M., Matsuura, S., Uchida, K., Kiuchi, H., Kaneda, N., Toya, S., Kohsaka, S. and Nagatsu, T. (1990). Effects of (6R)- and (6S)-tetrahydrobiopterin on L-3,4-dihydroxyphenylalanine (DOPA) formation in NRK fibroblasts transfected with human tyrosine hydroxylase type 2 cDNA. *Neurochem. Int.* **17**, 625–632.

Ishii, A., Kiuchi, K., Matsuyama, M., Satake, T. and Nagatsu, T. (1990). Ferrous ion activates the less active form of human adrenal tyrosine hydroxylase. *Neurochem. Int.* **16**, 59–64.

Kaneda, N., Kobayashi, K., Ichinose, H., Kishi, F., Nakazawa, A., Kurosawa, Y., Fujita, K. and Nagatsu, T. (1987). Isolation of novel cDNA clone for human tyrosine hydroxylase: alternative RNA splicing produces four kinds of mRNA from a single gene. *Biochem. Biophys. Res. Commun.* **146**, 971–975.

Kaneda, N., Kobayashi, K., Ichinose, H., Sasaoka, T., Ishii, A., Kiuchi, K., Kurosawa, Y., Fujita, K. and Nagatsu, T. (1990). Molecular biological approaches to catecholamine neurotransmitters and brain aging. In: *Aging of the Brain: Cellular and Molecular Aspect of Brain Aging and Alzheimer's Disease*. T. Nagatsu and O. Hayaishi (Eds). Japan Scientific Societies Press, Tokyo, pp. 53-66.

Kaneda, N., Sasaoka, T., Kobayashi, K., Kiuchi, K., Nagatsu, I., Kurosawa, Y., Fujita, K., Yokoyama, M., Nomura, T., Katsuki, M. and Nagatsu,T. (1990). Tissue-specific and high-level expression of the human tyrosine hydroxylase gene in transgenic mice. *Neuron* **6**, 583–594.

Kobayashi, K., Kaneda, N., Ichinose, H., Kishi, F., Nakazawa, A., Kurosawa, Y., Fujita, K. and Nagatsu, T. (1987). Isolation of a full-length cDNA clone encoding human tyrosine hydroxylase type 3. *Nucl. Acids Res.* **15**, 6733.

Kobayashi, K., Kaneda, N., Ichinose, H., Kishi, F., Nakazawa, Y., Kurosawa, Y., Fujita, K. and Nagatsu, T. (1988). Structures of the human tyrosine hydroxylase gene: alternative splicing from a single gene accounts for generation of four mRNA types. *J. Biochem.* **103**, 907–912.

Kobayashi, K., Kiuchi, K., Ishii, A., Kaneda, N., Kurosawa, Y., Fujita, K. and Nagatsu, T. (1988). Expression of four types of human tyrosine hydroxylase in COS cells. *FEBS Lett.* **238**, 431–434.

Le Bourdelles, B., Boularand, S., Boni, P., Horellou, P., Dumas, S., Grima, B. and Mallet, J. (1989). Analysis of the 5' region of the human tyrosine hydroxylase gene: combined patterns of exon splicing generate multiple regulated tyrosine hydroxylase iso-forms. *J. Neurochem.* **50**, 988–991.

Ledley, F. D., DiLella, A. G., Kwok, S. C. M. and Woo, S. L. C. (1985). Homology between phenylalanine and tyrosine hydroxylase reveals common structural and functional domains. *Biochemistry* **24**, 3389–3394.

Lewis, E. J., Harrington, C. A. and Chikaraishi, D. M. (1987). Transcriptional regulation of the tyrosine hydroxylase gene by glucocorticoid and cyclic AMP. *Proc. Natl. Acad. Sci. USA* **84**, 3550–3554.

Lloyd, K. G., Davidson, L. and Hornykiewicz, O. (1975). The neurochemistry of Parkinson's disease: effect of L-DOPA therapy. *J. Pharmacol. Exp. Therap.* **195**, 452–464.

McGeer, P. L. and McGeer, E. G. (1976). Enzymes associated with the metabolism of catecholamine, acetylcholine and GABA in human controls and patients with Parkinson's disease. *J. Neurochem.* **26**, 65–76.

Mogi, M., Harada, M., Kiuchi, K., Kojima, K., Kondo, T., Narabayashi, H., Rausch, D., Riederer, P., Jellinger, K. and Nagatsu, T. (1988). Homospecific activity (activity per enzyme protein) of tyrosine hydroxylase increases in Parkinsonian brain. *J. Neural Transm.* **72**, 77–81.

Mogi, M., Harada, M., Kojima, K., Kiuchi, K., Nagatsu, I. and Nagatsu, T. (1987). Effects of repeated systemic administration of 1-methyl-4-phenyl-1,2,3,6-tetrahydropyridine (MPTP) on striatal tyrosine hydroxylase activity *in vitro* and tyrosine hydroxylase content. *Neurosci. Lett.* **80**, 213–218.

Mogi, M., Kojima, K. and Nagatsu, T. (1984). Detection of inactive or less active forms of tyrosine hydroxylase in human brain and adrenal by a sandwich enzyme immunoassay. *Anal. Biochem.* **138**, 125–132.

Mogi, M., Kojima, K., Harada, M. and Nagatsu, T. (1986). Purification and immunological properties of tyrosine hydroxylase in human brain. *Neurochem. Int.* **8**, 423–428.

Mueller, R. A., Thoenen, H. and Axelrod, J. (1969). Adrenal tyrosine hydroxylase compensatory increase in activity after chemical sympathectomy. *Science* **163**, 468–469.

Nagatsu, I., Yamada, K., Karasawa, N., Sasaki, M., Takeuchi, T., Kaneda, N., Sasaoka, T., Kobayashi, K., Yokoyama, M., Nomura, T., Katsuki, M., Fujita, K. and Nagatsu,T (1991). Expression in brain sensory neurons of the transgene in transgenic mice carrying human tyrosine hydroxylase gene. *Neurosci. Lett.* **127**, 91–95.

Nagatsu, T., Kato, T., Numata (Sudo), Y., Ikuta, K., Sano, M., Nagatsu, I., Kondo, Y., Inagaki, S., Iizuka, R., Hori, A. and Narabayashi, H. (1977). Phenylethanolamine N-methyltransferase and other enzymes of catecholamine metabolism in human brain. *Clin. Clim. Acta* **75**, 221–232.

Nagatsu, T., Levitt, M. and Udenfriend, S. (1964). Tyrosine hydroxylase. The initial step in norepinephrine biosynthesis. *J. Biol. Chem.* **239**, 2910–2917.

Nagatsu, T. and Oka, K. (1987). Tyrosine 3-monooxygenase from bovine adrenal medulla. In: *Methods in Enzymology*. S. Kaufman (Ed.). Academic Press, New York, pp. 56–62.

Nagatsu, T., Yamaguchi, T., Kato, T., Sugimoto, T., Matsuura, S., Akino, M., Nagatsu, I., Iizuka, R. and Narabayashi, H. (1981). Biopterin in human brain and urine from controls and Parkinsonian patients: application of a new radioimmunoassay. *Clin. Chim. Acta* **109**, 305–311.

Nagatsu, T., Yamaguchi, T., Rahman, M. K., Trocewicz, J., Oka, K., Hirata, Y., Nagatsu, I., Narabayashi, H., Kondo, T. and Iizuka, R. (1984). Catecholamine-related enzymes and the biopterin cofactor in Parkinson's disease and related extrapyramidal diseases. In: *Advances in Neurology, vol. 40.* R. G. Hassler and J. F. Christ (Eds). Raven Press, New York, pp.467–473.

Neckameyer, W. S. and Quinn, W. G. (1989). Isolation and characterization of the gene for *Drosophila* tyrosine hydroxylase. *Neuron* **2**, 1167–1175.

O'Malley, K. L., Anhalt, M. J., Martin, B. M., Kalsoe, J. R., Winfield, S. L. and Ginns E. I. (1987). Isolation and characterization of the human tyrosine hydroxylase gene: identification of 5′ alternative splice sites responsible for multiple mRNAs. *Biochemistry* **26**, 6910–6914.

Tank, A. W. and Weiner, N. (1987). Tyrosine 3-monooxygenase from rat pheochromocytoma. In: *Methods in Enzymology*. S. Kaufman (Ed.). Academic Press, New York, pp. 71–82.

Uchida, K., Takamatsu, K., Kaneda, N., Toya, S., Tsukada, Y., Kurosawa, Y., Fujita, K., Nagatsu, T. and Kohsaka, S. (1988). Transfection of tyrosine hydroxylase cDNA into C6 cells. *Proc. Jpn. Acad.* **64**, Ser. B, 290–293.

Uchida, K., Takamatsu, K., Kaneda, N., Toya, S., Tsukada, Y., Kurosawa, Y., Fujita, K., Nagatsu, T. and Kohsaka, S. (1989). Synthesis of L-3,4-dihydroxyphenylalanine by tyrosine hydroxylase cDNA-transfected C6 cells: application for intracerebral grafting. *J. Neurochem.* **53**, 728–732.

Uchida, K., Ishii, A., Kaneda, N., Toya, S., Nagatsu, T. and Kohsaka, S. (1990). Tetrahydrobiopterin-dependent production of L-DOPA in NRK fibroblasts transfected with tyrosine hydroxylase cDNA: future use for intracerebral grafting. *Neurosci. Lett.* **109**, 282–286.

Udenfriend, S., Zaltzman-Nirenberg and Nagatsu, T. (1965). Inhibitors of purified beef adrenal tyrosine hydroxylase. *Biochem. Pharmacol.* **14**, 837–845.

Zigmond, R. E., Schwarzchild, M. A. and Rittenhouse, A. R. (1989). Acute regulation of tyrosine hydroxylase by nerve activity and by neurotransmitters via phosphorylation. *Ann. Rev. Neurosci.* **12**, 415–461.

Tyrosine Hydroxylase:
Its involvement to diseases

Tyrosine Hydroxylase, pp. 195–216
M. Naoi *et al.* (Eds)
© VSP 1993

CLINICAL EVALUATION OF TYROSINE HYDROXYLASE ACTIVITY IN BRAIN AND PERIPHERAL ORGANS

IRWIN J. KOPIN

National Institute of Neurological Disorders and Stroke,
National Institutes of Health, Bethesda, Maryland, USA

Abstract—Tyrosine hydroxylase was the last of the catecholamine biosynthetic enzymes to be discovered. The rate of tyrosine hydroxylation is regulated by several factors; the enzyme activity can be enhanced rapidly by activation so that there is not necessarily a correlation of the amount of enzyme and the rate of formation of DOPA. Although tyrosine hydroxylation is believed to be the rate limiting step in catecholamine biosynthesis, not all DOPA formed by this enzyme is decarboxylated; a significant portion escapes into the circulation. There is persuasive evidence that plasma levels of DOPA reflect tyrosine hydroxylation rates. Similarly, not all of the dopamine formed by decarboxylation of DOPA in the cytoplasm of noradrenergic neurons is β-hydroxylated. About half is deaminated to form dihydroxyphenylacetic acid (DOPAC). Both the DOPA and DOPAC which reach the circulation are metabolized mainly to homovanillic acid (HVA). Thus, HVA levels in plasma and excretion rates in urine reflect tyrosine hydroxylation in noradrenergic as well as dopaminergic neurons. Although only a small fraction of norepinephrine released from nerve terminals reaches the circulation, norepinephrine levels in plasma have been used as an index of adrenergic neuronal activity. Measurements of spillover appear to be a more reliable index, but reflect total body norepinephrine efflux. There are regional differences in norepinephrine spillover and to obtain an accurate picture of its turnover rate in any organ, it is important to assess overflow of catecholamine metabolites as well as of norepinephrine. It is possible also, by appropriate perturbation with debrisoquin of peripheral noradrenergic catecholamine metabolism, to estimate the rate of tyrosine hydroxylation in brain dopaminergic neurons. Thus, tyrosine hydroxylation rates can be assessed by measurements of catecholamine metabolites in plasma and urine.

Key words: tyrosine hydroxylase; norepinephrine; dopamine; DOPA; homovanillic acid; DOPAC.

INTRODUCTION

Tyrosine hydroxylase (TH), is the key enzyme in the synthesis of the three naturally occurring catecholamines, dopamine, norepinephrine (NE), and epinephrine

(EPI). The biosynthetic pathway for the catecholamines was deduced by Blaschko (1939) following the discovery by Holtz (1938) that dihydroxyphenylalanine (DOPA) could be decarboxylated rapidly by an enzyme found in mammalian kidney. The product was dopamine. Based on his observation that N-methyl-DOPA is not a substrate for the decarboxylating enzyme, Blaschko hypothesized correctly that dopamine was converted to NE by β-hydroxylation and that the final product, EPI, was the result of N-methylation of NE. The predicted sequence of EPI biosynthesis was confirmed with the discovery of dopamine in the adrenal medulla (Goodall, 1951) and the conversion of (^{14}C)-DOPA to norepinephrine in bovine adrenal medullary homogenates (Demis et al., 1956). It is interesting that the physiological importance of the catecholamines was discovered in the reverse order of their biosynthesis. EPI, the final product in the series, was first discovered (Abel, 1898) as the hormone released from the adrenal medulla. Its release during physiologically or environmentally provoked stress was the subject of great interest and extensive investigations (Cannon, 1929). A half century after discovery of EPI, its immediate precursor, NE, was identified as the neurotransmitter released from sympathetic nerve terminals (Euler, 1948; Peart, 1949). Because NE was found to be present in relatively high concentrations in certain regions of the brain (e.g. hypothalamus), Vogt (1954) suggested that NE also has a role as a neurotransmitter in the brain. At that time, because of its relative inactivity when injected intravenously, dopamine was regarded as only a precursor of NE and EPI. However, when dopamine was demonstrated in very high concentrations in the basal ganglia it was suggested that this catecholamine was an important brain neurotransmitter (Carlsson, 1959). In support of this view, Carlsson showed that pharmacological depletion of this catecholamine with reserpine produced a Parkinsonian syndrome in experimental animals and that this behavioral effect was reversed by DOPA administration.

Although in 1959, only the enzyme responsible for decarboxylation of DOPA had been characterized, there was an explosion of interest in catecholamine formation, release, and metabolism. The enzymes which convert dopamine to NE, and NE to EPI, dopamine-β-hydroxylase (DBH) and phenylethanolamine-N-methyl transferase (PNMT), respectively, were isolated and characterized during the next three years, (Levin et al., 1960; Axelrod, 1962) but the mechanism for conversion of tyrosine to DOPA was more elusive. There was even speculation that hydroxylation of tyrosine proceeded non-enzymatically since this oxidation was known to be catalyzed by ascorbic acid in vitro. The matter was settled, however, with the discovery of TH (Nagatsu et al., 1964a, b). In these papers, they showed that TH could be inhibited by DOPA and by NE, providing the first evidence that feedback control of this enzyme might regulate catecholamine synthesis rates. Spector et al. (1967) found that after inhibition of monoamine oxidase, NE accumulates in brain and heart until its levels reach two or three times those found in untreated animals. In the treated animals, conversion of radioactive tyrosine to NE was reduced, whereas formation of NE from DOPA was unaffected or increased. This indicated that at high NE content, formation of NE from tyrosine was diminished at the level of tyrosine hydroxylase. In an analogous experiment during sympathetic nerve stimulation, Sedvall *et al.* (1968) showed that the rate of ^{3}H-NE formation from ^{3}H-tyrosine was enhanced in the salivary gland, whereas formation of (^{14}C)-NE from (^{14}C)-DOPA was unaltered. Although there was a 5-fold greater conversion of tyrosine to NE in stimulated than in decentralized salivary glands, the tyrosine hydroxylase activity measured *in vitro* differed by only about 30%. That indicated that during nerve stimulation, the rate of NE formation is enhanced by increasing formation of DOPA from tyrosine, mainly by activation of TH and only slightly by induction of increased levels of TH. These findings suggested that a measure of tyrosine hydroxylation rates might provide an index of the activity of catecholaminergic neurons. Several approaches have been suggested to determine neuronal catecholaminergic activity based on assessment of biochemical parameters of catecholamine formation and release. Here we propose to review briefly the available methods to measure total body catecholaminergic activity and to separate peripheral sympathoadrenal medullary activity from brain catecholamine formation and metabolism as a means to assess brain catecholaminergic neuronal activity *in vivo* in intact animals or humans.

EARLY ATTEMPTS AT ASSESSING SYMPATHOADRENAL MEDULLARY ACTIVITY FROM PLASMA CATECHOLAMINES

Shortly after NE was demonstrated to be the sympathetic neurotransmitter, a fluorescence assay method for measurement of catecholamines in urine and plasma was developed (Lund, 1950). This assay, based on conversion of NE and EPI to highly fluorescent trihydroxyindole derivatives using strong alkali in the presence of an antioxidant, was subsequently modified to provide a clinically useful method to measure the catecholamines in plasma (although as much as 30 ml more blood was required) as well as in urine. Using the trihydroxyindole method, it was found that urinary excretion rates of EPI and NE were elevated during a variety of stressful situations, but the method was not sufficiently sensitive for routine assay of the catecholamines in small volumes of plasma or cerebrospinal fluid or for sequential analysis of multiple samples in humans.

The next advance made in the assay of plasma NE and EPI levels was based on enzymatic methods. In one method, which measures only NE, radiolabelled methyl groups from S-adenosyl methionine are transferred, using PNMT, to the nitrogen of NE to form EPI. In the other procedure, catechol-O-methyl transferase (COMT) is used to transfer the labelled methyl groups to the 3-hydroxyl position of the catechol moiety to form normetanephrine from NE and metanephrine from EPI. In both methods, the labelled products are isolated and the radioactivity assayed to provide sensitive specific measures of the unlabelled precursor molecules.

With the development of high performance liquid chromatography and sensitive electrochemical detection (Kissinger *et al.*, 1981), it became possible to assay simultaneously not only the catecholamines in plasma and cerebrospinal fluid, but also their precursor, L-DOPA, and their catechol deaminated products, dihydroxyphenylacetic acid (DOPAC) from dopamine and dihydroxyphenylglycol (DHPG) from the β-hydroxylated catecholamines, NE and EPI (Eisenhofer *et al.*, 1986). With the application of gas chromatography-mass spectroscopy, precise, sensitive methods were developed for the major O-methylated metabolites of the catecholamines. Homovanillic acid (HVA) formed from DOPAC is the major final metabolite of dopamine and is excreted in the urine. NE is converted to 3-methoxy-4-hydroxyphenylglycol (MHPG) via DHPG, and the oxidation product of MHPG, 3-methoxy-4-hydroxymandelic acid (VMA), is the major urinary metabolite of NE (see Kopin, 1985). Measurements of these metabolites (Fig. 1) in urine, plasma, or cerebrospinal fluid provide indices of the rate of tyrosine hydroxylation in the whole body, and with appropriate procedures, information about TH activity in selected individual organs can be obtained.

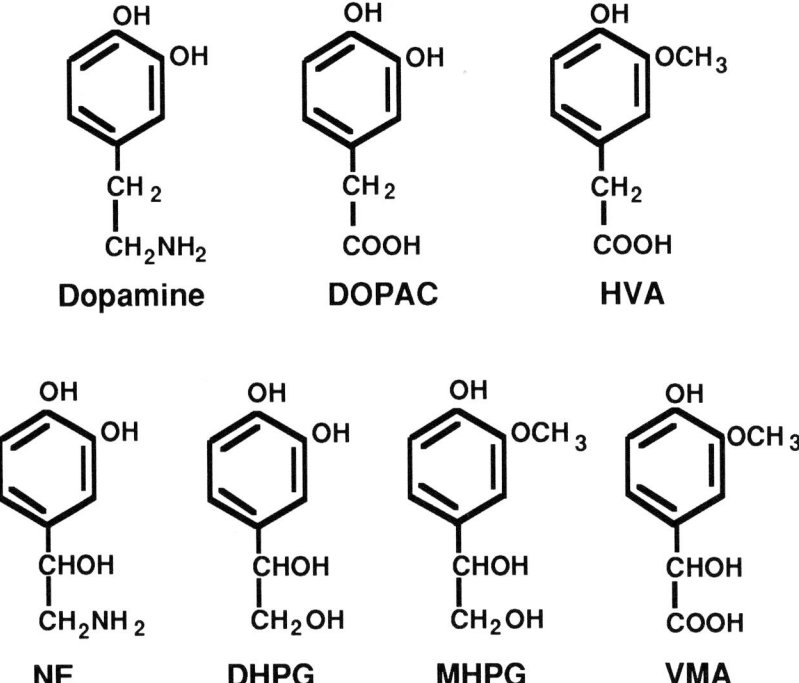

Figure 1. Norepinephrine, dopamine and their major urinary metabolites.

EPINEPHRINE FORMATION AND METABOLISM

Since most EPI is formed in the adrenal medulla and released directly into the circulation, secretion of this catecholamine can be monitored by direct measurements of its levels in plasma. After intravenous injection of ^{3}H-EPI, the catecholamine rapidly disappears from the plasma. The half-life of EPI is very short (about 1 min), most being destroyed by enzymatic O-methylation in the liver and kidney to form metanephrine. After intravenous injection of ^{3}H-EPI simultaneously with (^{14}C)-3-O-methyl-metanephrine, the ratio of ^{3}H(^{14}C) in excreted metanephrine and its conjugates was about 0.65 indicating that about 2/3 of the administered ^{3}H-EPI was converted to its O-methylated metabolite in humans (Kopin, 1960) as well as in rats (Kopin *et al.*, 1961). The metabolic fate of intravenously administered ^{3}H-EPI reflects precisely the disposition of EPI released from the adrenal medulla into the circulation. Thus, the urinary excretion rate of metanephrine, appropriately corrected for the fraction of ^{3}H-EPI which is not excreted as this metabolite, is a valid measure of the rate of EPI secretion. Urinary excretion rates of EPI or metanephrine provide an integrated view of release of the catecholamine over a reasonably long interval and have been used to examine effects of stressful situations on adrenal medullary activity. EPI is stored in the adrenal medulla; the released catecholamine is derived from the adrenal medullary stores, so that over a short interval there is no direct relationship between the rate of EPI synthesis and release. With prolonged increase in EPI release, however, the replenishment of its stores is attended by increased levels of TH. This occurs with repeated intervals of immobilization (Kvetňanský *et al.*, 1970), exposure to cold (Kvetňanský *et al.*, 1971a) or after hypoglycemic episodes (Kvetňanský *et al.*, 1971b).

Because of the short half-life of plasma EPI, its levels fluctuate rapidly and reflect, almost instantaneously, changes in the rate of EPI release from the adrenal. Excretion of EPI and its O-methylated metabolite provide useful indices of EPI release from the adrenal medulla, but since urine collections cannot be timed accurately for short intervals, changes in urinary excretion rates are useful only over relatively long intervals (over one hour).

The deaminated metabolites of EPI, DHPG and DHMA, and their O-methylated derivatives, MHPG and VMA, are formed also from NE. Since much more NE than EPI is formed in the body, these metabolites reflect mainly NE formation and metabolism. EPI is present also in the brain, but only in a few areas and in relatively low concentrations. At present there are no available strategies for assessing EPI synthesis, turnover, or metabolism in the brain of intact animals or humans.

NOREPINEPHRINE FORMATION, RELEASE, AND METABOLISM

As indicated earlier, measurements of plasma NE levels provide an index of release of NE from sympathetic nerve terminals. Because catecholamines are removed from the circulation rapidly, like EPI, NE levels may fluctuate rapidly and are sensitive to changes in the rate of NE release. What is actually measured in plasma is the net result of overflow of NE from the interstitial fluid into the plasma and the removal of NE from the plasma by re-entry into the tissues where it is largely O-methylated or recaptured into sympathetic nerve terminals. Although plasma NE levels correlate well with muscle sympathetic nerve activity (see Wallin *et al.*, 1988), for a more precise estimate of entry of NE into plasma, NE "total body spillover" into plasma or "overflow" from a selected tissue, must be measured by use of ^{3}H-NE. "Total body spillover" is estimated from the plasma levels of NE and its rate of removal as determined during a constant systemic infusion of ^{3}H-NE (Esler *et al.*, 1979). It is assumed that once removed into the tissues, the specific activity is so greatly diluted by the endogenously formed NE that only negligible amounts of the labelled compound are returned to the circulation. Although ^{3}H-NE recycling has been suggested on the basis of alumina extractable tritium (Henriksen and Christensen, 1989), the presence of ^{3}H-DHPG formed from ^{3}H-NE accounts for the excess tritium in plasma which led to the conclusion that ^{3}H-NE may be recycled (Eisenhofer *et al.*, in press). The regional rates of overflow of NE from specific regions is determined from the fractional removal of the ^{3}H-NE and endogenous NE content of the arterial and venous blood and the rate of blood flow through the region (see Esler *et al.*, 1990). The fraction (E) NE extracted from arterial blood can be estimated from the arterio-venous concentration difference in ^{3}H-NE:

$$E = (a^* - v^*)/a^*$$

divided by the arterial concentration of ^{3}H-NE (a^*). Spillover, S_o, of NE is equal to the blood flow, Q, multiplied by the sum of the veno-arterial difference in NE concentration, $V - A$ and the product of the arterial NE level, A, and the extraction fraction, E:

$$S_o = Q \cdot (V - A) + Q \cdot E \cdot A$$

Total body NE spillover and overflow rates of NE are related to sympathetic nerve activity (Wallin, 1988; Deka-Starosta *et al.*, 1989), but may not reflect precisely NE release rates. Several factors account for this lack of exact correspondence of NE entry into plasma and rate of NE release. First, there is a significant diffusion barrier to diffusion of NE across the capillary membrane (Fig. 2). The limited permeability of this barrier was measured in the capillaries of the dog heart using ^{131}I-albumin and (^{14}C)-sucrose (as diffusion-limited indicators) as well as ^{3}H-NE (Cousineau *et al.*, 1980). By appropriate kinetic analysis of the venous effluent from the heart after a bolus injection of the multiple isotopically labelled compounds, they were able to assess the capillary permeability of NE, to estimate the interstitial concentrations of NE and to evaluate the rate constant for NE uptake into sympathetic neurons. In dogs, capillary permeability surface products yield a constant of about 0.16 s^{-1} (ml/sec/ml interstitial fluid) (Cousineau *et al.*, 1980); in humans, the constant was found to be 0.10 s^{-1} (Rose *et al.*, 1985). Because of the resistance to diffusion of NE through the capillary membrane and the rapid transit time through the capillary relative to the capillary permeability surface constant, there is insufficient time for equilibration of NE between interstitial fluid and plasma. At more rapid blood flow rates, less ^{3}H-NE is removed from a given volume plasma than during slow flow rates. Similarly, at rapid flow rates, since the entry of NE into plasma is also diffusion-limited, the concentration increment in plasma NE seen in venous effluent plasma is lower than during slow flow rates (Grossman *et al.*, 1991). Although the NE concentration in the venous plasma may be lower during rapid flow, the net amount of NE removed (spillover per unit time) may be greater. These factors alter the relationship between NE release into the interstitial fluid and NE entry into plasma. Of course, although most NE released from the nerve terminals into the interstitial fluid is removed by reuptake into the neuron (see below), any change in uptake efficiency will alter the relationship of NE release and NE entry into plasma.

NE taken up into the sympathetic nerve terminal is either deaminated to form DHPG or is captured by vesicular uptake (see Kopin, 1985). In the isolated vas deferens preloaded with ^{3}H-NE, blockade of uptake with desipramine blocks completely the increased appearance of DHPG during electrical or potassium ion stimulation (Eisenhofer *et al.*, 1987). Release of DHPG is enhanced by reserpine, but this is unaffected by desipramine. These results indicate that in the rat vas deferens, DHPG formation is almost totally within the sympathetic nerve terminals.

In humans, 20 min after beginning a constant infusion of a mixture of ^{3}H-NE and ^{3}H-isoproterenol (^{3}H-ISO), when a reasonably steady state had been established, Goldstein *et al.* (1988), found that about 82% of ^{3}H-NE was cleared from arterial plasma during a single passage of blood through the coronary circulation, whereas only 14% of the ^{3}H-ISO had been lost (Table 1). Since ISO is not taken up into sympathetic nerve terminals but is taken up extraneuronally and O-methylated by COMT, it was concluded that 14% of the NE was removed by the same processes responsible for ISO clearance, and that the remainder of the ^{3}H-NE was removed by uptake into sympathetic neurons. Goldstein *et al.* (1988) calculated that 69% of the NE delivered to the heart was taken up into sympathetic nerves. This conclusion was supported by the marked diminution in cardiac removal of ^{3}H-NE after blockade of uptake with desipramine. In these studies, Goldstein *et al.* (1988) found that ^{3}H-DHPG outflow from the heart was only about 3–4% of the amount of ^{3}H-NE taken up into the sympathetic nerves, and that this was almost completely eliminated after treatment with desipramine. These observations indicate that as in the rat vas deferens, ^{3}H-DHPG was formed

Nerve terminal   **Cardiac or Skeletal Muscle**

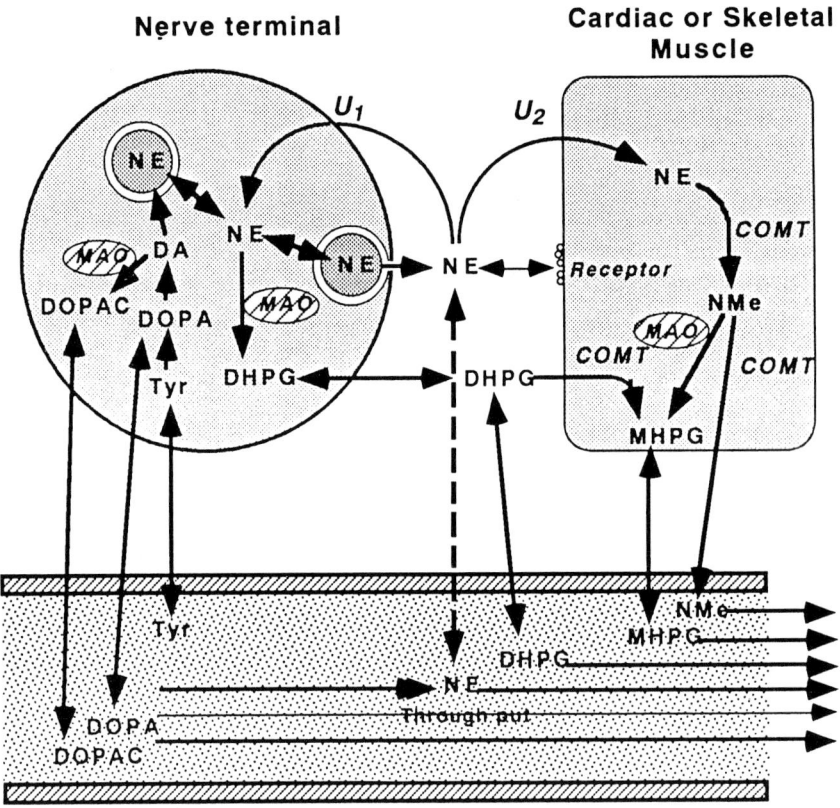

Figure 2.   Formation and disposition of norepinephrine, its precursors and its metabolites at sympathetic nerve terminals in cardiac or skeletal muscle (see text for explanations of the various processes for transfer and metabolism of administered and endogenously released norepinephrine).

almost exclusively from ^{3}H-NE taken up into the cardiac sympathetic nerve terminals; most of the tritiated catecholamine was retained, presumably in the storage vesicles. Endogenous cardiac DHPG formation exceeds, by far, the uptake of NE from the circulation in humans and in rabbits (Goldstein *et al.*, 1988; Eisenhofer *et al.*, 1991a). In humans (Table 1), the specific activity of ^{3}H-DHPG in the venous effluent from the heart (5.78 dpm/pmole) is only a small fraction of that of the venous plasma ^{3}H-NE (208 dpm/pmole). ^{3}H-DHPG appears to be formed from a pool of endogenous NE which dilutes the ^{3}H-NE entering from plasma; this pool has the same specific activity as NE released during sympathetic nerve stimulation. Figure 3 depicts the various compartments involved in the kinetic analysis of the disposition of ^{3}H-NE delivered via the blood stream to the heart. As indicated above, the diffusion barrier in myocardial capillaries does not permit complete equilibration between plasma and interstitial fluid NE. A portion of

NE delivered to the heart never enters the interstitial fluid (throughput) and may be considered to have been shunted or to flow through an impermeable vessel. The remainder of the afferent ^{3}H-NE enters and mixes with NE in the interstitial fluid.

Table 1.

Arterial entry and great cardiac venous output of endogenous and tritiated catechols during simultaneous infusion of tritiated *l*-norepinephrine and *dl*-isoproterenol[*]

Sampling site	Norepinephrine	DHPG	Isoproterenol
Radioactive Compound			
Arterial (dpm/min)	60750 ± 6120	3408 ± 710	44414 ± 3409
Venous (dpm/min)	12015 ± 1890	5254 ± 852	37615 ± 2513
$V - A$ (dpm/min)	−48735	1846	−6799
Endogenous Compound			
Arterial (pmole/min)	55.9 ± 5.06	284.8 ± 28.4	
Venous (pmole/min)	57.8 ± 6.65	623.5 ± 68.9	
$V - A$ (pmole/min)	1.9	338.7	
Specific Activity			
Arterial (dpm/pmole)	1087	8.22	
Venous (dpm/pmole)	208		
Increment (dpm/pmole)		5.45	

[*]Data calculated from Goldstein *et al.* (1988). DHPG, dihydroxyphenylglycol is calculated from blood flow whereas norepinephrine and isoproterenol are calculated using plasma flow.

Cousineau *et al.* (1980) predicted that during a steady state, the concentration of ^{3}H-NE in the interstitial fluid, I, would attain a level equal to $K_c/(K_c+K_n)$ that of venous plasma, which is equivalent to $K_2/(K_2+K_3+K_m)$ in Fig. 3. The mean value for this ratio, calculated from data in six humans reported by Rose *et al.*, (1985), was 0.252 ± 0.071. Since, under resting conditions the concentration of endogenous NE in the interstitial fluid was found to approximate that in the plasma (Rose *et al.* 1985), during a steady state, the specific activity of ^{3}H-NE in the interstitial fluid can be assumed also to be about 25% of that in the venous plasma. From the data on cardiac blood flow and arterial and great cardiac venous levels of catechols found during simultaneous infusion of ^{3}H-NE and ^{3}H-isoproterenol (Goldstein *et al.*, 1988), the rates of delivery to and removal from the heart of labelled and endogenous NE and DHPG were be determined (Table 1). This data and the results of the kinetic analyses by Cousineau *et al.* (1980) and by Rose *et al.* (1985) can now be used to estimate the various fluxes depicted in Fig. 3. The specific activity of ^{3}H-NE in the cardiac interstitial fluid, I, is calculated to be 52.0 dpm/pmole (i.e. 25% of the specific activity of NE in the great cardiac vein). From this specific activity and those of ^{3}H-NE in

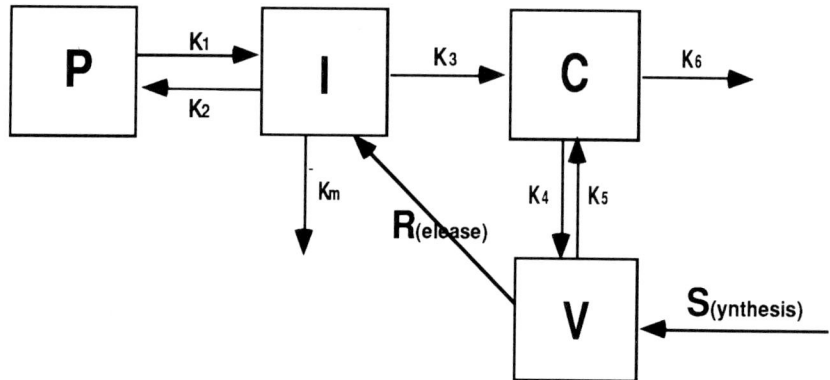

Figure 3. Compartmental model of the disposition and metabolism of norepinephrine shown diagrammatically in Fig. 2. P represents plasma; I, interstitial fluid; C, cytoplasm; and V, vesicles. The rate constants K_1 and K_2 are equal and represent diffusion across the capillary membrane, K_3 represents neuronal uptake and K_m extraneuronal uptake and metabolism (mostly O-methylation); k_6 is the rate of deamination to DHPG; K_4 and K_5 represent uptake into and release from the vesicles. In a steady state, the proportions of the source which enter the other compartments are each the ratio of the rate constant (K_i) for entry into that compartment to the sum of the rate constants for transfer to all the derivative compartments ($K_i + K_j + \ldots$).

the artery and in the great cardiac vein (Table 1), the proportions of endogenous NE in the venous blood derived from the artery and from the interstitial fluid can be calculated. If the amount of NE derived from the artery is A and the amount from interstitial fluid is B, then the specific activity of NE in the venous blood (SA_V) is:

$$SA_V = \frac{SA_A \cdot A + SA_I \cdot B}{A + B},$$

where SA_A and SA_I are the specific activities of ^{3}H-NE in the artery and interstitial fluid, respectively. From Table 1 we obtain

$$\frac{1087 \cdot A + 52.0 \cdot B}{A + B} = 208.$$

By simple algebra it can be shown that 15.1% (8.7 pmole/min and 9424 dpm/min) of the venous endogenous and tritiated NE was derived from the artery (throughput); 84.9% (49.1 pmole/min) of the venous endogenous NE was from the interstitial fluid. Since the specific activity of interstitial NE was 52 dpm/pmole, about 2548 dpm/min entered the venous blood from the interstitial fluid.

The net extraction of ^{3}H-NE from plasma into the myocardium ($A - V$) was 48735 dpm/min (Table 1). Since 2548 dpm/min was returned to the venous blood, the total entry of ^{3}H-NE into the myocardium was 51283 dpm/min (and 47.2 pmole NE/min). To maintain a specific activity of 52 dpm/pmole, a total of 1102 pmole/min must have been entering the interstitial fluid. Thus, about 1055 pmole/min was derived from the sympathetic neurons, presumably by release from the vesicular storage sites into the cytoplasm.

The proportion of ^{3}H-NE removed from the coronary circulation by uptake into sympathetic nerves was estimated from arterio-venous differences in the concentration of ^{3}H-NE and ^{3}H-isoproterenol (which is not taken up into sympathetic nerves but is taken up into cardiac muscle and O-methylated) 20 min after beginning a constant infusion of the labelled compounds (Goldstein *et al.*, 1988). About 69% (41,320 dpm/min) of the ^{3}H-NE in arterial blood appeared to have been removed by uptake into sympathetic nerves. The increment in plasma ^{3}H-DHPG (1846 dpm/min) accounted for only a small fraction (4.5%) of this ^{3}H-NE; this is an underestimate since about one third of the DHPG formed in the heart is converted to MHPG (Eisenhofer *et al.*, 1989). Only about 6.7% of the ^{3}H-NE formed DHPG; almost all (93.3%) of the catecholamine entering the cytoplasm was captured into the storage vesicles. Similarly the rate of myocardial DHPG efflux (Table 1) was 338.7 pmole/min. This is equivalent to an estimated production rate of about 508 pmole/min. Thus, of the total endogenous NE entering the cytoplasm only about 6.9% is deaminated and the remainder is recaptured into the storage vesicles.

The specific activity of ^{3}H-DHPG (5.45 dpm/pmole) produced by the heart (i.e. net entry into the plasma) is much lower than that of the interstitial fluid, indicating further substantial dilution of the tracer in the intraneuronal cytoplasm where deamination takes place. At sufficiently early times, before significant accumulation of ^{3}H-NE in the vesicular compartment, the specific activity of vesicular NE is negligible. To reduce the specific activity of the cytoplasmic ^{3}H-NE from 52 to 5.45 dpm/pmole, the rate of entry of unlabelled NE into the cytoplasm (from the vesicles) must be 6885 pmole/min. Almost all of this NE is recaptured into the vesicle, but a small portion is deaminated to form DHPG.

As indicated above, a portion (15.1%) of the ^{3}H-NE never enters the interstitial fluid (throughput). Of the 85% of ^{3}H-NE which does enter the interstitial fluid, 5.7% is returned to the plasma and 69% is taken up into the nerve terminal. About 13.3% enters extraneuronal tissue where it may be stored temporarily or metabolized. Most (76.3%) of the ^{3}H-NE entering the cytoplasm is sequestered in the storage vesicles but a small portion (5.4%), is deaminated to form ^{3}H-DHPG.

Similarly, the fate of endogenous NE released from the storage vesicles into the interstitial space and the exchange of vesicular and cytoplasmic NE can be quantified. A total of 940 pmole/min NE was released from the vesicles into the interstitial fluid, but only 49.1 pmole/min (less than 5%) was lost in the plasma, 806 pmole/min (about 82%) was taken up into the neuronal cytoplasm and the remainder was extraneuronally metabolized.

There is a rapid direct exchange of vesicular and cytoplasmic NE. Of a total of 7691 pmole NE/min entering the cytoplasm, nearly 90% (6885 pmole NE/min) was derived directly from the vesicles and about 93% of the cytoplasmic NE was taken up back into the cytoplasm. The deficit between total vesicular loss of NE (6885 pmole/min) to the cytoplasm and by release (940 pmole/min) into the interstitial fluid and recapture of NE from the cytoplasm (7183 pmole/min) must be satisfied by new synthesis of NE (642 pmole/min), if a steady state is to be maintained.

The efficient capture of ^{3}H-NE by sympathetic neurons in the heart contrasts with the relatively low clearances of ^{3}H-NE in the arm (arterio-brachial venous difference of only 14%) or leg (arterial-femoral venous difference of only 7%). This may be attributed to the relatively dense sympathetic innervation of the myocardium. The relationship of sympathetic nerve density and NE uptake was demonstrated in earlier studies of accumulation of (^{14}C)-l-NE in the tissues of rats during constant infusion of the isotopically labelled catecholamine (Kopin et al., 1965). Organs with dense sympathetic innervation and high blood flow rates concentrate NE from the plasma much more rapidly than those with sparse sympathetic terminals and/or relatively low blood flow rates. After constant intravenous infusion of (^{14}C)-l-NE for over 24 h, the specific activities of tissue NE reflect the proportions of NE derived from the circulation or from synthesis. In all organs, NE uptake from the circulation accounted for only a trivial proportion of the tissue NE. In the heart it was only about 5% and in other tissues less.

PLASMA DOPA LEVELS AND THE RATE OF TYROSINE HYDROXYLATION

Although hydroxylation of tyrosine to DOPA has been considered the rate-limiting in the formation of both DA and NE, not all of the DOPA formed in noradrenergic or dopaminergic neurons is decarboxylated immediately. Significant amounts of endogenously formed DOPA enter the circulation and plasma DOPA levels exceed those of the catecholamines (Eisenhofer et al., 1986). DOPA is present in some foods, but the predominant sources of DOPA in plasma are catecholaminergic neurons.

There is substantial evidence that plasma levels of DOPA may reflect the rate of tyrosine hydroxylation. In rats, pharmacological manipulations which would be expected to alter NE synthesis affect plasma DOPA levels (Eisenhofer et al., 1988). Administration of an inhibitor of tyrosine hydroxylase or of a ganglion blocking agent, chlorisondamine, diminishes plasma DOPA levels (Fig. 4). One half hour after administration of reserpine, when net release of NE into the cytoplasm from the synaptic vesicles is inhibiting TH, plasma levels of the norepinephrine metabolite, DHPG, rise (Fig. 5), whereas plasma levels of DOPA fall (Fig. 4). After 15 h, when NE stores have been depleted, plasma DHPG levels are low (Fig. 5) but plasma DOPA levels are increased (Fig. 4). This increase is prevented by clorgyline, which inhibits MAO–A in nerve terminals, presumably as a result of increased cytoplasmic levels of NE (and/or DA) which inhibit tyrosine hydroxylation. Similarly, infusion of NE in amounts sufficient to elevate cytoplasmic NE in nerve terminals diminishes DOPA levels in plasma. This effect is blocked by DMI-induced inhibition of NE uptake. Furthermore, forskolin, which activates TH via enhanced cyclic AMP (Abou-Donia *et al.*, 1986), also increases plasma DOPA levels.

The hypothesis that plasma DOPA reflects TH activity was supported in humans by the demonstration that in normal subjects, there are a significantly arterio-antecubital venous increments in plasma DOPA, whereas in sympathec-

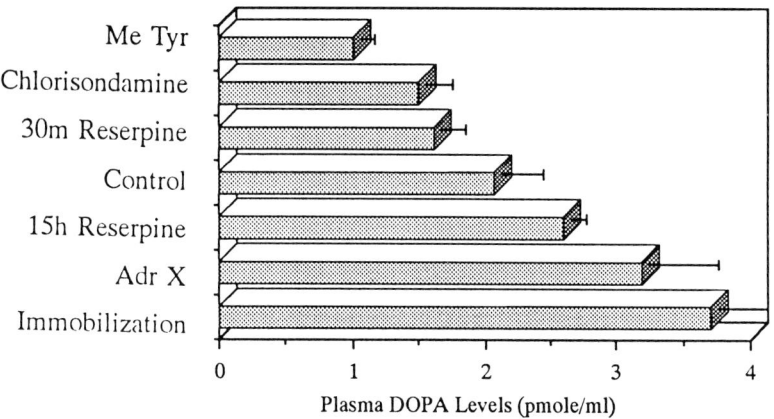

Figure 4. Plasma DOPA levels under various conditions (see text).

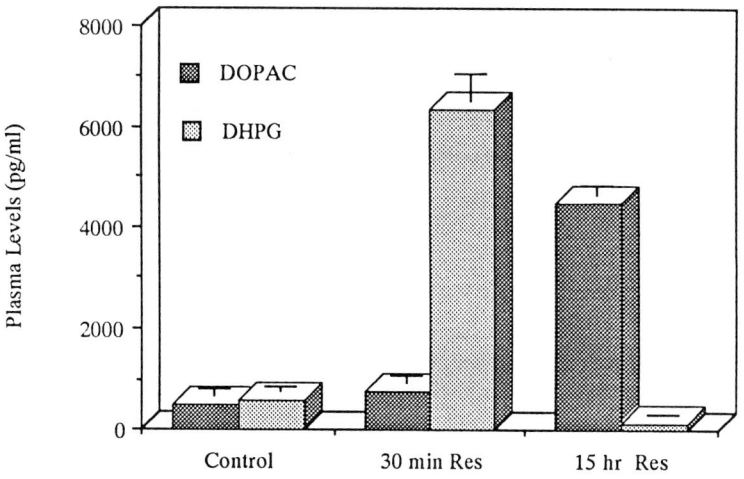

Figure 5. Time dependent effects of reserpine on plasma catechol levels.

tomized patients there are small (about 10%) decrements in the concentrations of this product of tyrosine hydroxylation (Goldstein *et al.*, 1987). Arterio-venous increments in plasma DOPA levels are correlated closely with arterio-venous increments in plasma levels of DHPG, MHPG, and NE (or their sum) across the vasculature of the brain, forearm, or heart. DOPA appears to account for about 20% of the total catechols and their metabolites which overflowed from these tissues (Eisenhofer *et al.*, 1989).

The correspondence of plasma DOPA levels with NE synthesis rates is further supported by studies in stressed animals. Even mild stressors, such as handling, elicit in rats an alerting response characterized by rapid and robust increased plasma levels of NE and EPI. This immediate rise in plasma catecholamines, particularly after a more severe stress such as immobilization, is followed within a few minutes by an increase in plasma levels of DOPA as well as of DOPAC, DHPG and MHPG. These findings are consistent with rapid enhancement of tyrosine hydroxylation when catecholamines are released and metabolized and support the view that plasma DOPA levels reflect rates of tyrosine hydroxylation in catecholaminergic neurons. Furthermore, immobilization stress-induced increments in plasma DOPA are reduced markedly by treatment with chlorisondamine to block sympathetic ganglion transmission or with α-methyl-paratyrosine to inhibit tyrosine hydroxylase. Other studies also confirm that sympathetic activation is attended by enhanced tyrosine hydroxylation and release of DOPA into the circulation. Thus Eisenhofer *et al.* (1991b) showed that in anesthetized dogs, electrical stimulation of the sympathetic nerves to the heart enhanced release of both norepinephrine and DOPA. The changes in cardiac DOPA production appeared to be related to the rate of release of both NE and its major deaminated metabolite, DHPG. Edrup *et al.* (1989a), however, reported that in humans, stimuli producing acute elevation of plasma NE levels did not increase levels of DOPA. Furthermore, they found that in rats, sympathectomy did not affect DOPA concentrations in skeletal muscle whereas there were striking decrements in tissue NE (Edrup *et al.*, 1989b). In contrast with these observations, Anden *et al.* (1989) showed that tissue DOPA concentrations increase when decarboxylation is inhibited, but that this increase is prevented by ganglionic blockade. These observations suggest that skeletal muscle takes up DOPA from the circulation and that, under some circumstances, this tissue store might be a source of plasma DOPA. To address this possibility, Szemeredi *et al.* (1991) examined the effects of curare and of chlorisondamine on stimulation-induced increments in plasma DOPA, NE and DHPG in pithed rats. Stimulation of the spinal nerves of pithed rats evoked striking increases in plasma NE, DHPG, and DOPA which were almost completely blocked by ganglionic blockade. Curare blockade of skeletal muscle contraction had no significant effect on increments in NE and DHPG during stimulation, but attenuated by about 50% the increment in plasma DOPA. This is consistent with the notion that skeletal muscle provides a reservoir for

DOPA, which, under some conditions can be returned to the circulation, but that the origin of DOPA is mostly in catecholaminergic neurons.

CATECHOLAMINE METABOLITES AS INDICES OF TYROSINE HYDROXYLATION RATES

Clearly, in a steady state the sum of the excretion rates of all DOPA metabolites, including O-methyl-DOPA, dopamine, norepinephrine and epinephrine and their metabolites, must equal the rate of tyrosine hydroxylation in the whole body, but not all dopamine metabolites are derived from dopaminergic neurons. In rodent brain, the predominant dopamine metabolite is the initial product of its deamination, 3,4-dihydroxyphenylacetic acid (DOPAC). In primate brain and in cerebrospinal fluid, the O-methylated derivative of this compound, homovanillic acid (HVA) is the major metabolite, (see Kopin, 1985). HVA levels in plasma and its excretion rates in urine provide a measure of the rate of dopamine synthesis, but it must be remembered that DOPA and dopamine are formed in noradrenergic neurons (as precursors to norepinephrine) as well as in dopaminergic neurons. Also, DOPA released into plasma can be metabolized to HVA. There is considerable evidence that not all dopamine formed in noradrenergic neurons is converted to norepinephrine. Anden and coworkers (Anden and Grabowska-Anden, 1983; Anden *et al.*, 1985) showed that in NE-predominant brain regions, dopamine turnover was more rapid than NE turnover and dopamine metabolite levels parallel the rate of dopamine formation. Furthermore, Scatton *et al.* (1984) showed that electrical stimulation of ascending noradrenergic fibres increased hippocampal levels of DOPAC as well as those of MHPG. Similarly, Curet *et al.* (1985), found increments in DOPAC and HVA in the locus coeruleus after antidromic stimulation of noradrenergic neurons. Drug-induced enhancement of noradrenergic activity also increases both HVA and MHPG in the cerebrospinal fluid of rats (Sheinen, 1986). These observations indicate that a considerable portion of dopamine formed in brain noradrenergic neurons escapes capture by the dopamine-β-hydroxylase-containing storage vesicles and, instead of being converted to norepinephrine, is deaminated to form DOPAC and subsequently HVA.

Similarly, in the peripheral sympathetic neurons, deamination of cytoplasmic dopamine which is incompletely captured by the NE storage vesicles is deaminated to form DOPAC; this metabolite is O-methylated to HVA or enters the plasma. In experimental animals, procedures known to accelerate norepinephrine release and to accelerate tyrosine hydroxylation, such as immobilization stress or adrenalectomy, elevate plasma DOPA levels whereas inhibition of TH by administration of α-methyltyrosine or ganglionic blockade with chlorisondamine diminish plasma DOPAC levels.

In humans, patients with neurogenic orthostatic hypotension may be divided into those with multiple system atrophy (MSA) and those with primary autonomic failure (PAF). Patients with MSA have normal levels of plasma NE when reclining, but because of a central nervous system abnormality, which usually produces other symptoms as well as autonomic insufficiency, they are unable to activate an otherwise intact peripheral sympathetic nervous system; plasma NE levels fail to increase when the patients assume an erect posture. These patients have normal basal plasma DOPAC levels (Goldstein *et al.*, 1989). Patients with PAF differ from those with MSA in that lesions appear to be localized to the peripheral sympathetic neurons; they have no central nervous system deficits and plasma levels of NE are low even when they recline, and, of course, fail to increase when the patients are erect. These patients have low plasma DOPAC levels (Goldstein *et al.*, 1989). Thus plasma DOPAC appears to be largely derived from sympathetic neurons in humans as well as in animals.

RATES OF TYROSINE HYDROXYLATION IN BRAIN

One method for measuring the brain contribution to total tyrosine hydroxylation and catecholamine production is by estimating the rate of catecholamine metabolite outflow from the brain. This is determined from the cerebral blood flow and the jugular venous-arterial differences in concentrations of HVA (for dopamine) or MHPG (for norepinephrine). Using this method, Maas *et al.* (1980) calculated the rate of HVA production by human brain was about 6.5 mmole/min using blood obtained from the right jugular vein. This represented about one third (390 mg/h) of the total urinary HVA excretion (1176 mmole h^{-1}). In a similar study, Lambert *et al..* (1991) found that overflow of HVA was 36% greater into the left jugular vein than from the right, whereas the reverse was the case for NE and MHPG. They estimated that HVA production by the brain was only about half that found by Maas *et al.* (1980). The difference was attributed to use of plasma flow by Lambert *et al.* and blood flow by Maas *et al.* the latter assuming (incorrectly) that HVA enters the red blood cells rapidly. Measurements of HVA production, as indicated above, includes metabolism of dopamine in noradrenergic neurons. MHPG arteriovenous differences across the cerebral circulation also have been measured in monkeys (Maas *et al.*, 1977) and humans (Maas *et al.*, 1979). By this means norepinephrine turnover in noradrenergic neurons can be assessed, but HVA is produced by both noradrenergic and dopaminergic neurons.

An alternative method for assessing brain production of catecholamine metabolites involves use of a peripheral monoamine oxidase inhibitor. Debrisoquin has been used to block selectively HVA (and MHPG) production only in peripheral tissues, since it does not penetrate the blood-brain barrier, and is concentrated in noradrenergic neurons where it blocks MAO. Elevated cytoplasmic catecholamine levels inhibiting tyrosine hydroxylation as well as the blockade of deamination results in lowered plasma levels of MHPG and HVA. If peripheral tissue blockade of formation of these metabolites were complete, then urinary excretion would represent solely brain metabolism.

Whereas VMA and MHPG are derived solely from noradrenergic neurons, as indicated earlier, HVA may be derived from both noradrenergic and dopaminergic neurons as well as from DOPA which escapes decarboxylation in the tissues and reaches plasma (Fig. 2). Noradrenergic neurons in the peripheral sympathetic nervous system are quantitatively important for the production of dopamine metabolites, most of which is HVA, as well as for production of the metabolites of norepinephrine, VMA and MHPG. In the brain, dopaminergic neurons predominate, but brain HVA is derived from noradrenergic as well as dopaminergic neurons. If it is assumed that the fraction of dopamine which escapes capture into the vesicles of central noradrenergic neurons is similar to the fraction which is deaminated to DOPAC in peripheral noradrenergic neurons, then even with incomplete inhibition of peripheral tyrosine hydroxylation, it is possible to deduce the proportion of urinary HVA derived from brain dopaminergic neurons. The scheme upon which this is based is depicted in Fig. 6. The rates of tyrosine hydroxylation in noradrenergic and dopaminergic neurons are represented as N and D, respectively. The proportions of DOPA which enter plasma from noradrenergic and dopaminergic neurons are designated p_n and p_d, respectively. There are then three sites of dopamine formation (Fig. 6) from DOPA: in noradrenergic neurons ($d_n \cdot N$), in peripheral tissues ($p_n \cdot N + p_d \cdot D$) and in dopaminergic neurons ($d_d \cdot D$). A portion, (f_β), of dopamine formed in noradrenergic neurons is β-hydroxylated ($f_\beta \cdot d_n \cdot N$) and the remainder, [$(1 - f_\beta) \cdot d_n \cdot N$], is deaminated to form DOPAC and then HVA. The HVA excretion rate is the sum of these three sources:

$$HVA = (1 - f_\beta) \cdot d_n \cdot N + (p_n \cdot N + p_d \cdot D) + d_d \cdot D.$$

Since $p_d + d_d = 1$, this equation becomes:

$$HVA = [(1 - f_\beta) \cdot d_n + p_n] \cdot N + D.$$

The sum of the rates of production and excretion of the norepinephrine metabolites, VMA and MHPG approximates the rate of norepinephrine formation in noradrenergic neurons:

$$[VMA + MHPG] = f_\beta \cdot d_n \cdot N.$$

From these equations it is evident that

$$HVA = m \cdot [VMA + MHPG] + D,$$

where

$$m = \frac{(1 - f_\beta) \cdot d_n + p_n}{f_\beta \cdot d_n}$$

Dopaminergic Neurons

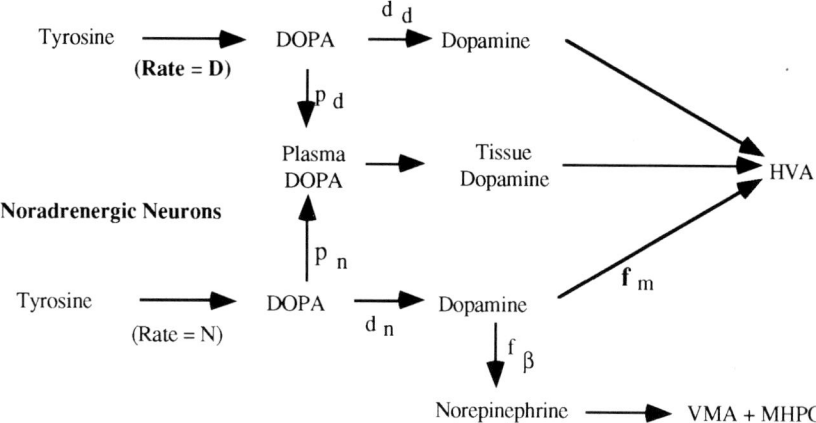

Figure 6.   Schematic representation of the origins of homovanillic acid (HVA), 3-methoxy-4-hydroxyphenylethylglycol (MHPG) and vanillylmandelic acid (VMA).

Thus, the rate of HVA excretion is the rate of tyrosine hydroxylation in dopaminergic neurons in brain, D, and a constant (m) times the rate of excretion of MHPG plus VMA. This is a straight line with slope m and intercept D on the y-axis.

Such a relationship can be demonstrated by examining the urinary excretion rates of HVA, VMA and MHPG before and during the administration of varying doses of debrisoquin to diminish peripheral production of these metabolites. If the urinary excretion rates of HVA are plotted (ordinate) against the sum of the urinary excretion rates of VMA and MHPG (abscissa), a straight line would be expected (Fig. 7). The slope of this line represents the ratio of the portion of dopamine formed in noradrenergic neurons which is ultimately excreted as HVA and the portion converted to norepinephrine, reflected by the sum of VMA plus MHPG excretion. The intercept on the y-axis is an estimate of the rate of HVA produced in brain dopaminergic neurons. In monkeys, it was found that about 25% of urinary HVA is derived from the brain dopaminergic neurons (Kopin *et al.*, 1988). Brain-derived HVA was decreased by about 75% after treatment with MPTP to destroy brain dopaminergic neurons. Debrisoquin is being used currently in studies of humans and preliminary results suggest that a similar, small fraction of urinary HVA reflects tyrosine hydroxylation in brain dopaminergic neurons.

CONCLUSION

Tyrosine hydroxylation is rate-limiting in the production of catecholamines. The rate of tyrosine hydroxylation varies with demand and can be accelerated promptly

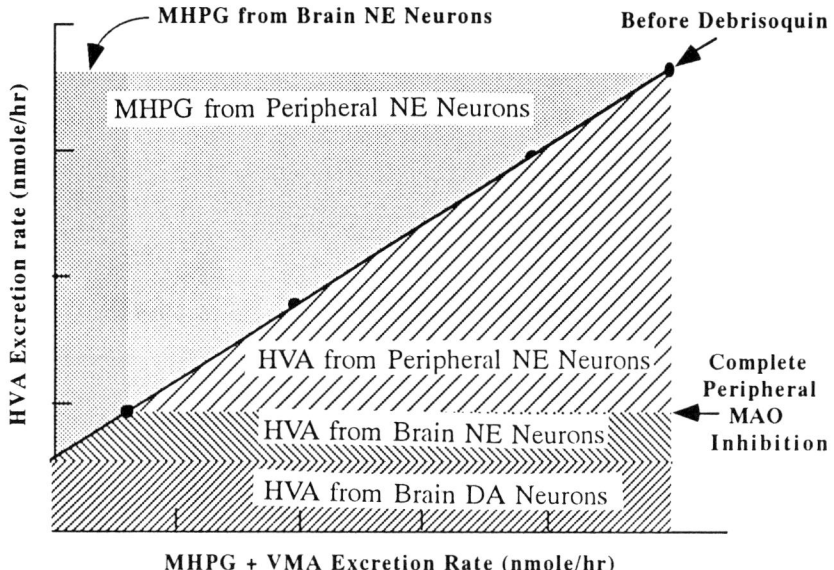

Figure 7. Hypothesized relationship between urinary excretion of catecholamine metabolites before and during varying degrees of peripheral MAO inhibition by debrisoquin.

by activation of TH; induction of increased tissue levels of TH increase the capacity of the tissue to respond to demands for enhanced catecholamine synthesis, but do not necessarily reflect the rate of tyrosine hydroxylation. Plasma levels and spillover into plasma of DOPA as well as DOPAC and DHPG reflect tyrosine hydroxylation rates whereas those of NE reflect stimulation-induced release of the neurotransmitter. Frequently, but not always, these are parallel in their increments and decrements. Arterio-venous increments and blood flow measurements permit assessment of regional, including brain, formation of catecholamine metabolites and NE overflow. By appropriate use of debrisoquin to perturb peripheral catecholamine metabolite formation and excretion, it is possible to assess the rate of tyrosine hydroxylation in brain dopaminergic neurons.

REFERENCES

Abel, J. J. and Crawford, S. C. (1897). On the blood pressure raising constituent of the suprarenal capsule. *Bull. Johns Hopk. Hosp.* **8**, 151–157.

Abou-Donia, M. M., Wilson, S. P., Zimmerman, T. P., Nichol, C. A. and Viveros, O. H. (1986). Regulation of guanosine triphosphate cyclohydralase and tetrahydrobiopterin levels and the role of the cofactor in tyrosine hydroxylation in primary culture of adrenalmedullary chromaffin cells. *J. Neurochem.* **46**, 1190–1199.

Anden, N. E. and Grabowska-Anden, M. (1983). Formation of deaminated metabolites of dopamine in noradrenaline neurons. *Naunyn-Schmiedebergs Arch. Pharmacol.* **324**, 1–6.

Anden, N. E., Grabowska-Anden, M., Lindgren, S. and Oweling, M. (1985). Very rapid turnover of dopamine in noradrenaline cell body regions. *Naunyn-Schmiedebergs Arch. Pharmacol.* **329**, 258–263.

Axelrod, J. (1962). Purification and properties of phenylethanolamine-N-methyl transferase. *J. Biol. Chem.* **237**, 1657–1660.

Blaschko, H. (1939). The specific action of L-dopa Decarboxylase. *J. Physiol. (London)* **96**, 50P–51P.

Cannon, W. B. (1929). *Bodily Changes in Pain, Hunger, Fear and Rage. Appleton-Century-Crofts, 2nd Edition (New York)*.

Carlsson, A. (1959). The occurrence, distribution and physiological role of catecholamines in the nervous system. *Pharm. Rev.* **12**, 490–493.

Cousineau, D., Rose, C. P. and Goresky, C. A. (1980). Labeled catecholamine uptake in the dog heart: interactions between capillary wall and sympathetic nerve uptake. *Circ. Res.* **47**, 329–338.

Curet, O., Dennis, T. and Scatton, B. (1985). The formation of deaminated metabolites of dopamine in the locus coeruleus depends upon noradrenergic neuronal activity. *Brain Res.* **335**, 297–301.

Deka-Starosta, A., Garty, M., Zukowska-Grojec, Z., Keiser, H. R., Kopin, I. J. and Goldstein, D. S. (1989). Renal sympathetic nerve activity and norepinephrine release in rats. *Am. J. Physiol.* **257**, R229–R236.

Demis, D. J., Blaschko, H. and Welch, A. D. (1956). The conversion of dihydroxyphenylalanine-2-(C^{14})(dopa) to norepinephrine by bovine adrenal medullary homogenate. *J. Pharmacol. Exp. Ther.* **117**, 208–212.

Edrup, E., Christensen, N. J., Andreasen, J. and Hilsted. (1989a). Plasma dihydroxyphenylalanine (DOPA) is independent of sympathetic activity in humans. *Eur. J. Clin. Invest.* **19**(6), 514–517.

Edrup, E., Richter, E. A. and Christensen, N. J. (1989b). DOPA, norepinephrine, and dopamine in rat tissues: no effect of sympathectomy on muscle DOPA. *Am. J. Physiol.* **256**(2 pt 1), E284–E287.

Eisenhofer, G., Goldstein, D. S., Stull, R., Keiser, H. R., Sunderland, T., Murphy, D. L. and Kopin, I. J. (1986). Simultaneous liquid-chromatographic determination of 3,4-dihydroxyphenylglycol, catecholamines, and 3,4-dihydroxyphenylalanine in plasma, and their responses to inhibition of monoamine oxidase. *Clin. Chem.* **32**(11), 2030–2033.

Eisenhofer, G., Ropchak, T. G., Stull, R. W., Goldstein, D. S., Keiser, H. R. and Kopin, I. J. (1987). Dihydroxyphenylglycol and intraneuronal metabolism of endogenous and exogenous norepinephrine in the rat vas deferens. *J. Pharmacol. Exp. Ther.* **241**, 547–553.

Eisenhofer, G., Ropchak, T., Nguyen, H., Keiser, H. R., Kopin, I. J. and Goldstein, D. S. (1988). Source and physiological significance of plasma 3,4-dihydroxyphenylalanine in the rat. *J. Neurochem.* **51**, 1204–1213.

Eisenhofer, G., Brush, J. E., Cannon, R. O., Stull, R., Kopin, I. J. and Goldstein, D. S. (1989). Plasma dihydroxyphenylalanine and total body and regional noradrenergic activity in humans. *J. Clin. Endocrin. Metab.* **68**, 247–255.

Eisenhofer, G., Esler, M. D., Goldstein, D. S. and Kopin, I. J. (1991). Neuronal uptake, metabolism, and release of tritium-labeled norepinephrine during assessment of its plasma kinetics. *Am. J. Physiol.*, (in press).

Eisenhofer, G., Smolich, J. J. and Esler, M. D. (1991). Neuronal reuptake of norepinephrine and production of dihydroxyphenylglycol by cardiac sympathetic nerves in the anesthetized dog. *Circulation*, (in press).

Esler, M., Jackman, G., Bobik, A., Kelleher, D., Jennings, G., Leonard, P., Skews, H. and Korner, P. (1979). Determination of norepinephrine apparent release rate and clearance in humans. *Life Sci.* **25**, 1461–1470.

Esler, M., Jennings, G., Lambert, G., Meredith, I., Horne, M. and Eisenhofer, G. (1990). Overflow of catecholamine neurotransmitters to the circulation: source, fate, and functions. *Physiol. Rev.* **70**, 963–985.

von Euler, U. S. (1948). Identification of the sympathomimetic ergone in adrenergic nerve fibres (sympathin) with laevo-noradrenaline. *Acta Physiol. Scan.* **16**, 63–74.

Goldstein, D. S., Brush, J. E., Eisenhofer, G., Stull, R., Esler, M. (1988). *In vivo* measurement of neuronal uptake of norepinephrine in the human heart. *Circulation* **78**, 41–48.

Goldstein, D. S., Udelsman, R., Eisenhofer, G., Keiser, H. R. and Kopin, I. J. (1987). Neuronal source of plasma dihydroxyphenylalanine. *J. Clin. Endocrinol. Metab.* **64**, 856–861.

Goldstein, D. S., Polinsky, R. J, Garty, M., Robertson, D., Brown, R. T., Biaggioni, I., Stull, R. and Kopin, I. J. (1989). Patterns of plasma levels of catechols in neurogenic orthostatic hypotension. *Ann. Neurol.* **26**, 558–563.

Goodall, Mc. C. (1951). Studies of adrenaline and noradrenaline in the mammalian heart and suprarenals. *Acta. Physiol. Scand.* **24** (Suppl. 85), 7–51.

Grossman, E., Chang, P. C., Hoffman, A., Tamrat, M., Kopin, I. J. and Goldstein, D. S. (1991). Tracer norepinephrine kinetics: dependence on regional blood flow and the site of infusion. *Am. J. Physiol.* **260**, R946–952.

Henriksen, J. H. and Christensen, N. J. (1989). Plasma norepinephrine in humans: limitations in assessment of whole body norepinephrine kinetics and plasma clearance. *Am. J. Physiol.* **257**, E743–E750.

Holtz, P., Heise, R. and Ludtke, K. (1938). Fermentativer Abbau von L-dioxphenylalanin (dopa) durch Niere. *Naunyn-Schmeideberg's Arch. Exp. Path. Pharmak.* **191**, 87–118.

Kissinger, P. T., Bruntlett, C. S. and Shoup, R. E. (1981). Neurochemical application of liquid chromatography with electrochemical detection. *Life Sci.* **28**, 455–465.

Kopin, I. J. (1960). Technique for the study of alternate metabolic pathways: Epinephrine metabolism in man. *Science* **131**, 1372–1374.

Kopin, I. J., Axelrod, J. and Gordon, E. (1961). The metabolic fate of ^{3}H-epinephrine and (^{14}C)-metanephrine in the rat. *J. Biol. Chem.* **236**, 2109–2113.

Kopin, I. J., Gordon, E. K. and Horst, W. D. (1965). Studies of uptake of I-norepinephrine-(^{14}C). *Biochem. Pharmacol.* **14**, 753–759.

Kopin, I. J. (1985). Catecholamine metabolism: Basic aspects and clinical significance. *Pharmacol. Rev.* **37**, 333–364.

Kopin, I. J., Bankiewicz, K. S. and Harvey-White, J. (1988). Assessment of brain dopamine metabolism from plasma HVA and MHPG during debrisoquin treatment: Validation of monkeys treated with MPTP. *Neuropsychopharmacology* **1**, 119–125.

Kvetňanský, R., Weise, V. K. and Kopin, I. J. (1970). Elevation of adrenal tyrosine hydroxylase and phenylethanolamine-N-methyltransferase by repeated immobilization of rats. *Endocrinology* **87**, 744–749.

Kvetňanský, R., Gewirtz, G. P., Weise, V. K. and Kopin, I. J. (1971a). Catecholamine-synthesizing enzymes in the rat adrenal gland during exposure to cold. *Am. J. Psychiatry* **220**, 928–931.

Kvetňanský, R., Silbergeld, S., Weise, V. K. and Kopin, I. J. (1971b). Effects of restraint on rat adrenomedullary response to 2-deoxyglucose. *Psychopharmacologia* **20**, 22–31.

Lambert, G. W., Eisenhofer, G., Cox, H. S., Horne, C. M., Kalff, V., Kelly, M., Jennings, G. L. and Esler, M. D. (1991). Direct determination of homovanillic acid release from the human brain, an indicator of central dopaminergic activity. *Life Sciences* , (in press).

Levin, E. Y., Levenberg, B. and Kaufman, S. (1960). The enzymatic conversion of 3,4-dihydroxyphenylethylamine to norepinephrine. *J. Biol. Chem.* **235**, 2080–2086.

Lund, A. (1949). Fluorimetric determination of adrenaline in blood III. A new sensitive and specific method. *Acta Pharmacol. (Kbh.)* **5**, 231–247.

Maas, J. W., Hattox, S. E., Landis, D. H. and Roth, R. H. (1977). A direct method for studying 3-methoxy-4-hydroxy-phenethyleneglycol (MHPG) production by brain in awake animals. *Eur. J. Pharmacol.* **46**, 221–228.

Maas, J. W., Hattox, S. E., Greene, N. M. and Landis, D. H. (1979). 3-Methoxy-4-hydroxyphenylethyleneglycol production by human brain *in vivo*. *Science* **205**, 1025–1027.

Maas, J. W., Hattox, S. F., Greene, N. M. and Landis, D. H. (1980). Estimates of dopamine and serotonin synthesis by the awake human brain. *J. Neurochem.* **34** (6), 1547–1549.

Maas, J. W., Contreras, S. A. Bowden, C. L. and Weintraub, S. E. (1985). Effects of debrisoquin on CSF and plasma HVA concentrations in man. *Life Sci.* **36**, 2163–2170.

Nagatsu, T., Levitt, M. and Undenfriend, S. (1964a). Conversion of L-tyrosine to 3,4-dihydroxyphenylalanine by cell-free preparations of brain and sympathetically innervated tissues. *Biochem. Biophys. Res. Commun.* **14**, 543–549.

Nagatsu, T., Levitt, M. and Undenfriend, S. (1964b). Tyrosine hydroxylase: the initial step in norepinephrine biosynthesis. *J. Biol. Chem.* **239**, 2910–2917.

Peart, W. S. (1959). The nature of splenic sympathin. *J. Physiol. (London)* **108**, 491–501.

Rose, C. P., Burgess, J. H. and Cousineau, D. (1985). Tracer norepinephrine kinetics in coronary circulation of patients with heart failure secondary to chronic pressure and volume overload. *J. Clin. Invest.* **76**, 1740–1747.

Scatton, B., Dennis, T. and Cruet, O. (1984). Increase in dopamine and DOPAC levels in noradrenergic terminals after electrical stimulation of the ascending noradrenergic pathways. *Brain Res.* **298**, 193–196.

Scheinen, H. (1986). Enhanced noradrenergic neuronal activity increases homovanillic acid levels in cerebrospinal fluid. *J. Neurochem.* **47**, 665–667.

Sedvall, G. C., Weise, V. K. and Kopin, I. J. (1968). The rate of norepinephrine synthesis measured *in vivo* during short intervals: influence of adrenergic impulse activity. *J. Pharmacol. Exp. Ther.* **159**, 274–282.

Spector, S. Gordon, R., Zaltzman-Nirenberg, P., Levitt, M. and Udenfriend, S. (1963). Norepinephrine synthesis from tyrosine-(C^{14}) in isolated perfused guinea pig heart. *Science* **139**, 1299–1301.

Szemeredi, K., Pacak, K., Kopin, I. J., Goldstein, D. S. (1991). Sympathoneural and skeletal muscle contributions to plasma DOPA responses in pithed rats. *J. Autonom. Nervous Sys.*, (in press).

Vogt, M. (1954). The concentration of sympathin in different parts of the central nervous system under normal conditions and after the administration of drugs. *J. Physiol. (London)* **123**, 451–481.

Wallin, B. G., Sundolg, G., Eriksson, B. M., Dominiak, P., Grobecker, H. and Lidbland, L. E. (1981). Plasma noradrenaline correlates to sympathetic muscle nerve activity in normotensive man. *Acta Physiol. Scand.* **111**, 69–73.

Wallin, B. G. (1988). Relationship between sympathetic nerve traffic and plasma concentrations of noradrenaline in man. *Pharmacol. Toxicol.* **63** (Suppl.), 19–11.

Tyrosine Hydroxylase, pp. 217–224
M. Naoi *et al.* (Eds)
© VSP 1993

DECREASE OF TYROSINE HYDROXYLASE ACTIVITY AND UNDERSTANDING OF PARKINSON'S DISEASE

HIROTARO NARABAYASHI

Neurological Clinic, 5-12-8 Nakameguro, Meguro-ku, Tokyo 153, Japan

Abstract—Nigrostriatal dopamine deficiency, which is the axial feature of Parkinson's disease, is due to decrease of tyrosine hydroxylase activity as established by Nagatsu *et al.* This opened the way of rapid development of pharmacological treatment today, although the reason for decrease of activity is still not fully explained. In the long-term experiences of various treatment, including levodopa, dopamine-agonist, L-*threo*-DOPS and microelectrode stereotaxic surgery, it was gradually established that in the chronic stage of the long-term course of Parkinson's disease, deficiency of norepinephrine seems to be added to that of dopamine and become important in producing the complex clinical pictures in the later stage.

Key words: rigidity; tremor; akinesia; DA; NE; DβH.

BASIC RESEARCH: NIGROSTRIATAL DA DEFICIENCY

The discovery of a dopamine (DA) deficiency in the brains of patients with Parkinson's disease (PD) is one of the most remarkable milestones in both clinical and basic neurosciences. The finding of a marked beneficial effect of levodopa therapy on Parkinsonism, which aims to compensate DA deficiency by administration of levodopa, the precursor for DA, opened up a new era of pharmacological treatment for the disease and supports the hypothesis that this deficiency is the essential mechanism of the disease. Understanding of neurodegenerative disorders as conditions based on neurotransmitter deficiency and the idea of treating the disease by compensating for the deficiency in transmitter by administering its precursor, became a fashion in clinical research and treatment of degenerative diseases, although many of the trials thus proposed failed to prove clinically useful.

It is well recognized today that the reason for the DA deficiency is the lowering of tyrosine hydroxylase (TH) activity and not that of dopa-decarboxylase (DC) activity, which point was established by Nagatsu *et al.* (1977; 1979). They measured the activities of a series of catecholamine (CA) enzymes and found that the activity of enzymes at each step of biosynthesis of CAs, such as DA, norepinephrine (NE) and epinephrine (E) was lower in the brains of PD patients than in normal brains and, more importantly, that TH, the rate-limiting enzyme of CA biosynthesis, showed the greatest drop in activity. These discoveries on DA and Parkinsonism provoked research interest in the detailed mechanism of DA transmission at the synaptic site and characterization of the DA receptor, which soon led to the molecular biological investigation. The ensuing increased knowledge about the DA receptors then facilitated development of various DA agonists for use in clinical treatment.

As a reason for nigrostriatal DA deficiency or lowered TH activity, many possible explanations have been proposed. In neuropathology, selective degeneration of the melanin-containing neurons in the compact zone of the substantia nigra (SNc) has long been known as an axial pathological change in PD. This degenerative change was interpreted as causing DA deficiency in the striatum via a possible nigrostriatal projection, the existence of which was not known when the DA deficiency in PD was found but was later demonstrated by Anden et al. (1964). However, no explanation was given as to the reason for neuronal death within the SNc. Encephalitis of unknown etiology, viral infections, role of repeated minor head trauma and of vascular changes, and others were proposed in the literature, but almost all of these hypotheses have disappeared without substantial proof.

The breakthrough discovery by Langston et al. (1983) that MPTP (1-methyl-4-phenyl-1,2,3,6-tetrahydropyridine) as an exogenous toxin, can selectively destroy the nigrostriatal DA system and produce symptoms mimicking Parkinsonism, presented the possibility that MPTP itself or some similar substance(s) might cause selective degenerative change in the nigrostriatum finally leading to neuronal death. This raised the question as to whether chronic exposure to some environmental substance might be responsible for the disease, a question that has still not been answered. The possibility also exists that some specific endogenously produced substance(s) may be toxic to the nigrostriatal DA system. In fact, tetrahydroisoquinoline (TIQ) has been proposed by Nagatsu and Yoshida (1988) and Yoshida and Nagatsu (1990) as one of such candidates.

Another question is whether the nigral neuronal death comes first, followed inevitably by lowering of intraneuronal TH, or vice versa. Regarding the former possibility, some error of intracellular DA metabolism has been hypothesized to cause neuronal death through production of superoxide during oxidation of DA (Cohen, 1983), although no increase in monoamine oxidase (MAO) activity in PD brains has been found.

Shortage of tetrahydrobiopterin (BH4), the co-factor of TH, in PD was studied by Nagatsu et al. (1981). Relative improvement following administration of BH4 to younger-starting PD cases was reported by Narabayashi et al. (1982). A decreased level of TH protein in the nigrostriatum and a compensatory increase of TH homospecific activity were described by Mogi *et al.* (1988), which may indicate that the main and axial disturbance was in the change of TH enzyme itself and not of the co-factor.

One of the challenging topics for the future might be the role of possible growth factor dysfunction in the mechanism of neuronal degeneration, the idea having been introduced during the trials of tissue transplantation aimed at compensating for the DA deficiency.

CLINICAL OBSERVATIONS AND ANALYSES

Neural pathway for rigidity, tremor and akinesia

The author's interest has been on the analysis of the morphological and physiological network underlying each axial motor symptom, rigidity, tremor, and akinesia of the disease. With rapid advancement of basic pharmacological studies, clinical data on levodopa therapy have been accumulated for more than twenty years now and provide for a more precise analysis of these symptoms of PD.

Levodopa therapy is generally accepted as most effective on rigidity and akinesia, but less so on tremor. This may indicate that the nigrostriatal DA deficiency is mainly concerned with neuronal mechanisms producing rigidity and akinesia and not primarily concerned with those for tremor production.

Neurophysiological data obtained by a microelectrode recording technique during stereotaxic thalamotomy, which the author has performed for more than forty years indicate that the morphological structure producing rigidity is different from that for tremor (Narabayashi, 1986b, 1988, 1990a). The former is the pallidal outflow to the VL (ventrolateral nucleus) of the thalamus, which is under the control of the striatum, and the latter is the Vim (ventral intermediate nucleus) of the thalamus, which does not receive the pallidal projection and, therefore, is not under direct striatal influence. The Vim is located posteriorly to the VL and receives mainly the cerebellar afferents and the proprioceptive sense from the periphery.

DA agonists such as bromocriptine or pergolide mesylate show a dopaminergic effect, which is not so strong enough as that of levodopa, and display a similar divergence of effects between rigidity and tremor. BH4, the co-factor of TH, is more effective on rigidity and akinesia than on tremor, when it works in cases of early-starting Parkinsonism (Furukawa *et al.*, 1991)

It is, therefore,understood that rigidity is a symptom due to release of pallidal neurons from striatal, perhaps putaminal, inhibition, but tremor seems to be different in its mechanism of generation.

Levodopa-induced dyskinesia is also a member of the striatopallidal symptom group, since it is abolished well by surgery on the VL, but less so by surgery on the Vim (Narabayashi, 1984). Rigidity and levodopa-induced dyskinesia are interpreted as reverse phenomena depending on the same anatomical structures but under the different striatal DA content. The neural mechanism underlying akinesia is also of interest (Barbeau, 1972). But differently from rigidity and tremor, the term "akinesia"is not clearly defined even in its daily clinical use, and for detailed analysis and interpretation of the symptom the reader should refer to other papers by the author (Narabayashi, 1983, 1990c, in preparation)

Long-term observation of clinical course of PD

PD is known as a slowly progressive degenerative disease occurring in the later period of life, mostly after the age of fifty or sixty. The grade of symptoms in the course of disease progression, i.e. the grade of difficulties in movement and in activity in daily living (ADL), is well described by the five stage gradings, I to V, proposed by Hoehn and Yahr (1967). It has generally been accepted that PD is a disorder of lowered DA metabolism in the nigrostriatum. However, when the progressive changes and worsening of clinical symptoms from the onset of the disease and also the change in efficacy of drugs such as levodopa and DA agonist, especially in the cases with long-standing disease, are carefully analyzed, the degenerative process that transpired during the course of the disease appears not to be explainable by a disorder of DA metabolism alone.

Early onset cases. The author and his colleagues were interested in the specific group of patients in whom the disease started at a relatively young age, such as below the age of forty (Yokochi, 1979a, b; Narabayashi, 1986a, 1990b, c; Gibb *et al.*, 1991). These younger-onset cases occupy 7 to 10% of the whole Parkinsonian population, which percentage slightly differs depending on the statistics in different countries.

Cases of younger onset present the following specific clinical features when compared with the classical cases commencing at a later age: These cases present mainly rigidity and akinesia with less resting tremor. The speed of progressive worsening of clinical pictures is observed to be relatively slower than in the classical PD cases. They usually do not present autonomic signs or psychological changes such as depressive mood, bradyphrenia or intellectual lowering, even in the chronic phase of the illness, which symptoms are common in the long-standing classical PD cases. The response to levodopa therapy is usually very marked, but a much higher incidence of levodopa-induced dyskinesia of the extremities and the trunk is observed than in the classical PD cases. But when the pharmacological therapy with levodopa and/or DA agonist with/without surgical treatment is carefully and smoothly performed without side effect, these cases can maintain almost normalized activity in jobs and home life quite satisfactorily for as long as ten to twenty years, or sometimes even longer. These observations and experiences seem to indicate that the axial metabolic disturbance in these younger-onset cases is DA deficiency, which is assumed to be localized within the nigrostriatum and responds to levodopa quite satisfactorily.

(2) *Later-onset cases.* On the other hand, the classical PD patients respond to levodopa also well but not so markedly or dramatically as do the younger-onset cases. In the later stage of the disease, the response to levodopa is usually milder, less and limited. Difficulty of maintenance of posture, gait disturbances, especially freezing of gait and psychological changes such as depressive mood and bradyphrenia are symptoms that become commonly observable in the later stage in these patients under long-term levodopa therapy, and these symptoms are not modified or benefited even by a further increase of levodopa or DA agonist dose (Narabayashi and Nakamura, 1981a). In other words, progressive worsening in general ADL, even under the long-term levodopa therapy, is common in these cases, which suggests that factors other than DA deficiency alone might also be participating in symptomatology of the later stage.

The author postulated that the role of NE deficiency might be important in this aspect. Degeneration of the *locus ceruleus* (LC) has been described as one of the main neuropathological findings together with that of the SNc in PD, although its role in the emergence of clinical symptoms is not sufficiently clear. The lowered activity of dopamine-β-hydroxylase (DβH), the enzyme for biosynthesis of NE, in the brain tissue (Nagatsu *et al.*, 1977) and also in the CSF (Nagatsu *et al.*, 1982) of classical PD patients has been described by Nagatsu and others, which suggests a lowered level of NE in the entire CNS.

Administration of L-*threo*-3,4-dihydroxyphenylserine (L-*threo*-DOPS), which is the industrial precursor of NE and is converted to NE by decarboxylation in the brain tissue, was tried with beneficial results. Details of the results, including those of the double-blind trial, have been reported elsewhere (Narabayashi *et al.*, 1981b, 1986c, 1987). The moderate effect was seen mostly on the motor symptoms, such as disturbances of postural balancing and gait, especially on gait freezing. It is interesting that favorable changes in depressive mood and bradyphrenia were also noted, which, along with motor difficulties, are the basic symptoms in PD, especially in the chronic stage (Narabayashi *et al.*, 1991). The author considers that the classical PD in its chronic stage or occurring in the elderly is a disorder not only of DA deficiency but also of NE deficiency. NE deficiency is considered to arise several years after the first start of symptoms due to DA deficiency in these cases (Narabayashi, 1990c). It seems that the frequency of NE deficiency in the younger-starting cases is much less and the progression is slower than in classical PD. But in the elderly patients with disease onset after the age of seventy or eighty, clinical pictures often become worse much faster but respond well to L-*threo*-DOPS, suggesting that more widespread pharmacological changes may start faster in these cases.

Therefore, the speed and course of progressive worsening in presenting a wider spectrum of symptoms assumed to be dependent on the extent of aging of the brain at the onset of the disease process.

PROGRESSIVE WORSENING OF CLINICAL SYMPTOMS

There is no question that PD, at the beginning and in the initial several years, is a disorder of DA biosynthesis due to lowered TH activity. But over the slowly progressive course, metabolic disturbances other than those stemming from DA deficiency, especially those attributable to NE deficiency, must be considered. Therefore, based on careful follow-up observation over the long term, PD is to be considered a disorder of CA's and not of DA alone. Progression of a single symptom, such as rigidity becoming severer during the disease course, can be explained by gradual degeneration of nerve cell within the certain functional system concerned with the symptom. The widening of symptoms by addition of other symptoms such as psychological changes will be explained only by additional pathology in the different functional systems, such as the NE-driven structures.

Furthermore, progression of pathology is also considered within the dopaminergic projection, which are described in detail in other papers by the author. Within the nigrosriatum, the start of pathology was analyzed and interpreted to start from the ventrolateral part of SNc, which is known to project mainly to the putamen and is correlated with generation of rigidity. Within a few months or years, this will be followed by additional pathology involving the medial part of SNc, i.e. nigrocaudal projection, which may correlate to development of akinesia in clinical picture. Therefore, the progressive widening of pathology seems also to occur within the dopaminergic structures (Narabayashi, 1990c, in preparation; Gibb *et al.*, 1991).

The reason for such progressive involvement of another functional system in PD pathology has not been explained. Since DA and NE belong to the same metabolic chain of monoamines, interrelation between enzymes in each step of CA biosynthesis should be investigated in the future. For example, the question should be answered as to through which mechanism does the lowering of TH activity cause or interfere with the later-occurring lowering of $D\beta H$ activity.

Note

This paper is dedicated to Prof Nagatsu to celebrate his 60th birthday with my deep appreciation and gratitude to his guidance and teaching in the field of catecholamine research.

REFERENCES

Anden, N. E, Carlsson, A., Dahlstrom, A., Fuxe, K., Hillarp, N. A. and Larsson, K. (1964). Demonstration and mapping out of nigro-neostriatal dopamine neurons. *Life Sci.* **33**, 523–530.

Barbeau, A. (1972). Contributions of levodopa therapy to the neuropharmacology of akinesia. In: *Parkinson's Disease, Vol. 1.* J. Siegfried (Ed.). Hans Huber Publisher, Bern, pp.151–174.

Cohen, G. (1983). The pathology of Parkinson's disease: biochemical aspects of dopamine neuron senescence. *J. Neural Transm.* **19** (Suppl.), 89–103.

Gibb, W. R. G., Narabayashi, H., Yokochi, M., Iizuka, R. and Less, A. J. (1991). New pathologic observations in juvenile onset Parkinsonism with dystonia. *Neurology* **41**, 820–822.

Furukawa, Y., Nishi, K., Kondo, T., Mizuno, Y. and Narabayashi, H. (1991). Juvenile Parkinsonism: ventricular CSF biopterin levels and clinical features. *J. Neurol. Sci.,* (in press).

Hoehn, M. M. and Yahr, M. D. (1967). Parkinsonism: onset, progression, and mortality. *Neurology* **17**, 427–442.

Langston, J. W., Ballard, P. A., Tetrud, J. W. and Irwin, I. (1983). Chronic Parkinsonism in human due to a product of meperidine-analog synthesis. *Science* **219**, 979–980.

Mogi, M., Harada, M., Kiuchi, K., Kojima, K., Kondo, T. Narabayashi, H., Rausch, D., Riederer, P., Jellinger, K. and Nagatsu, T. (1988). Homospecific activity (activity per enzyme protein) of tyrosine hydroxylase increases in Parkinsonian brain. *J. Neural Transm.* **72**, 77–81.

Nagatsu, T., Kato, T., Numata (Sudo), Y., Ikuta, K., Sano, M., Nagatsu, I., Kondo, Y., Inagaki, S., Iizuka, R., Hori, A. and Narabayashi, H. (1977). Phenylethanolamine N-methyltransferase and other enzymes of catecholamine metabolism in human brain. *Clin. Chim. Acta* **75**, 221–232.

Nagatsu, T., Kato, T., Nagatsu, I., Kondo, Y., Ingaki, S., Iizuka, R. and Narabayashi, H. (1979). Catecholamine-related enzymes in the brain of patients with Parkinsonism and Wilson's disease. In: *Advances in Neurology, Vol. 24.* L. J. Poirier, T. L. Sourkes and P. J. Bedard (Eds). Raven Press, New York, pp. 283–292.

Nagatsu, T., Yamaguchi, T., Kato, T., Sugimoto, T., Matsuura, S., Akino, M., Nagatsu, I., Iizuka, R. and Narabayashi, H. (1981). Biopterin in human brain and urine from controls and Parkinsonian patients: Application of a new radioimmunoassay. *Clin. Chim. Acta* **109**, 305–311.

Nagatsu, T., Wakui, Y., Kato, T., Fujita, K., Kondo, T., Yokochi, F. and Narabayashi, H. (1982). Dopamine beta-hydroxylase activity in cerebrospinal fluid of Parkinsonian patients. *Biomed. Res.* **3**, 95–98.

Nagatsu, T. and Yoshida, M. (1988). An endogenous substance of the brain, tetrahydroisoquinoline, produces Parkinsonism in primates with decreased dopamine, tyrosine hydroxylase and biopterin in the nigrostriatal regions. *Neurosci. Lett.* **87**, 178–182.

Narabayashi, H. and Nakamura, R. (1981a). The freezing phenomenon. Problems in long-term L-dopa treatment for Parkinsonism. In: *Research Progress in Parkinson's Disease.* F. C. Rose and R. Capildeo (Eds). Pitman Medical, Tunbridge Wells, Kent, pp. 248–253.

Narabayashi, H., Kondo, T., Hayashi, A., Suzuki, T. and Nagatsu, T. (1981b). L-threo-3,4-dihydroxyphenylserine treatment for akinesia and freezing of Parkinsonism. *Proc. Jap. Acad.* **57-B**, 351–354.

Narabayashi, H., Kondo, T., Nagatsu, T., Sugimoto, T. and Matuura, S. (1982). Tetrahydrobiopterin administration for Parkinsonian symptoms. *Proc. Jap. Acad.* **58-B**, 283–287.

Narabayashi, H. (1983). Pharmacological basis of akinesia in Parkinson's disease. *J. Neural Transm.* **19** (Suppl.), 143–151.

Narabayashi, H., Yokochi, F. and Nakajima, Y. (1984). Levodopa-induced dyskinesia and thalamotomy. *J. Neurol. Neurosurg. Psychiat.* **47**, 831–839.

Narabayashi, H., Yokochi, M., Iizuka, R. and Nagatsu, T. (1986a). Juvenile Parkinsonism. In: *Handbook of Clinical Neurology, Vol. 5 (49).* P. J. Vinken, G. W. Bruyn and H. L. Klawans (Eds). Elsevier Science Publisher, Amsterdam, pp. 153–165.

Narabayashi, H. (1986b). Tremor: its generating mechanism and treatment. In: *Handbook of Clinical Neurology, Vol. 5 (49).* P. J. Vinken, G. W. Bruyn and H. L. Klawans (Eds). Elsevier Science Publisher, Amsterdam, pp. 597–607.

Narabayashi, H., Kondo, T., Yokochi, F. and Nagatsu, T. (1986c). Clinical effects of L-*threo*-3,4-dihydroxyphenylserine in cases of Parkinsonism and pure akinesia. In: *Advances in Neurology, Vol. 45.* M. D. Yahr and K. J. Bergmann (Eds). Raven Press, New York, pp. 593–602.

Narabayashi, H. and Kondo, T. (1987). Results of a double-blind study of L-*threo*-DOPS in Parkinsonism. In: *Recent Developments in Parkinson's Disease, Vol. II.* S. Fahn, C. D. Marsden, D. B. Calne and M. Goldstein (Eds). Macmillan Healthcare Information, New Jersey, pp. 279–291.

Narabayashi, H. (1988). Lessons from stereotaxic surgery using microelectrode techniques in understanding Parkinsonism. *Mt. Sinai J. Med.* **55**, 50–57.

Narabayashi, H. (1990a). Surgical treatment in the levodopa era. In: *Parkinson's Disease.* G. Stern (Ed.). Chapmann and Hall, London, pp. 597–646.

Narabayashi, H. (1990b). Juvenile or early-starting Parkinsonism as a prototype of nigrostriatal dopamine deficiency. In: *Basic, Clinical, and Therapeutic Aspects of Alzheimer's and Parkinson's Disease, Vol. 2.* T. Nagatsu, A. Fisher and M. Yoshida (Eds). Plenum Press, New York, pp. 181–186.

Narabayashi, H. (1990c). Clinical analysis of juvenile and classical Parkinsonism and underlying pathophysiological mechanisms. In: *Proceedings of the XIth International Congress of Neuropathology.* The Japanese Society of Neuropathology, Kyoto, pp. 43–47.

Narabayashi, H., Yokochi, F., Ogawa, T. and Igakura, T. (1991). Analysis of L-*threo*-3,4,-dihydroxyphenylserine effect on motor and psychological symptoms in Parkinson's disease. *Brain Nerve* **43**, 263–268.

Narabayashi, H. Three types of akinesia in progressive course of Parkinson's disease. In: *Advances in Neurology.* Raven Press, New York, (in preparation).

Yokochi, M. (1979a). Juvenile Parkinson's disease. 1. Clinical aspects. *Shinkei Shinpo* **23**, 1048–1059.

Yokochi, M. (1979b). Juvenile Parkinson's disease. 2. Pharmaco-kinetic study. *Shinkei Shinpo* **23**, 1060–1073.

Yoshida, M., Niwa, T. and Nagatsu, T. (1990). Parkinsonism in monkeys produced by chronic administration of an endogenous substance of the brain, tetrahydroisoquinoline: the behavioral and biochemical change. *Neurosci. Lett.* **119**, 109–113.

Tyrosine Hydroxylase, pp. 225–235
M. Naoi *et al.* (Eds)
© VSP 1993

TYROSINE HYDROXYLASE IN PARKINSON'S DISEASE

WOLF-DIETER RAUSCH and PETER RIEDERER

*Institute for Medical Chemistry, Vet. Med. Univ. Vienna, A-1030, Vienna,
Austria and Clinical Neurochemistry, Department of Psychiatry,
University of Würzburg, D-8700 Würzburg, Germany*

Abstract—The particular situation for tyrosine hydroxylase (TH) in Parkinson's disease
(PD) is that the enzyme is confronted with a multitude of pathological processes. A
lower amount of enzyme protein may be produced in the neuronal cell body. Part of the
enzyme decoded could be structurally altered by age and disease related changes in the
milieu interieur and its influence on the genetic apparatus. Enzyme may be lost from
defective or leaky nerve terminals in acute (inflammatory) processes of degeneration.
Inside the diseased neuron cofactor and oxygen supply appear diminished (exceeding the
compensatory affinity changes). Periods of (or permanent) hypoxia may prevail. Toxic
compounds (endogenously formed or derived from exogenous origin) accumulate over
time and either inhibit, or in higher concentrations destroy the TH enzyme system. Due
to low surrounding catecholamine concentrations, feedback inhibition may be abolished
and this and cellular autoreceptor signals force the remaining enzyme into a permanently
activated, possibly phosphorylated state where it is more prone to degenerative external
processes. These are present in the form of active oxygen species or other radicals, as
hydrogen peroxide (due to increased deamination of dopamine and a decrease in detox-
ication of radicals) and iron are elevated, while antioxidant concentrations are lowered.
All these events can create cellular and membrane damage involving lipid peroxidation.
It is rather likely that a concert of these events finally will lead not only to functional
disintegration of TH but also to dopaminergic cell death.

Key words: tyrosine hydroxylase; Parkinson's disease; phosphorylation; radicals; iron; lipid peroxidation.

INTRODUCTION

The particular susceptibility of the neuronal enzyme tyrosine hydroxylase (TH,
tyrosine 3-monooxygenase E. C. 1. 14. 3. a) makes this enzyme a primary target in neurodegenerative disorders in which the dopaminergic system is involved
(Fig. 1). The resulting deficit produces characteristic neurological symptoms as
expressed in Parkinson's disease (PD, Ehringer and Hornykiewicz, 1960). Biological investigations of the causative factors in neurodegenerative diseases rely

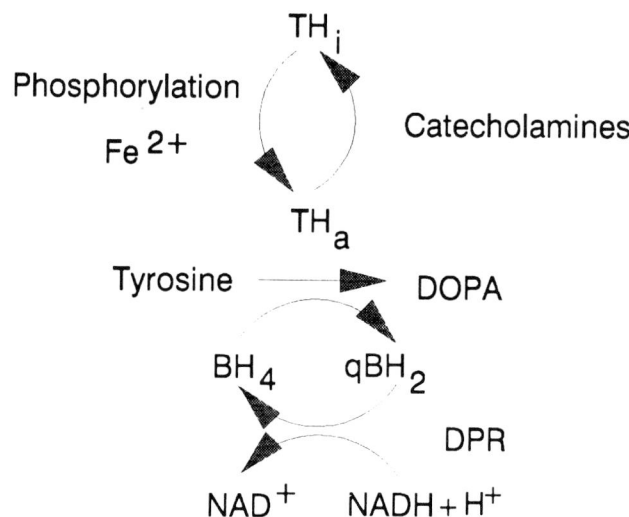

Figure 1. Schematic regulation of TH in human brain. Abbreviations used: TH_i inactive TH; TH_a active TH; BH_4 (6R)-L-erythro-5,6,7,8-tetrahydrobiopterin, qBH_2 quinonoid-L-erythro-7,8-dihydrobiopterin; DPR, dihydropteridine reductase; NAD^+ ($NADH + H^+$) nicotinamide adenine dinucleotide oxidized (reduced) form.

on the investigation of human post mortem brain material. Apparently from this fact there is an open field for artefacts, pre and post mortem secondary changes, overlying secondary diseases, drug interactions or effects all superimposing the investigated neurodegenerative phenomenon (Riederer and Wuketich, 1976). Thus, particular attention has also focussed on the field of basic research trying to understand the initial triggering mechanisms. Profound investigation in the field of protein and particularly enzyme research has characterized TH biology as enzyme structure, kinetics, cofactor and effector requirements, its regulation, phosphorylation and even down to its molecular properties as gene location, encoding, transcription and translation, enzyme induction, metabolism and turnover rate.

In this paper it is attempted to elucidate the role of TH in PD in the view of current knowledge. Evidences are compared to understand whether TH represents merely a marker of a surrounding degenerative process, whether this enzyme may be partially involved or differentially regulated or even represents the major target of neurotoxic agents in PD.

LOSS OF TH ACTIVITY IN PD

The most general symptom of PD is the degeneration of melanin containing cells in the substantia nigra (Hassler, 1938). These cells represent a particular group of dopamine synthesizing neurons.

Table 1.
Changes of TH activity in PD

| | TH activity in different brain areas | | | | | | | | |
| Caudate nucleus | | | Putamen | | | Substantia nigra | | | References |
C	PD	% C	C	PD	% C	C	PD	% C	
18.7	3.2	17.1	17.4	3.1	17.8	17.4	6.1	35.0	Lloyd *et al.* (1975)[a]
60.6	6.6	10.9	76.7	3.2	4.2	99.7	12.6	12.6	Nagatsu *et al.* (1984)[b]
27.8	3.5	12.6	16.2	1.2	7.4	19.4	4.9	25.2	Riederer *et al.* (1978)[c]
26.5	1.8	6.8	18.1	1.6	8.8	6.8	1.1	16.2	Mogi *et al.* (1988)[d]

C controls, PD Parkinson's disease, % C Parkinson's disease data as % of controls. Different assay conditions and units were used. *a*: trapping of $^{14}CO_2$, nmol CO_2/ 0.5 h /100 mg protein, *b*: HPLC-technique, nmol/g/h, *c*: isolation of $_{11}C$ -DOPA, nmol/g/h, *d*: HPLC technique, pmol/min/mg protein.

TH, the key enzyme in the synthesis of the catecholamines, is particularly affected in PD. An extensive loss of TH activity in the brains of PD when compared to age matched controls has been observed (Table 1), in good correlation with a reduction of the dopamine content in these particular areas (Bernheimer *et al.*, 1973; Birkmayer and Riederer, 1975). Regional differences exist as the reduction of TH activity is most pronounced in the corpus striatum (particularly in the putamen), equally or less reduced in the substantia nigra, whereas other brain stem nuclei such as locus coeruleus, nucleus ruber and the raphe reticularis formation are frequently less or not affected (Riederer *et al.*, 1978). These results are consistent with data from other authors (Lloyd *et al.*, 1975; McGeer and McGeer, 1976; Nagatsu *et al.*, 1977) and they indicate a varying extent of enzyme degeneration in different areas of the Parkinsonian brain due to particular local conditions. In dopaminergic tracts of the ventral tegmental area TH activity was not reduced (Javoy-Agid *et al.*, 1981).

For instance such a selective vulnerability may be connected to the neuromelanin content of a subpopulation of dopaminergic cells of the mesencephalon (Hirsch *et al.*, 1989). Interestingly neurotoxins such as 1-methyl-4-phenyl-1,2,3,6-tetrahydropyridine (MPTP) similarly affect particularly the nigrostriatal tract and only exert toxic effects on other dopaminergic systems in higher concentrations. Its neurotoxic action may be enhanced by interaction with melanin, serving as an intracellular pool where it can be gradually released (D'Amato *et al.*, 1987).

An age dependent decrease of TH activity was found in human brain (McGeer *et al.*, 1971). The decrease of TH activity in PD is too extensive and variable to calculate age or duration of disease correlations. A correlation of dopaminergic cell loss and TH activity, though not explicitly stated in literature, could be concluded from various pathological studies.

TH activity is also significantly reduced in the adrenal gland in PD (Riederer *et al.*, 1978) suggesting that PD has traits of a generalized disease.

Moreover the loss of activity in PD is not absolutely confined to TH, as also the activities of other monooxygenases (dopamine-β-hydroxylase, phenylethanolamin-N-methyltransferase) are decreased (Nagatsu et al., 1984).

The loss and fate of TH protein has been the subject of recent studies. Inactive ("ghost") TH protein could remain within the dopaminergic neuron (Mogi et al., 1984), as it has been found for the adrenals. The actual amount of protein when measured with a sandwich immunoassay is lower in brain areas of PD (Mogi et al., 1988), indicating that the decrease of TH activity found in PD brains is due to the disappearance of TH protein as a result of cell death. Part of its fate may be the integration into Lewy bodies in catecholaminergic regions, which have been shown to stain positively for TH (Nakashima and Ikuta, 1984), even so this phenomenon does not represent a particular PD event.

TH — AN INDICATOR OF EARLY CHANGES IN PD

Post mortem brain analysis depicts final changes in PD without giving too much information about the onset of this disease. Interestingly TH has little value as a peripheral marker. Unlike all the other catecholamine synthesizing enzymes TH presence or activity has not been detected in human blood. Thus its long term disappearance may only be extrapolated from post mortem studies. Early changes at onset of PD could in part be deduced from clinical observations related to the catecholamine systems e.g. early symptoms of PD, orthostatic hypotension together with PD or clinical efficacy of L-Dopa treatment. Further diagnostic tools represent the apomorphine test for the neuronal system (Barker et al., 1989) and the visualization of dopamine synthesis in peripheral ganglia and adrenal medulla by imaging techniques.

REGULATION OF TH ACTIVITY IN PD

Various mechanisms of regulation have been described for the activity of TH. Long term and short term regulation through nerve stimulation, stress, cofactor, drug treatment and allosteric modulation (e.g., phosphorylation) were differentiated (Masserano and Weiner, 1983). Different forms of stress induce mRNA translation and will delay an increase in TH protein and activity. Whether these complex regulatory systems have any function in PD is unknown. A total cell loss of about 70-80 % is necessary before PD becomes clinically manifested thus the system as a whole comprises tremendous compensatory ability.

TH together with tryptophan hydroxylase have a particular demand for a pterin cofactor, L-erythro-5,6,7,8-tetrahydrobiopterin (Kaufman and Fisher, 1974). Cofactor requirements also involve an active reducing system in order to replenish oxidized pterins. The pterin cofactor levels in the brain are normally not saturating TH in vivo. Thus the pterin cofactor in itself could be an important regulatory factor for TH activity. The cofactor shows a similar distribution and good correlation with TH (Nagatsu *et al.*, 1981). Indeed in PD the pterin concentrations have been found to be lower when compared to age-matched controls (Nagatsu *et al.*, 1981). As the cofactor can pass the blood-brain barrier peripheral levels may indicate central changes. Lowered blood levels were described when assayed in PD. However pilot trials have shown that cofactor therapy with tetrahydrobiopterin has little beneficial effect in PD (Birkmayer and Riederer, 1985). In contrast, therapy with NADH to replenish the biopterin cofactor is claimed to be effective at times as an additional therapy of PD (Birkmayer *et al.*, 1990).

AVAILABILITY OF TYROSINE

Peripheral tyrosine concentrations can influence TH activity in the brain. However, tyrosine available in different brain areas is not different in PD and age-matched controls (Riederer *et al.*, 1975). Tyrosine is competing with biopterin for the enzyme and promotes biopterin production. A stimulation of serum levels of biopterin due to an oral tyrosine load is less pronounced in PD than in healthy controls (Yamaguchi *et al.*, 1981). Thus its interaction with biopterin synthesis may be disturbed.

A ROLE OF OXYGEN SUPPLY IN THE REGULATION OF TH IN PD?

The oxygen dependence of TH suggests that brain oxygen availability should influence catecholamine synthesis in the human brain (Fisher and Kaufman, 1972). Even so data and theories on oxidative or free radical damage in PD prevail, interestingly reports on the oxygen supply for TH in PD are lacking. Definitely TH may change in its activity as a response to hypoxic conditions, yet a certain compliance to hypoxia and stress may be assumed as the enzyme can undergo conformational changes to increase its affinity for oxygen (Carlsson, 1978). An oxygen deficit as in the case of carbon monoxide poisoning may finally lead to a characteristic form of Parkinsonism (Jellinger, 1989).

PHOSPHORYLATION AND DEPHOSPHORYLATION OF TH IN PD

Among a variety of factors known to affect the activity of tyrosine hydroxylase catecholamines, cyclic AMP-dependent protein kinase (Joh *et al.*, 1978), calmodulin-dependent protein kinase II and polyanions result in the reversible modulation of the enzyme activity (Fig. 1).

Different sites of TH may be phosphorylated as could be shown for tryptic fragments of the TH molecule (Haycock *et al.*, 1982). *In vitro* experiments suggested that the inactivation of the enzyme by the catecholamines, the end products of the enzyme and the activation by the cyclic AMP-dependent protein kinase might be most important in controlling the enzyme. Incubation of the enzyme with catecholamines at a concentration as low as 10–100 nM resulted in a rapid inactivation. The inactive/stable form exhibited no activity under physiological conditions. TH may usually exist as the catecholamine induced inactive/stable form in the nervous system which can be markedly activated by a cyclic AMP-dependent protein kinase (Fujisawa and Okuno, 1989).

Most of the studies on this regulatory mechanism for TH so far reported have been done with rat enzyme. Human TH from pheochromocytoma was found to be remarkably activated by cyclic AMP-dependent protein kinase. Calmodulin-dependent protein kinase and polyanion also modulated the activity of the enzyme. These results suggest that TH may be regulated in a similar manner in the human as in the rat (Okuno et al., 1989).

Human brain post mortem samples when assayed for TH activity in the presence of cyclic AMP and ATP did not lead to TH stimulation. The addition of exogenous protein kinase catalytic subunit led to increases in TH activity particularly in PD samples, indicating deranged kinase systems. These tissues, however, did not respond to conditions promoting both Ca-calmodulin dependent and Ca-acetylcholine dependent phosphorylation (Rausch et al., 1988, Table 2).

Table 2.
Stimulation of TH activity in human caudate nucleus under various phosphorylating conditions

Stimulating agent	basal	stimulated	% of basal value	$p <$
cAMP (100 μM) and ATP (1 mM)				
Controls (9)	33.3 ± 4.9	33.9 ± 5.2	102 ± 4	NS
PD (9)	19.3 ± 4.3^a	19.6 ± 5.6^a	115 ± 11	NS
Protein kinase (3 μg)				
Controls (13)	40.3 ± 4.1	48.2 ± 6.0	118 ± 26	NS
PD (14)	18.6 ± 3.7^b	24.9 ± 3.7^b	148 ± 14	0.05
Ca^{2+}-Calmodulin (0.3 μg)				
Controls (9)	38.9 ± 7.6	39.8 ± 8.4	108 ± 6	NS
PD (10)	21.0 ± 4.6^a	19.6 ± 4.6^a	107 ± 8	NS

Data taken from Rausch et al. (1988). Homogenates from human brain caudate nucleus were incubated together with or without the stimulating agent. Values are given as the means ± SEM of nanomoles DOPA/g/h. Numbers in parentheses indicate number of brain areas. NS not significant.
[a] $p < 0.05$ compared to corresponding control levels.
[b] $p < 0.01$ compared to corresponding control levels.

Thus labile kinases have to be assumed along with a possibly higher sensitivity of the phosphorylated enzyme towards degeneration. The remaining TH molecules show particularly high activity when compared to control enzyme. This so called "homospecific activity" (activity per enzyme protein) suggests molecular changes in the TH enzyme as a compensatory reaction (Mogi et al., 1988). To date it is unknown which factors (e.g. phosphorylation) bring about this activation.

THE ROLE OF IRON FOR TH

TH is an iron containing enzyme and ferrous but not ferric irons were found to stimulate TH activity (Nagatsu *et al.*, 1964). The iron content determined for adrenal TH amounts to 0.5–0.75 mol/mol of enzyme (Hoeldtke and Kaufman, 1979). *In vitro* Fe^{2+} ions are capable of stimulating the enzyme. Particular dependence on Fe^{2+} for TH activity has been demonstrated for the human brain (Rausch *et al.*, 1988). More than 13-fold activation has been demonstrated for post mortem human brain samples. However, even with lower basal values and corresponding lower stimulated values, PD samples did not respond differently from controls. *In vitro* preincubation of TH enzyme equally leads to loss of activity together with increased iron dependence, indicating that general degenerative phenomena may account for these changes.

Ferrous iron participates in the hydroxylation reaction as well as it is capable of removing excess H_2O_2 (Shiman *et al.*, 1971). Equally *in vitro* addition of catalase, by decomposing H_2O_2 to molecular oxygen (Kaufman, 1977), leads to increased TH activity. *In vitro*, when both catalase and iron are present, H_2O_2 reduces TH activity at high concentrations only (Rausch *et al.*, 1988).

Through the Fenton reaction Fe^{2+} contributes to the formation of free radicals. H_2O_2 and H_2O_2-derived radicals (O_2^- and $\cdot OH$) and iron seem to accumulate during aging (Hallgren and Sourander, 1958). It is unclear if an interaction of endogenous free radicals (such as O_2^-, $\cdot OH$, peroxides) and aldehydes or quinolines can damage TH or its catalytic structures, or if increased lipid peroxidation in the substantia nigra of PD (Dexter *et al.*, 1986) also affects the TH molecule.

Total iron is also increased in the substantia nigra of PD (Dexter *et al.*, 1987, Sofic *et al.*, 1988). In any case aging and particularly PD is accompanied by a dramatic shift of the Fe^{2+}/Fe^{3+} ratio in the substantia nigra from 2.45:1 to 1:1 indicating a deposition and/or increased oxidative processes in the progress of the disease (Riederer *et al.*, 1989). It can be assumed that these changes also include TH protein, TH being a prominent iron-containing protein in this area. Together with a decreased availability of the antioxidants glutathionine and ascorbate in the substantia nigra of PD, this makes this area even more likely to be affected in any kind of oxidative damage (Riederer *et al.*, 1989).

There is no evidence that *in vivo* iron can stimulate TH activity since neither iron-deficiency nor iron overload alters TH activity *in vivo* (Youdim, 1989). Nevertheless beneficial effects of iron therapy have been demonstrated in individuals with PD (Birkmayer and Birkmayer, 1989). Peripheral actions of iron may exist, since iron crosses the blood-brain barrier only at very low speed and the blood-brain barrier in itself appears intact in PD. *In vitro* experiments have shown that a stimulation of adrenal TH activity by iron could result in an increased formation of L-Dopa (Ishii *et al.*, 1990).

PD-SPECIFIC GENETIC DERANGEMENT OF TH?

A large body of evidence regarding the mRNA decoding TH has been accumulated over the last years. Four types of human TH have been found by cDNA cloning (Kobayashi *et al.*, 1988). This protein diversity comes about through alternative splicing from a single gene. Tissue and stage specific expression of different TH molecules has been reported suggesting that alternative splicing plays a regulatory role in the functioning of TH. Indeed different kinetic properties and homospecific activities were described (Kobayashi *et al.*, 1988). Conclusive findings to prove that such changes occur in PD have not been made so far.

The expression of the TH gene is regulated in a tissue specific manner during neuronal development and differentiation and is modulated by a variety of factors such as cyclic AMP, glucocorticoids, nerve growth factor and transsynaptic stimuli. The amount of mRNA for TH, presumably due to cell loss, is decreased in PD (Dumas *et al.*, 1990). Furthermore, proof has been gained that a decreased TH mRNA resides in surviving dopaminergic neurons in the substantia nigra of PD subjects by an *in situ* hybridization study. The subnormal mRNA content may reflect changes in the levels of gene subscription in the diseased neurons (Javoy-Agid, 1990).

FUTURE PERSPECTIVES

New therapeutic approaches in PD may aim at preservation of TH in PD. Provided general cell toxicity originates from the formation of free radicals, antioxidant treatment with α-tocopherol or the MAO-inhibitor L-deprenyl may also act protectively on TH as it may delay the progress of PD (Factor *et al.*, 1990). Over the last years great progress has been made in the field of dopaminergic cell culture techniques, where toxic events can be studied directly and TH serves as an important marker of cellular integrity (Michel *et al.*, 1990). Grafting or transplantation techniques of dopaminergic cells still appear not to be a practical treatment of PD. Thus the gene expression of TH in nonneuronal cells as Schwann cells (Owens *et al.*, 1991), in fibroblasts (Wolff *et al.*, 1989), neuroblastoma and neuroendocrine cell lines (Horellou *et al.*, 1989; Uchida and Kohsaka, 1989) may offer new perspectives replacing implantation of fetal or adrenal tissue. Irrespectively of any future approach, the presence of metabolically active TH will serve as a main indicator for assessing cellular function.

REFERENCES

Barker, R., Duncan, J. and Lees, A. (1989). Subcutaneous apomorphine as a diagnostic test for dopaminergic responsiveness in Parkinsonian syndrome. *Lancet* **I**, 675.

Bernheimer, H., Birkmayer, W., Hornykiewicz, O., Jellinger, K. and Seitelberger, F. (1973). Brain dopamine and the syndromes of Parkinson and Huntington. *J. Neurol. Sci.* **20**, 415–455.

Birkmayer, W. and Birkmayer, J. G. D. (1989). Iron therapy in Parkinson's disease. Stimulation of endogenous presynaptic L-DOPA biosynthesis by the iron compound oxyferriscorbone. In: *Early Diagnosis and Preventive Therapy in Parkinson's Disease.* H. Przuntek and P. Riederer (Eds). Springer Verlag, Vienna, New York, pp. 323–327.

Birkmayer, W. and Riederer, P (1975). Responsibility of extrastriatal areas for the appearance of psychotic symptoms. *J. Neural Transm.* **37**, 175–182.

Birkmayer, W. and Riederer, P. (1985). *Die Parkinson-Krankheit* Springer Verlag, Wien-New York.

Birkmayer, W., Birkmayer, J. G. D., Vrecko, K. and Paletta, B. (1990). The clinical benefit of NADH as stimulator of endogenous L-DOPA biosynthesis in Parkinsonian patients. In: *Advances in Neurology* Vol. **53**, *Parkinson's Disease: Anatomy, Pathology, and Therapy.* M. B. Streifler, A. D. Korczyn, E. Melamed and M. B. H. Youdim (Eds). Raven Press, New York, pp. 545–549.

Carlsson, A. (1978). The *in vivo* estimation of rates of tryptophan and tyrosine hydroxylation: effects of alterations in enzyme environment and neuronal activity. In: *Aromatic Amino Acids in the Brain.* G. E. W. Wolstenholme and W. Fitzsimmons (Eds). Elsevier, Holland, pp. 117–134.

D'Amato, D., Alexander, G. M., Schwartzmann, R., Kill, C. A., Price, D. L. and Snyder, S. H. (1987). Molecular mechanisms of MPTP-induced toxicity II. Neuromelanin: A role in MPTP-induced neurotoxicity. *Life Sci.* **40**, 705–712.

Dexter, D. T., Carter, C., Agid, F., Agid, Y., Lees, A. J., Jenner, P. and Marsden, C. D. (1986). Lipid peroxidation is a cause of nigral death in Parkinson's disease. *Lancet* **2**, 639–640.

Dexter, D. T., Wells, F. R., Agid, F., Agid, Y., Lees A. J., Jenner, P. and Marsden, C. D. (1987). Increased nigral iron content in post mortem Parkinsonian brain. *Lancet* **2**, 1219–1220.

Dumas, S., Javoy-Agid, F., Hirsch, E., Agid, Y. and Mallet, J. (1990). Tyrosine hydroxylase gene expression in human ventral mesencephalon: detection of tyrosine hydroxylase messenger RNA in neurites. *J. Neurosci. Res.* **25**, 569–575.

Ehringer, H. and Hornykiewicz, O. (1960). Verteilung von Noradrenalin und Dopamin im Gehirn des Menschen und ihr Verhalten bei Erkrankungen des extrapyramidalen Systems. *Klin. Wschr.* **38**, 1236–1239.

Factor, S. A., Sanchez-Ramos J. R. and Weiner W. J. (1990). Vitamine E therapy in Parkinson's disease. In: *Advances in Neurology* Vol. **53**, *Parkinson's Disease: Anatomy, Pathology, and Therapy.* M. B. Streifler, A. D. Korczyn, E. Melamed and M. B. H. Youdim (Eds). Raven Press, New York, pp. 457–461.

Fisher, D. B. and Kaufman, S. (1972). The inhibition of phenylalanine and tyrosine hydroxylase by high oxygen levels. *J. Neurochem.* **19**, 1359–1365.

Fujisawa, H. and Okuno, S. (1989). Regulation of the activity of tyrosine hydroxylase in the central nervous system. *Adv. Enzyme Regul.* **28**, 93–110.

Hallgren, B. and Sourander, P. (1958). The effect of age on the non-heme iron in human brain. *J. Neurochem.* **3**, 41–51.

Hassler, R. (1938). Zur Pathologie der Paralysis agitans und des postenzephalitischen Parkinsonismus. *J. Psychol. Neurol.* **48**, 387–476.

Haycock, J. W., Bennett, W. F., George, R. J. and Waymire, J. C. (1982). Multiple site phosphorylation of tyrosine hydroxylase. *J. Biol. Chem.* **257**, 13699–13703.

Hirsch, E. C., Graybiel, A. M., Agid, Y. (1989). Selective vulnerability of pigmented dopaminergic neurons in Parkinson's disease. *Acta. Neurol. Scand.* **126** (Suppl.), 19–22.

Hoeldtke, R. and Kaufman S. (1979). Bovine adrenal tyrosine hydroxylase: purification and properties. *J. Biol. Chem.* **252**, 3160–3169.

Horellou, P., Guibert, B., Leviel, V. and Mallet, J. (1989). Retroviral transfer of a human tyrosine hydroxylase cDNA in various cell lines: regulated release of dopamine in mouse anterior pituitary AtT-20 cells. *Proc. Natl. Acad. Sci. USA* **86**, 7233–7237.

Ishii, A., Kiuchi K., Matsuyama, M., Satake, T. and Nagatsu, T. (1990). Ferrous ion activates the less active form of human tyrosine hydroxylase. *Neurochem. Int.* **16**, 59–64.

Javoy-Agid, F., Hirsch, E. C., Dumas, S., Duyckaerts, C., Mallet, J. and Agid, Y. (1990). Decreased tyrosine hydroxylase messenger RNA in the surviving dopamine neurons of the substantia nigra in Parkinson's disease: an *in situ* hybridization study. *Neuroscience* **38**, 245–253.

Javoy-Agid, F., Ploska, A. and Agid, Y. (1981). Microtopography of tyrosine hydroxylase, glutamic acid decarboxylase, and choline acetyltransferase in the substantia nigra and ventral tegmental area of control and Parkinsonian brains. *J. Neurochem.* **37**, 1218–1227.

Jellinger, K. (1989). Pathology of Parkinson syndrome. In: *Handbook of Experimental Pharmacology* **Vol. 88**. D. B. Calne (Ed.). Springer-Verlag, Berlin, pp. 47–112.

Joh, T. H., Park, D. H. and Reis, D. J. (1978). Direct phosphorylation of brain tyrosine hydroxylase by c-AMP-dependent protein kinase: mechanism of enzyme activation. *Proc. Natl. Acad. Sci. USA* **75**, 4744–4748.

Kaufman, S. and Fisher, D. B. (1974). Pterin-requiring aromatic amino acid hydroxylases. In: *Molecular Mechanisms of Oxygen Activation.* O. Hayaishi (Ed.). Academic Press, New York, pp. 285–369.

Kobayashi, K., Kiuchi, K., Ishii, A., Kaneda, N., Kurosawa, Y., Fujita, K. and Nagatsu, T. (1988). Expression of four types of human tyrosine hydroxylase in COS cells. *FEBS Lett.* **238**, 431–434.

Lloyd, K. G., Davidson, L. and Hornykiewicz, O. (1975). The neurochemistry of Parkinson's disease: effect of L-dopa therapy. *J. Pharmacol. Exp. Ther.* **195**, 453–464.

Masserano, J. M. and Weiner, N. (1983). Tyrosine hydroxylase regulation in the central nervous system. In: *Molecular and Cellular Biochemistry* **Vols 53/54**. Martinus Nijhoff, Boston.

McGeer, E. G., McGeer, P. L. and Wada, J. A. (1971). Distribution of tyrosine hydroxylase in human and animal brain. *J. Neurochem.* **18**, 1647–1658.

McGeer, P. L. and McGeer, E. G. (1976). Enzymes associated with the metabolism of catecholamines, acetylcholine and GABA in human controls and patients with Parkinson's disease and Huntington's chorea. *J. Neurochem.* **26**, 65–76.

McGeer, P. L., McGeer, E. G. and Suzuki. J. S. (1977). Aging and extrapyramidal function. *Arch. Neurol.* **34**, 33–35.

Michel, P. P., Dandapani, P. K., Krusel, B., Sanchez-Ramos, J., Hefti, F. (1990). Toxicity of 1-methyl-4-phenylpyridinium for rat dopaminergic cells in culture. Selectivity and irreversibility. *J. Neurochem.* **54**, 1102–1109.

Mogi, M., Harada, M., Kiuchi, K., Kojima, K., Kondo, T., Narabayashi, H., Rausch, W.-D., Riederer, P., Jellinger, K. and Nagatsu, T. (1988). Homospecific activity (activity per enzyme protein) of tyrosine hydroxylase increases in Parkinsonian brain. *J. Neural Transm.* **72**, 77–82.

Mogi, M., Kojima, K. and Nagatsu, T. (1984). Detection of inactive or less active form of tyrosine hydroxylase in human adrenals by a sandwich immunoassay. *Anal. Biochem.* **138**, 125–132.

Nagatsu, T., Kato,T., Numata, Y., Ikuta, K. and Sano, M. (1977). Phenylethanolamine N-methyltransferase and other enzymes of catecholamine metabolism in human brain. *Clin. Chim. Acta* **75**, 221–232.

Nagatsu, T., Levitt, M. and Udenfriend S. (1964). Tyrosine hydroxylase: the initial step in norepinephrine synthesis. *J. Biol. Chem.* **239**, 2910–2917.

Nagatsu, T., Yamaguchi, T., Kato, T., Sugimoto, T., Matsuura, S., Akino, M., Nagatsu, I., Iizuka, R. and Narabayashi, H. (1981). Biopterin in human brain and urine from controls and Parkinsonian patients: application of a new radioimmunoassay. *Clin. Chim. Acta* **109**, 305–311.

Nagatsu, T., Yamaguchi, T., Rahman, M.K., Trocewicz, J., Oka. K., Hirata, Y, Nagatsu, I., Narabayashi, H., Kondo, T. and Iizuka, R. (1984). Catecholamine-related enzymes and the biopterin cofactor in Parkinson's disease and related extrapyramidal diseases. *Adv. Neurol.* **40**, 467–473.

Nakashima, S. and Ikuta, F. (1984). Tyrosine hydroxylase protein in Lewy bodies of Parkinsonian and senile brains. *J. Neurol. Sci.* **66**, 91–96.

Okuno, S., Kanayama, Y. and Fujisawa, H. (1989). Regulation of human tyrosine hydroxylase activity. Effects of cyclic AMP-dependent protein kinase, calmodulin-dependent protein kinase II and polyanion. *FEBS Lett.* **253**, 52–54.

Owens, G. C., Johnson, R., Bunge, R. P. and O'Malley, K. L. (1991). L-3,4-dihydroxyphenylalanine synthesis by genetically modified Schwann cells. *J. Neurochem.* **56**, 1030–1036.

Rausch, W.-D., Hirata, Y., Nagatsu, T., Riederer, P. and Jellinger, K. (1988). Tyrosine hydroxylase activity in caudate nucleus from Parkinson's disease: effects of iron and phosphorylating agents. *J. Neurochem.* **50**, 202–208.

Riederer, P. and Wuketich, S. (1976). Time course of nigrostriatal degeneration in Parkinson's disease. *J. Neural Transm.* **38**, 277–301.

Riederer, P., Rausch, W.-D., Birkmayer, W., Jellinger, K. and Seemann, D. (1978). CNS Modulation of adrenal tyrosine hydroxylase in Parkinson's disease and metabolic encephalopathies. *J. Neural Transm.* **14** (Suppl.), 121–131.

Riederer, P., Sofic E., Rausch, W.-D., Schmidt, B., Reynolds, G. P., Jellinger, K. and Youdim, M. B. H. (1989). Transition metals, ferritin, glutathionine and ascorbic acid in Parkinsonian brains. *J. Neurochem.* **52**, 515–520.

Shiman, R., Akino, M. and Kaufman, S. (1971). Solubilization and partial purification of tyrosine hydroxylase from bovine adrenal medulla. *J. Biol. Chem.* **246**, 1330–1340.

Sofic, E., Riederer, P., Heinsen, H., Beckmann H., Reynolds, G. P., Hebenstreit, G. and Youdim, M. B. H. (1988). Increased iron (III) and total iron content in post mortem substantia nigra of Parkinsonian brain. *J. Neural Transm.* **74**, 199–205.

Uchida, K. and Kohsaka, S. (1989). Current advances in neural transplantation. *Hum. Cell.* **2**, 150–155.

Wolff, J. A., Fisher, L. J., Xu, L., Jinnah, H. A., Langlais, P. J., Iuvone, P. M., O'Malley, K. L., Rosenberg, M. B., Shimohama, S. and Friedmann, T. (1989). Grafting fibroblasts genetically modified to produce L-dopa in a rat model of Parkinson disease. *Proc. Natl. Acad. Sci. USA* **86**, 9011–9014.

Yamaguchi, T., Nagatsu, T., Kojima, K., Iizuka, R., Kondo, T. and Narabayashi, H. (1981). Short-term regulation of tissue biopterin concentration. Effect of tyrosine, nomifensine and gamma-butyrolactone. *Bull. Jap. Neurochem. Soc.* **20**, 412–415.

Youdim, M. B. H. (1989). Dopaminergic neurotransmission and status of brain iron. In: *Early Diagnosis and Preventive Therapy in Parkinson's Disease.* H. Przuntek and P. Riederer (Eds). Springer-Verlag, Vienna, New York. pp. 151–160.

Tyrosine Hydroxylase, pp. 237–251
M. Naoi *et al.* (Eds)
© VSP 1993

BIOGENIC AMINES IN CEREBRAL ISCHEMIA AND TRAUMA

MARIA SPATZ,[1] HANNA M. PAPPIUS[2] and BOGUMIR B. MRSULJA[3]

[1]*Stroke Branch, National Institute of Neurological Disorders and Stroke, National Institutes of Health, Bethesda, MD, USA*
[2]*Experimental Neurochemistry, Montreal Neurological Institute, Montreal, Canada*
[3]*Department of Biochemistry, School of Medicine, Belgrade, Yugoslavia*

Abstract—The review summarizes the changes in biogenic amine systems induced by cerebral ischemia and trauma. It focuses on the findings related to the role of monoamines in the complex pathomechanisms of ischemic and traumatic brain injury. The available data suggest that the ischemic alteration of the dopaminergic system is associated with both early changes in inhibitory and excitatory neurotransmission as well as late neuronal damage. The disturbance in the serotoninergic system is related to the development of ischemic brain edema. In brain trauma, the changes in both the noradrenergic and serotoninergic systems are involved in functional disturbances, e.g. cortical depression associated with injury.

Key words: brain; ischemia; trauma; biogenic amines.

INTRODUCTION

Ischemic and traumatic injury to the brain leads to a variety of changes, including breakdown of the blood-brain barrier and development of edema as well as disturbances of biogenic amines and other neurotransmitters systems. The altered metabolism of the monoamines (dopamine (DA), norepinephrine (NE) and serotonin (5-HT)) neurotransmitters has been attributed to either energy failure, limited glucose supply or anoxia, all of which have been documented to take place when cerebral blood flow is markedly compromised. However, many studies have clearly shown that neither ischemic nor traumatic changes in the monoaminergic systems of the brain are directly related to the transient loss of oxygen or energy (Mrsulja *et al.*, 1989). Moreover, evidence has been accumulated that following brain trauma (induced by a superficial cortical freezing lesion) the involvement of the serotonergic and noradrenergic (but not dopaminergic) system in functional depression is not associated with alteration of blood flow or edema (Pappius, 1991). In support of monoaminergic involvement in the pathomechanisms of cerebral ischemia and trauma are recent data, from both animal and clinical studies, indicating an improvement of functional brain recovery after pharmacologic, modulation of neurotransmitter activity (Feeney and Sutton, 1987; Miyazaki *et al.*, 1989; Papius, 1991).

In this chapter, we will review the available data pertaining to the role of monoamines in the complex pathomechanisms of brain ischemia and trauma.

BRAIN ISCHEMIA

Three principal factors are thought to be involved in the ischemic breakdown of brain function: (1) exhaustion of the high-energy phosphate pool; (2) impairment of the ATPase-dependent sodium-potassium-calcium gradient; and (3) excessive liberation of neurotransmitters such as glutamate, monoamines, and others.

Cerebral ischemia interrupts the supply of nutrients to an area of tissue which contains nerve endings and axons among other cellular elements. The circulatory disturbance leads not only to a release of neurotransmitters and but also to changes in the properties of their receptors. This in turn contributes to the exacerbation of the consequences of the initial insult since neurotransmitter synthesis, degradation and release are in constant flux *in vivo*. Many of the secondary processes occurring in the damaged tissue and in the regions distant to the injury are complex in nature. In the severely affected areas, the neurons degenerate and eventually disappear while in the penumbra distant viable neurons may be capable of biochemical and functional recovery. Therefore, the possibility exists that neurotransmitters may play a role in both degenerative and regenerative processes as well as in functional disturbances not associated with cell death.

The effect of focal and global ischemia on biogenic amines has been studied in a variety of experimental models but the exact significance of the observed changes induced by ischemia in the monoaminergic systems is unclear. Most extensive investigations concerned with the changes in content and turnover of monoamines (5-HT, NE and DA) and their metabolites during and after ischemia were performed in Mongolian gerbils (Mrsulja *et al.*, 1975, 1976, 1977, 1978; Zervas *et al.*, 1975; Moskowitz and Wurtman, 1976; Welch *et al.*, 1976; Cvejic *et al.*, 1980). These and our later studies (Cvejic *et al.*, 1984; Spatz *et al.*, 1985; Kumami *et al.*, 1989; Mrsulja *et al.*, 1989a, b; Spatz and Mrsulja, 1990) have shown that ischemia alters the content of cerebral amines, their precursor metabolites, and the activities of synthesizing and degrading enzymes. The occurrence and extent of regional monoamine changes depend on the age of the animals, duration of cerebral deprivation of blood supply, and degree of subsequent reestablishment of blood circulation. For example, the same ischemic insult which reduced the content of 5-HT and NE in the cortex of adult gerbils delayed or failed to induce these amines in the cortex of young animals (Spatz *et al.*, 1985; Kumami *et al.*, 1989). The reduction in the content of biogenic amines observed after long-term ischemia was not seen following a short ischemic period unless circulation was reestablished to the brain (Mrsulja *et al.*, 1989a). In addition, the appearance of attenuated metabolism of biogenic amines was manifested by a pattern of slow development of lesions described as "maturation" and coincided with neuronal morphologic deterioration (Mrsulja *et al.*, 1976). All these observations indicated that the progression of ischemic brain injury continued during reperfusion of the previously ischemic brain. Moreover, most importantly, these investigations draw attention to: (1) the excessive release of DA, NE and 5-HT; (2) the lack of correlation between the ischemic changes in energy-related metabolites and biogenic amines both during and after ischemia; and (3) the persistence of disturbances in monoamine metabolism during the postischemic period which contradicted the previous notion that postischemic restoration of energy metabolism in the brain signifies a reversal of all consequences of ischemia to the preischemic state. Generally, there are two postischemic phases: the period immediately after ischemia characterized by brain hypermetabolism followed by a period of reduced cerebral metabolism. In reflow, increased turnover rates and decreased turnover times are found for NE, DA and 5-HT in the cerebral cortex; in the striatum, the turnover rate for DA was also increased while NE was reduced (Cvejic *et al.*, 1981). These observations represent the first biochemical evidence for regional differences in the response of biogenic amine pathways to ischemia in adult animals.

Noradrenergic system

The changes in NE metabolism are a part of general disturbances of neurotransmitter pathways induced by an ischemic insult. Recent investigation of NE turnover in rats subjected to incomplete forebrain ischemia indicated an increased turnover of NE in the early postischemic period (24 h) but not after a longer period of recirculation whereas hippocampal (CA1) neuronal damage was only seen at the later time of (48 and 96 h). Based on these data, it was suggested that the increased activity of the locus coeruleus (LC) modulates the effects of glutaminergic receptors since the stimulation of the NE system takes place during and immediately following marked glutamate release in the hippocampus (Miyazaki *et al.*, 1989).

The significance of the activation of the NE system is controversial. An increased neuronal damage of the hippocampus (CA1 region) was observed after bilateral LC lesions in ischemia in rats (Blomquist *et al.*, 1985). However, other studies have demonstrated an improved recovery of energy metabolism in the NE-depleted hemisphere of the unilaterally lesioned LC in rats (Busto *et al.*, 1985). Amelioration of ischemic damage and of catecholamine alteration was reported by increasing the levels of circulating NE or preischemic treatment with clonidine (α_2-adrenergic agonist, which is an inhibitor of catecholamines release, especially NE in the brain), respectively (Koide *et al.*, 1986; Miyazaki *et al.*, 1989).

Dopaminergic system

Recently, the potential role of neurotransmitters in the pathogenesis of cerebral dysfunction resulting from ischemia was explored with the application of *in vivo* microdialysis and electrochemical chronoamperometry. These investigations demonstrated massive increase in extracellular concentrations of DA, NE, 5-Ht and their metabolites, and other neurotransmitters (both excitatory and inhibitory) in the striatum, cortex, and hippocampus shortly after circulatory arrest in various animal models (Phebus *et al.*, 1986; Brannan *et al.*, 1987; Globus, 1988, 1989; Ogura *et al.*, 1989; Obrenovich *et al.*, 1990; Ahn *et al.*, 1991).

The observed ischemic increases in monoamine levels in the extracellular space probably reflect both their increased release as well as decreased removal. It is possible that an increased activity of tyrosine hydroxylase (Calderini *et al.*, 1978) and alteration of the neuronal membrane (Villacara *et al.*, 1989) contribute to the synthesis and release of catecholamines in the postischemic period. The synaptic release of amines can be driven by increased extracellular potassium or intracellular calcium. The homeostasis of both electrolytes is disturbed during ischemia; namely, there is an increased extracellular potassium and increased intracellular calcium due to the failure of the NA-K-Ca-ATPase mechanisms, which require ATP. Removal of biogenic amines from the extracellular space is largely due to neuronal reuptake, which is also ATP-dependent. The mechanism(s) involved in the postischemic release of biogenic amines are unclear. Recently, biphasic striatal DA release during transient ischemia and reperfusion in gerbils was reported (Ahn *et al.*, 1991). The first increase in extracellular DA was observed after ischemia lasting longer than 2 min, and the second, even larger release of biogenic amines, was seen during recirculation. Since striatal ATP is exhausted within 1 min and restored within 15 min after 5 min of ischemia (Lust *et al.*, 1986; Mrsulja *et al.* 1986, 1989a), we believe that factors other than ATP depletion must contribute to changes which occur in biogenic amines during the postischemic period.

The striatum, known to be more vulnerable than the cortex to ischemia, is innervated by both the corticostriatal glutaminergic pathway and the dopaminergic nigrostriatal tract which interact with each other (Cheramy *et al.*, 1986; Chiodo and Berger, 1986; Kornhuber and Kornhuber, 1986; Romo *et al.*, 1986). Both DA and glutamate release have been implicated as playing a major role in the selective striatal vulnerability to ischemia due to the excitatory activity of these substances (Globus *et al.*, 1988). Even though it was generally thought that the nigrostriatal pathway has an inhibitory action, recently it was also shown to have an excitatory component (Ohno *et al.*, 1985). Lately, we focused on further elucidation of regional vulnerability to ischemia by correlating the DA release with changes in dopaminergic receptors in the striatum and cortex in the gerbil model of bilateral brain ischemia (15 min) alone or with 1–2 h recirculation (Chang *et al.*, 1992). These studies showed that the ischemic release of

DA from the striatum was greater (about 400-fold over preischemic levels) than that from the cortex (12-fold over preischemic levels). A significant decrease in affinity (= increased K_D) without any changes in the number of binding sites for D_1-receptors was found in striatal membranes after ischemia alone. A decreased number (= B_{max}) of binding sites for D_2-receptor ligand was detected in the striatal but not in the cortical membranes after ischemia and reflow.

The results indicate that DA release into the striatal but into the cortical extracellular space is associated with changes in kinetic properties of D_1-(inhibitory) and D_2- (excitatory) receptors. The findings are most likely related to the amount of released DA which depends on the regional innervation and/or the presence of the receptor. The observed decreased affinity for D_1-binding sites and the significantly reduced number of D_2-binding sites strongly suggest desensitization and down-regulation of DA receptors, respectively.

The release of DA in the striatum is controlled by presynaptic D_2-receptors localized in the substantia nigra (SN) (Altar *et al.*, 1987; Boyar and Altar, 1987; Filloux *et al.*, 1987). D_1-receptors do not mediate the DA release but affect the DA metabolism (Saller and Salama, 1985). It is generally accepted that the striatal D_1- and D_2-receptors are postsynaptic in nature although some studies suggested the presence of both presynaptic and postsynaptic dopaminergic receptors. It has also been shown that both the D_1- and D_2-receptors are localized or colocalized on nondopaminergic neurons (Joyce and Marshall, 1987; Ohno *et al.*, 1987). Hence, the observed down-regulation of D_2-receptors in the striatum appears to be the result of excessive ischemic release of DA rather than a cause of that release. It is possible that the release of DA from the nigrostriatal nerve terminals in the striatum could be due to a similar ischemic alteration of D_2-receptors present in the *SN*. In view of the striatal localization of postsynaptic D_1- and D_2-receptors on nondopaminergic neurons and the demonstrated interaction between DA and cholinergic (Scatton, 1982; Stoof *et al.*, 1982) or GABA systems (Waszczak and Walters, 1983) it is possible that the desensitization and down-regulation of DA receptors may adversely affect transmission in either system and in this way contribute further to ischemic sequelae. Thus, the findings strongly suggest that the ischemic release of DA in the striatum is associated with early transient changes in both inhibitory and stimulatory neurotransmission which may also influence other neurotransmitter systems.

Studies concerned with the late effect of ischemia demonstrated a reduction of D_1- and D_2-receptor binding sites associated with cellular necrosis in the caudate-putamen (Przedborski, 1991) and disappearance of D_1- (but not D_2-) receptor and forskolin binding sites with loss of intracytoplasmic immunoreactivity for cAMP-regulated phosphoprotein (DARPP) related to neuronal necrosis in the dorsolateral striatum (Benefenati *et al.*, 1987). The causative involvement of the DA system in neuronal necrosis is supported by the fact that depletion of DA either by inhibition of synthesis or release prevented striatal necrosis (Clements and Phebus; 1988; Globus *et al.*, 1987) or ameliorated the damage of nerve terminals (Weinberger *et al.*, 1985). Moreover, the observed decreased incidence of ischemic stroke in patients with Parkinson's disease might be related to the disturbed production of DA (Struck *et al.*, 1990). The question then arises as to the mechanism by which DA causes neuronal degeneration. It has been suggested that the formation of free radicals and subsequent lipid peroxidation induced by the released DA among other mechanisms may contribute to cell damage.

During reperfusion, DA may undergo oxidation when exposed to molecular oxygen. The products of DA oxidation can bind to glutathione or sulfhydryl groups on proteins which may inactivate cellular enzymes and/or contribute to structural derangement of the membrane. It has been known that superoxide ions produced during catecholamine oxidation are potentially dangerous free radicals (Graham, 1984). Moreover, the catecholamine (DA) metabolized by mitochondrial monoamine oxidase produces hydrogen peroxide, a free radical precursor and catecholamine can reduce transitional metals to an oxidative state that catalyzes free radical production from hydrogen peroxide (Lovstad, 1984). This event could occur upon reperfusion, when oxygen is available as a cofactor for monoamine oxidase. Furthermore, DA can react with hydroxyl radicals and form 6-hydroxydopamine, a neurotoxin (Slivka and Cohen, 1985). Products of catecholamine oxidation and the monoamine oxidase-generated H_2O_2 can only induce neuronal injury when the antioxidative cellular systems (such as superoxide dismutase, catalase, glutathione peroxidase, etc.) are overwhelmed. The reduction of striatal glutathione reductase (Stanimirovic *et al.*, 1988) at 1 h after 15 min of ischemia in gerbils, at a time of the observed persistent increase of dopamine turnover (Chang *et al.*, 1992; Cvejic *et al.*, 1981) and increased superoxide anion formation (Mrsulja *et al.*, 1990) as well as enhanced index of lipid peroxidation of cell membranes (Villacara *et al.*, 1989) are consistent with the postulated role of DA involvement in the processes leading to brain injury following ischemia.

Serotoninergic system

The association of 5-HT changes with ischemic cerebral edema (accumulation of water in the tissues) was extensively reviewed previously (Spatz and Mrsulja, 1990). Briefly, we would like to reiterate that the participation of 5-HT in the development and progression of both cytotoxic and vasogenic cerebral edema (as classified by Klatzo, 1967) has been controversial (Fenske *et al.*, 1976) and a subject of investigation for many years (Spatz and Mrsulja, 1990). Vasogenic edema may occur as a result of brain trauma. The edema caused by ischemia has both cytotoxic and vasogenic components; hence, it should be considered as a separate entity. The formation of ischemic edema basically represents an osmotic phenomenon which is triggered by biochemical dysfunction of the membranes. A significant correlation between maximal accumulation of water, inhibition of Na-K-ATPase, and 5-HT release suggest that the changes in the activity of Na-K-ATPase and the 5-HT system contribute to the development of ischemic edema.

Pharmacologic investigations of 5-HT turnover provided additional support for 5-HT involvement in ischemic edema. Substances that either inhibit 5-HT synthesis (α-paraphenylalanine (PCPA)) or release (propentyphylline (a methylxanthine derivative)) were shown to reduce the brain swelling. Moreover, methysergide an antagonist of presynaptic 5-HT receptor, increased 5-HT turnover and formation of ischemic brain edema (Cvejic *et al.*, 1984; Mrsulja *et al.*, 1984). The reported ischemic modulation of kinetic properties for 5-HT$_1$ due to 5-HT$_{1B}$ (but not 5-HT$_{1A}$) as well as 5-HT$_2$ receptor binding sites in the cortex was consistent with observed disturbances in the serotonergic system (Villacara *et al.*, 1989; Kumami *et al.*, 1990). These changes suggested a down-regulation of 5-HT$_{1B}$ receptor binding sites by ischemia alone while ischemia with recirculation additionally resulted in an increased affinity (low K_D) for 5-HT$_{1B}$, the appearance of two binding sites for 5-HT$_2$ ligands, and simultaneous reduction in synaptosomal membrane fluidity.

The ischemically altered kinetic properties of 5-HT$_{1B}$ receptor sites associated with presynaptic autoreceptors which modulate 5-HT release may be responsible for the observed 5-HT release and the subsequent changes in affinity for the 5-HT$_2$ (S$_2$) postsynaptic receptors. Taken together these changes strongly suggest that the neuronal 5-HT$_1$-mediated inhibitory function is altered in the cortex at the onset of ischemic edema and reinforce our conclusion that serotonin is involved in the formation of ischemic cerebral edema.

BRAIN TRAUMA

Recent evidence suggests that perturbations of monoaminergic transmission also play a role in functional disturbances which develop in response to brain trauma and is unrelated to reduction of cerebral blood flow (Pappius, 1991).

Model of brain injury and assessment of cerebral function

A focal freezing lesion in the rat was used as a model of cerebral injury and the functional changes in the lesioned brain were delineated with the deoxyglucose (DG) method of Sokoloff *et al.* (1977) to measure local cerebral glucose utilization (LCGU) (Papius, 1981). These studies showed that, with time after a focal freezing lesion, the most marked effect on LCGU was a widespread depression developing throughout the entire cortex of the traumatized hemisphere. A decrease in the average cortical utilization was measurable 4 h after injury,

fell to about 50% of normal 3 days following the lesion, and remained somewhat depressed even at 10 days post-lesion (Pappius, 1988). The effects of the
lesion were unrelated to its location (Colle et al., 1986) and heat lesions had a
similar effect (Pappius, 1981). Since there was no evidence that the changes
in the cortical LCGU represented a primary metabolic effect of traumatization,
and in keeping with the hypothesis that the functional state of cerebral tissue
is closely coupled to metabolism (Sokoloff, 1981), the demonstrated depression
of cortical glucose utilization was interpreted to be a manifestation of cerebral
dysfunction. In other words, the energy needs of cortical areas of the injured
brain are envisaged as being diminished because of a functional depression in
the affected brain regions. Decisive support for this interpretation was obtained
recently when it was demonstrated that the energy status and substrate (glucose)
supply had not been compromised in the cortex of the traumatized hemisphere
at the time of the greatest decrease in cortical LCGU (Buczek et al., 1991). The
demonstration of increased ATP, CrP, and glucose levels in the affected cortical
regions, thus of an enriched metabolic profile, is consistent with the hypothesis
that the decreased cortical glucose use in traumatized brain is due to a diminished
need for energy rather than to its decreased supply.

In search of mechanisms underlying the functional depression resulting from
injury, the potential role of both the NE and 5-HT neurotransmitter systems was
investigated (Pappius and Wolfe, 1983a, b, 1984; Pappius, 1991).

Noradrenergic system

Both inhibition of biogenic amine synthesis with α-methyltyrosine and α_1-norad-
renergic receptor blockage with prazosin (PZ) ameliorated the cortical depression
of LCGU in the traumatized hemisphere in rats with a focal freezing lesion
(Pappius and Wolfe, 1983b; Inoue et al., 1991), suggesting that the NE system is
involved in functional disturbances associated with brain injury. PZ was effective
when given as a single dose of 1 mg/kg 30 min before lesioning and when given
before the lesion and for 2 days thereafter (3 mg/kg/day in 3 divided doses).
The latter regimen normalized the LCGU in areas where it was depressed by
lesioning despite the fact that it induced globally decreased LCGU in normal
animals. Yohimbine (YOH), mainly an α_2-noradrenergic receptor blocker, was
much less effective and only when given for 3 days. Whether the effects of
YOH were mediated by α_2-adrenergic receptors known to be present in both the
locus coeruleus and cortical regions (Fallon and Loughlin, 1987; Nakamura et al.,
1988) or at other adrenergic receptors or were due to its 5-HT-mimetic properties
(Pettibone et al., 1985) is unclear. Surprisingly, cortical NE levels decreased
bilaterally in unilaterally lesioned rat brain between days 1 and 6 after lesioning
(Pappius and Wolfe, 1983b), and this was accompanied by bilateral increases on
days 1 and 3 when it was measured in the amount of the principal metabolite

of NE in rat brain, 3-methoxy-4-hydroxyphenylglycol sulfate (Pappius, 1991). At the same time in this model of brain injury, DA and its metabolites were unchanged (Pappius and Wolfe, 1983b).

Further evidence for the involvement of the NE system in functional disturbance in injured brain has come from studies on the *in vivo* binding of a specific, selective α_1-adrenoreceptor ligand [^{125}I]HEAT ([^{125}I]-iodo-2-[β-(4-hydroxyphenyl)-ethylaminomethyl]tetralone) (Pappius *et al.*, 1991). In normal animals, the *in vivo* binding of [^{125}I]HEAT occurred at two different sites with different affinities (Gjedde *et al.*, 1991) similar to this reported for α_{1A} (high-affinity) and α_{1B} (low-affinity) adrenoreceptor subtypes (Minneman, 1988). Three days following a freezing lesion, at the time of the most marked depression of cortical glucose utilization, the density (B_{max}) of both the high- and low-affinity binding sites of [^{125}I]HEAT was altered as compared to normal (Pappius *et al.*, 1991). B_{max} of the low-affinity binding was significantly increased in three cortical areas of the lesioned hemisphere examined, as compared to both the normal and nonlesioned side. No significant effect of injury was seen in subcortical structures of the lesioned hemisphere nor on any area of the nonlesioned side. The areas of increased low-affinity (α_{1B}) binding correlated closely with areas of diminished cortical glucose utilization, suggesting that the demonstrated changes in the density of α_{1B}-adrenoreceptors are of functional importance in injured brain. In contrast, the B_{max} of high-affinity binding of [^{125}I]HEAT was significantly decreased bilaterally in most cortical and subcortical structures in which it was determined. This down-regulation may be related to the bilateral activation of the NE metabolism in injured brain, referred to above. It was clearly not correlated with the unilateral, mostly cortical, decrease of glucose utilization. Thus, there is no evidence that decreased global α_{1A}-binding contributes to the cortical dysfunction resulting from injury as delineated by unilateral decrease in cortical energy metabolism. The general conclusion to be drawn from these studies, as well as those on the effect of PZ in injured brain, is that the NE system plays a role in he functional consequence of brain injury and that mediation by α_{1B}-adrenoreceptors is most likely involved.

Serotoninergic system

There is also evidence that the 5-HT system contributes to the functional cortical depression induced by injury. It appears to be unilaterally activated in the cortex of the injured hemisphere where the 5-HT content was significantly decreased 24 h after a freezing lesion, while the level of the metabolite, 5-hydroxyindoleacetic acid (5-HIAA), was increased up to day 6 post-lesion (Pappius and Dadoun, 1987). That these changes are correlated with changes in cortical glucose utilization and hence the postulated functional depression in injured brain was demonstrated when inhibition of 5-HT synthesis with p-chlorophenylalanine was shown to result in significant increases in cortical LCGU in the lesioned hemisphere where it was decreased by injury in untreated animals (Pappius *et al.*, 1988). Preliminary results indicate that the 5-HT$_2$ receptor blocker, ketanserin, also ameliorated the depression of cortical utilization in the traumatized brain (Pappius, 1991), implying that the effects of activation of the 5-HT system in traumatized brain may be mediated by 5-HT$_2$ receptors.

The conclusion that both the NE and 5-HT systems are involved in cortical depression associated with injury is compatible with what is known regarding their function. In the cortex, both are thought to be inhibitory (Bloom *et al.*, 1972; Phillis and Kostopoulos, 1977; Aston-Jones and Bloom, 1991a, b) and it has been postulated that both affect cortical information processing (Fallon and Laughlin, 1987; Foote and Morrison, 1987).

COMMENTS

Both cerebral ischemia and trauma induce disturbances in the inhibitory noradrenergic and serotoninergic systems. Ischemia but not trauma also causes a change in the excitatory dopaminergic system. Thus, the monoaminergic alterations disrupt the balance between inhibitory and excitatory neurotransmission essential for normal function and cellular integrity.

Acknowledgment

The editorial assistance of D. G. Schoenberg is greatly appreciated.

REFERENCES

Ahn, S. S., Blaha, C. D., Alkire, M. T., Wood, E., Gray-Allan, P., Marrocco, R. T. and Moore, W. S. (1991). Biphasic striatal dopamine release during transient ischemia and reperfusion in gerbils. *Stroke* **22**, 674–679.

Altar, C. A. and Hauser, K. (1987). Topography of substantia nigra innervation by D1 receptor-containing striatal neurons. *Brain Res.* **410**, 1–11.

Altar, C. A., Boyar, W. C., Oei, E. and Wood, P. L. (1987). Dopamine autoreceptors modulate the *in vivo* release of dopamine in the frontal, cingulate, and entorhinal cortices. *J. Pharmacol. Exp. Therap.* **242**, 115–120.

Aston-Jones, G. and Bloom, F. E. (1981a). Norepinephrine-containing locus coeruleus neurons in behaving rats exhibit pronounced responses to non-noxious environmental stimuli. *J. Neurosci.* **1**, 887–900.

Aston-Jones, G. and Bloom, F. E. (1981b). Activity in epinephrine-containing locus coeruleus neurons in behaving rats anticipates fluctuations in the sleep-waking cycle. *J. Neurosci.* **1**, 876–886.

Bareggi, S. R., Porta, M., Selenati, A., Assael, B. M., Calderini, G., Collice, M., Rossandra, M. and Morselli, P. L. (1975). Homovanillic acid and 5-hydroxyindole-acetic acid in the CSF of patients after severe head injury. *Eur. Neurol.* **13**, 528.

Benefenati, F., Pich, E. M., Grimaldi, R., Zoli, M., Fuxe, K., Toffano, G. and Agnati, L. F. (1989). Transient forebrain ischemia produces multiple deficits in dopamine D_1 transmission in the lateral neostriatum of the rat. *Brain Res.* **498**, 376–380.

Blomqvist, P., Lindvall, O. and Wieloch, T. (1985). Lesions of the locus coeruleus system aggravate ischemic damage in the rat brain. *Neurosci. Lett.* **58**, 353–358.

Bloom, F. E., Hoffer, B. J., Siggins, G. R., Barker, J. L. and Nicoll, R. A. (1972). Effects of serotonin on central neurons: microiontophoretic administration. *Fed. Proc.* **31**, 97–106.

Boyar, W. C. and Altar, C. A. (1987). Modulation of *in vivo* dopamine release by D_2 but not D_1 receptor agonists and antagonists. *J. Neurochem.* **48**, 824–831.

Brannan, T., Weinberger, J., Knott, P., Taff, I., Kaufmann, H., Togasaki, D., Nieves-Rosa, J. and Marker, H. (1987). Direct evidence of acute, massive striatal dopamine release in gerbils with unilateral stroke. *Stroke* **18**, 108–110.

Buczek, M., Ratcheson, R. A., Lust, W. D., McHugh, M. and Pappius, H. M. (1991). The effects of focal cortical freezing lesion on regional energy metabolism. *J. Cereb. Blood Flow Metab.* **11**, 845–851.

Busto, R., Harik, S. I., Yoshida, S., Scheinberg, P. and Ginsberg, M. D. (1985). Cerebral norepinephrine depletion enhances recovery after brain ischemia. *Ann. Neurol.* **18**, 329–336.

Calderini, G., Carlson, A. and Nordström, C.-H. (1978). Influence of transient ischemia on monoamine metabolism in the rat brain during nitrous oxide and phenobarbitone anaesthesia. *Brain Res.* **157**, 303–310.

Chang, C. J., Ishii, H., Yamamoto, T., Yamamoto, M. and Spatz, M. (1992). The effect of cerebral ischemia on regional dopamine release and D_1- and D_2-receptors. *J. Neurochem.*, in press.

Cheramy, A., Romo, R., Godeheu, G., Baruch, P. and Glowinski, J. (1986). *In vivo* presynaptic control of dopamine release in the caudate nucleus: facilitatory or inhibitory influence of L-glutamate. *Neuroscience* **19**, 1081–1090.

Chiodo, L. A. and Berger, T. W. (1986). Interactions between dopamine and amino acid-induced excitation and inhibition in the striatum. *Brain Res.* **357**, 198–203.

Clements, J. A. and Phebus, L. A. (1988). Dopamine depletion protects striatal neurons from ischemia-induced cell death. *Life Sci.* **42**, 707–711.

Colle, L. M., Holmes, L. J. and Pappius, H. M. (1986). Correlation between behavioral status and cerebral glucose utilization in rats following freezing lesion. *Brain Res.* **397**, 27–36.

Cvejic, V., Micic, D. V. and Mrsulja, B. B. (1981). Catecholamine turnover in cerebral cortex and caudate during long-term reflow following transient ischemia in gerbils. In: *Cerebral Vascular Disease,* **Vol. 13**. J. S. Meyer *et al.* (Eds). Excerpta Medica, Amsterdam, pp. 261–264.

Cvejic, V., Micic, D. V. and Mrsulja, B. B. (1984). Depression of catecholamine metabolism following temporary brain ischemia in gerbils. *Jugoslav. Physiol. Pharmacol. Acta* **20**, 109–114.

Cvejic, V., Micic, D. V., Djurivic, B. M., Mrsulja, B. J. and Mrsulja, B. B. (1980). Monoamines and related enzymes in cerebral cortex and basal ganglia following transient ischemia in gerbils. *Acta Neuropathol. (Berlin)* **51**, 71–77.

Fallon, J. H. and Loughlin, S. E. (1987). Monoamine innervation of cerebral cortex and a theory of the role of monoamines in cerebral cortex and basal ganglia. In: *Cerebral Cortex,* **Vol. 6:** *Further Aspects of Cortical Function, Including Hippocampus.* E. G. Jones and A. Peters (Eds). Plenum Press, New York, pp. 41–127.

Feeney, D. M. and Sutton, R. L. (1987). Pharmacotherapy for recovery of function after brain injury. *Crit. Rev. Neurobiol.* **13**, 135–197.

Fenske, A., Sinterhauf, K. and Reulen, H. J. (1976). The role of monoamines in the development of cold-induced edema. In: *Dynamics of Brain Edema*. H. M. Pappius and W. Feindel (Eds). Springer-Verlag, Heidelberg, pp. 150–154.

Filloux, F. M., Wamsley, J. K. and Dawson, T. D. (1987). Dopamine D_2 auto- and postsynaptic receptors in the nigrostriatal system of the rat brain: localization by quantitative autoradiography with [^{3}H] sulpiride. *Eur. J. Pharmacol.* **128**, 61–68.

Foote, S. L. and Morrison, J. H. (1987). Extrathalamic modulation of cortical function. *Annu. Rev. Neurosci.* **10**, 67–95.

Gjedde, A., Dyve, S., Yang, Y.-J., McHugh, M. and Pappius, H. M. (1991). Bi-affinity of α_1-adrenoreceptor binding in normal rat brain *in vivo*. *Synapse* **9**, 1–6.

Globus, M. Y.-T., Busto, R., Dietrich, W. D., Martinez, E., Valdes, I. and Ginsberg, M. D. (1988). Effect of ischemia on the *in vivo* release of striatal dopamine, glutamate, and γ-aminobutyric acid studied by intracerebral microdialysis. *J. Neurochem.* **51**, 1455–1464.

Globus, M. Y.-T., Busto, R., Dietrich, W. D., Martinez, E., Valdes, I. and Ginsberg, M. D. (1989). Direct evidence for acute and massive norepinephrine release in the hippocampus during transient ischemia. *J. Cereb. Blood Flow Metab.* **9**, 829–896.

Globus, M. Y.-T., Ginsberg, M. D., Dietrich, W. D., Busto, R. and Scheinberg, P. (1987). Substantia nigra lesion protects against ischemic damage in the striatum. *Neurosci. Lett.* **80**, 251–256.

Graham, D. D. (1984). Catecholamine toxicity: A proposal for the molecular pathogenesis of manganese neurotoxicity and Parkinson's disease. *Neurotoxicology* **5**, 83–96.

Inoue, M., McHugh, M. and Pappius, H. M. (1991). The effect of α_1-adrenergic receptor blockers prazosin and yohimbine on cerebral metabolism and biogenic amine content of traumatized brain. *J. Cereb. Blood Flow Metab.* **11**, 242–252.

Joyce, J. N. and Marshall, J. F. (1987). Quantitative autoradiography of dopamine D_2 sites in rat caudate-putamen: Localization to intrinsic neurons and not to neocortical afferents. *Neuroscience* **20**, 773–795.

Klatzo, I. (1967). Neuropathological aspects of brain edema. *J. Neuropath. Exp. Neurol.* **26**, 1–14.

Koide, T., Wieloch, T. W. and Siesjö, B. K. (1986). Circulating catecholamines modulate ischemic brain damage. *J. Cereb. Blood Flow Metab.* **6**, 559–565.

Kornhuber, J. and Kornhuber, M. E. (1986). Presynaptic dopaminergic modulation of cortical input to the striatum. *Life Sci.* **39**, 669–674.

Kumami, K., Mrsulja, B. B., Ueki, Y., Djurivic, B. M. and Spatz, M. (1989). Effect of ischemia on noradrenergic and energy-related metabolites in the cerebral cortex of young and adult gerbils. *Metab. Brain Dis.* **3**, 273–277.

Kumami, K., Yamamoto, T., Villcara, A., Mrsulja, B. B. and Spatz, M. (1990). Ischemic brain edema: 5-Hydroxytryptamine receptors and the physical state of synaptosomal membranes. *Adv. Neurol.* **52**, 47–56.

Lovstad, R. A. (1984). Catecholamine stimulation of copper-dependent haemolysis: Protective action of superoxide dismutase, catalase, hydroxyl radicals and scavengers and serum proteins (ceruloplasmin, albumin and apotransferrin). *Acta Pharmacol. Toxicol.* **54**, 340–345.

Lust, W. D., Ueki, Y. and Mrsulja, B. B. (1986). Metabolic profile of regional ischemia in the gerbil brain. II. Dynamics of the first minute after 5-minute ischemia. *Jugoslav. Physiol. Pharmacol. Acta* **22**, 195–196.

Minneman, K. P. (1988). Alpha 1-adrenergic receptor subtypes, inositol phosphates, and sources of cellular Ca^{2+}. *Pharmacol. Rev.* **40**, 87–119.

Miyazaki, M., Nazarali, A. J., Boisvert, D. P., Bayens-Simmonds, J. and Baker, G. B. (1989). Inhibition of ischemia-induced brain catecholamine alterations by clonidine. *Brain Res. Bull.* **22**, 207–211.

Moskowitz, M. A. and Wurtman, R. J. (1976). Acute stroke and brain monoamines. In: *Cerebrovascular Disease*. P. Scheinberg (Ed.). Raven Press, New York, pp. 156–166.

Mrsulja, B. B., Ueki, Y. and Lust, W. D. (1986). Regional metabolic profile in early stage of global ischemia in the gerbil. *Metab. Brain Dis.* **1**, 205–220.

Mrsulja, B. B., Djuricic, B. M., Ueki, Y., Lust, W. D. and Spatz, M. (1989a). Cerebral ischemia: changes in monoamines are independent of energy metabolism. *Neurochem. Res.* **14**, 1–7.

Mrsulja, B. B., Djuricic, B. M., Cvejic, V. and Micic, D. V. (1984). Pharmacological approach to postischemic brain edema. In: *Recent Progress in the Study and Therapy of Brain Edema*. K. G. Go and A. Baethmann (Eds). Plenum Press, New York, pp. 683–689.

Mrsulja, B. B., Djuricic, B. M., Micic, D. V., Cvejic, V., Mrsulja, B. J., Kostic, V., Stojanovic, T. and Maletic-Savatic, M. (1989b). Biochemistry of cerebral ischemia: pathophysiological considerations. *Jugoslav. Physiol. Pharmacol. Acta* **25**, 89–106.

Mrsulja, B. B., Lust, W. D., Mrsulja, B. J. and Passonneau, J. V. (1977). Effect of repeated cerebral ischemia on metabolites and metabolic rates in gerbil cortex. *Brain Res.* **119**, 480–488.

Mrsulja, B. B., Mrsulja, B. J., Cvejic, V., Djuricic, B. M. and Rogac, Lj. (1978). Alterations of putative neurotransmitters and enzymes during ischemia in gerbil cerebral cortex. *J. Neurol. Trans.* **14** (Suppl.), 23–38.

Mrsulja, B. B., Mrsulja, B. J., Spatz, M. and Klatzo, I. (1975). Action of cerebral ischemia on decreased levels of 3-methoxy-4-hydroxy-phenylglycol sulfate, homovanillic acid, and 5-hydroxyindoleacetic acid. *Brain Res.* **98**, 400–404.

Mrsulja, B. B., Mrsulja, B. J., Spatz, M. and Klatzo, I. (1976). Catecholamines in brain ischemia — effect of alpha-methyl-p-tyrosine and pargyline. *Brain Res.* **104**, 373–378.

Mrsulja, B. B., Mrsulja, B. J., Spatz, M., Ito, U., Walter, J. T., Jr and Klatzo, I. (1976). Experimental cerebral ischemia in Mongolian gerbils. IV. Behaviour of biogenic amines. *Acta Neuropathol. (Berlin)* **36**, 1–8.

Mrsulja, B. B., Stanimirovic, D., Micic, D. V. and Spatz, M. (1990). Excitatory amino acid receptors, oxido-reductive process and brain oedema following transient ischaemia in gerbils. *Acta Neurochir.* **51** (Suppl.), 180–182.

Nakamura, S., Sakaguchi, T., Kimura, F. G. and Aoki, F. (1988). The role of alpha$_1$-adrenoreceptor-mediated collateral excitation in the regulation of the electrical activity of locus coeruleus neurons. *Neuroscience* **27**, 921–929.

Obrenovich, T. P., Sarna, G. S., Matsumoto, T. and Symon, L. (1990). Extracellular striatal dopamine and its metabolites during transient cerebral ischaemia. *J. Neurochem.* **54**, 1526–1532.

Ogura, K., Shibuya, M., Suzuki, Y., Kanamori, M. and Ikegaki, I. (1989). Changes in striatal dopamine metabolism measured by *in vivo* voltammetry during transient brain ischemia in rats. *Stroke* **20**, 783–787.

Ohno, Y., Sasa, M. and Takaori, S. (1985). Dopamine D$_2$-receptor mediated-excitation of caudate nucleus neurons from the substantia nigra. *Life Sci.* **37**, 1515–1521.

Pappius, H. M. (1981). Local cerebral glucose utilization in thermally traumatized rat brain. *Ann. Neurol.* **9**, 484–491.

Papius, H. M. (1991). Brain injury: New insights in neurotransmitter and receptor mechanisms. *Neurochem. Res.* **16**, 919–949.

Pappius, H. M. and Dadoun, E. R. (1987). Effects of injury on the indoleamines in cerebral cortex. *J. Neurochem.* **49**, 321–325.

Pappius, H. M. and Wolfe, L. S. (1983a). Functional disturbances in brain following injury: Search for underlying mechanisms. *Neurochem. Res.* **8**, 63–72.

Pappius, H. M. and Wolfe, L. S. (1983b). Involvement of serotonin and catecholamines in functional depression of traumatized brain. *J. Cereb. Blood Flow Metab.* **3(1)**, S226–S227.

Pappius, H. M. and Wolfe, L. S. (1984). Effects of drugs on local cerebral glucose utilization in traumatized brain: mechanism of action of steroids revisited. In: *Recent Progress in the Study and Therapy of Brain Edema.* K. G. Go and A. Baethmann (Eds). Plenum Press, New York, pp. 11–26.

Pappius, H. M., Dadoun, R. and McHugh, M. (1988). The effect of p-clorophenyl-alanine on cerebral metabolism and biogenic amine content of traumatized brain. *J. Cereb. Blood Flow Metab.* **8**, 324–334.

Pappius, H. M., Dyve, S., McHugh, M. and Gjedde, A. (1992). Effect on α_1-adrenoreceptors in rat brain *in vivo.* In: *Proceedings of Satellite Symposium: The Role of Neurotransmitters in Brain Injury.* Y.-T. Mordecai, M. D. Globus and W. D. Dietrich (Eds). Plenum Publishing Corp., New York.

Pettibone, D. J., Pfeuger, A. B. and Totaro, J. A. (1985). Comparison of the effects of recently developed α_2-adrenergic antagonists with yohimbine and rauwolscine on monoamine synthesis in rat brain. *Biochem. Pharmacol.* **34**, 1093–1097.

Phebus, L. A. and Clemens, J. A. (1989). Effects of transient, global cerebral ischemia on striatal extracellular dopamine, serotonin, and their metabolites. *Life Sci.* **44**, 1335–1342.

Phebus, L. A., Perry, K. W., Clemens, J. A. and Fuller, R. W. (1986). Brain anoxia releases striatal dopamine in rats. *Life Sci.* **38**, 2447–2453.

Phillis, J. W. and Kostopoulos, G. K. (1977). Activation of a noradrenergic pathway from the brain stem to rat cerebral cortex. *Gen. Pharmacol.* **8**, 207–211.

Przedborski, S., Kostic, V., Jackson-Lewis, V., Cadet, J. L. and Burke, R. E. (1991). Effect of unilateral perinatal hypoxic-ischemic brain injury in the rat on dopamine D_1 and D_2 receptors and uptake sites: A quantitative autoradiography study. *J. Neurochem.* **57**, 1951–1961.

Romo, R., Cheramy, A., Godeheu, G. and Glowinski, J. (1986). *In vivo* presynaptic control of dopamine release in the cat caudate nucleus: further evidence for the implication of corticostriatal glutamatergic neurons. *Neuroscience* **19**, 1091–1099.

Saller, C. F. and Salama, A. I. (1985). Dopamine receptor subtypes: *In vivo* biochemical evidence for functional interaction. *Eur. J. Pharmacol.* **109**, 297–300.

Scatton, B. (1982). Further evidence for the involvement of D_2, but not D_1 receptors in dopminergic control of striatal cholinergic transmission. *Life Sci.* **31**, 2883–2890.

Slivka, A. K. and Cohen, G. (1985). Hydroxyl radical attack on dopamine. *J. Biol. Chem.* **260**, 15466–15472.

Sokoloff, L. (1981). Localization of functional activity in the central nervous system by measurement of glucose utilization with radioactive deoxyglucose. *J. Cereb. Blood Flow Metab.* **1**, 7–36.

Sokoloff, L., Reivich, M., Kennedy, C., Des Rosiers, M. H., Patlak, C. S., Pettigrew, K. D., Sakurada, O. and Shinohara, M. (1977). The [^{14}C]-deoxyglucose method for the measurement of local glucose utilization: theory, procedure, and normal values in the conscious and anesthetized albino rat. *J. Neurochem.* **28**, 897–916.

Spatz, M. and Mrsulja, B. B. (1990). Monoamines and cerebral ischemia. In: *Cerebral Ischemia and Resuscitation.* A. Schurr and B. M. Rigor (Eds). CRC Press, Boca Raton, Florida, pp. 179–189.

Spatz, M., Cvejic, M., Kumami, K., Ueki, Y., Djuricic, B. M., Wrobleska, B. and Mrsulja, B. B. (1986). Neurotransmitters in ischemia with particular reference to 5-hydroxytryptamine. In: *Pharmacology of Cerebral Ischemia.* J. Krigelstein (Ed.). Elsevier, Amsterdam, pp. 181–187.

Spatz, M., Ueki, Y., Djuricic, B. M. and Mrsulja, B. B. (1985). 5-Hydroxytryptamine in cerebral cortex of adult and young gerbils. In: *Cerebral Vascular Disease*. J. S. Meyer, H. Lechner, M. Reivich and E. O. Ott (Eds). Excerpta Medica, Amsterdam, pp. 254–259.

Stanimirovic, D., Djuricic, B. M. and Mrsulja, B. B. (1988). Glutathione reductase during and after brain ischemia in gerbils. *Metab. Brain Dis.* 3, 293–296.

Stoof, J. C., DeBoer, T., Sminia, P. and Mulder, A. H. (1982). Stimulation of D_2-dopamine receptors in rat neostriatum inhibits the release of acetylcholine and dopamine but does not affect the release of γ-aminobutyric acid, glutamate, or serotonin. *Eur. J. Pharmacol.* **84**, 211–214.

Struck, L. K., Rodnitzky, R. L. and Dobson, J. D. (1990). Stroke and its modification in Parkinson's disease. *Stroke* **21**, 1395–1399.

Villacara, A., Kumami, K., Yamamoto, T., Mrsulja, B. B. and Spatz, M. (1989). Ischemia modification of cerebrocortical membranes: 5-hydroxytryptamine receptors, fluidity, and inducible *in vivo* lipid peroxidation. *J. Neurochem.* **53**, 595–601.

Waszczak, B. L. and Walters, J. R. (1982). Dopamine modulation of the effects of γ-aminobutyric acid on substantia nigra pars reticulata neurons. *Science* **220**, 218–221.

Weinberger, J., Nieves-Rosa, J. and Cohen, G. (1985). Nerve terminal damage in cerebral ischemia: Protective effect of alpha-methyl-para-tyrosine. *Stroke* **16**, 864–870.

Welch, K. M. A., Chabi, E., Dodson, R. F., Wang, T.-P., Nell, J. and Bergin, B. (1976). The role of biogenic amines in the progression of cerebral ischemia and edema: modification of p-chlorphenylalanine, methysergide and pentoxyfylline. In: *Dynamics of Brain Edema*. H. M. Pappius and W. Feindel (Eds). Springer–Verlag, Berlin, New York, pp. 193–202.

Zervas, N. T., Hori, H., Negora, M., Wurtman, R., Larin, F. and Lavyne, M. (1975). Reduction in brain dopamine following experimental ischemia. *Nature (London)* **247**, 283–284.

Tyrosine Hydroxylase, pp. 253–281
M. Naoi *et al.* (Eds)
© VSP 1993

STRESS-INDUCED CHANGES IN TYROSINE HYDROXYLASE AND OTHER CATECHOLAMINE BIOSYNTHETIC ENZYMES

RICHARD KVETŇANSKÝ[1,2] and ESTHER L. SABBAN[3]

[1] *Institute of Experimental Endocrinology, Slovak Academy of Sciences, Bratislava, Czechoslovakia*
[2] *National Institute of Neurological Disorders and Stroke, NIH, Bethesda, Maryland, USA*
[3] *Department of Biochemistry and Molecular Biology, New York Medical College, Valhalla, New York, USA*

Abstract—Studies on stress indicate that some stressors rapidly increase tyrosine hydroxylase (TH) activity, whereas others, such as cold or immobilization stress, lead to long-term changes in TH activity, characterized by increased levels of the enzyme protein. Cold stress increases TH mRNA rapidly, whereas there is a lag in the appearance of increased protein and a further delay in increased TH activity. A single, as well as repeated immobilization stress, raises TH mRNA levels, although at least 2 immobilizations are required for sustained effects. DOPA decarboxylase (DDC) was not found to be changed by stress. Dopamine-β-hydroxylase (DBH) activity is also elevated by cold and immobilization stress. DBH mRNA levels do not significantly rise in response to a single immobilization but are elevated following two or more repeated immobilization stresses. Phenylethanolamine N-methyltransferase (PNMT) activity and mRNA levels showed a smaller but consistent rise in response to cold stress. The neural inputs to the adrenal medulla are especially important for regulation of TH and DBH by stress, and the humoral input is particularly important for PNMT regulation. The information suggesting the mechanisms of stress-induced changes in the catecholamine biosynthetic enzymes is discussed.

INTRODUCTION

Chronic or repeated exposure to various stressors is known to increase the synthesis and release of catecholamines in the sympathetic-adrenomedullary system (Gordon *et al.*, 1966; Kvetňanský and Mikulaj, 1970; Kvetňanský *et al.*, 1971d, 1984; Stone and McCarty, 1983; McCarty and Stone, 1984). The elevated synthesis of catecholamines is accompanied by an increase in activity of tyrosine hydroxylase (TH, EC 1.14.16.2), the first enzyme in the catecholamine biosynthetic pathway (Nagatsu *et al.*, 1964; Levitt *et al.*, 1965). Alterations in catecholamine-synthesizing enzyme activities have been observed in response to different stressors (Usdin *et al.*, 1976, 1980, 1984; Van Loon *et al.*, 1989; Kvetňanský *et al.*, 1992).

The results of early studies indicated that alterations in tyrosine hydroxylase activity induced by cold or immobilization stress resulted from increases in the amount of enzyme protein and its synthetic rate (Kvetňanský *et al.*, 1970b, 1971d; Chuang and Costa, 1974; Hoeldtke *et al.*, 1974; Chuang *et al.*, 1975). Later it was shown that chronic cold exposure or repeated immobilization induced not only elevation of adrenal TH activity, but also elevation of TH mRNA concentrations (Fluharty *et al.*, 1985a; Stachowiak *et al.*, 1985, 1986; Tank *et al.*, 1985; Baruchin *et al.*, 1990; McMahon *et al.*, 1992; Sabban *et al.*, 1992).

Stress was also found to increase the other catecholamine biosynthetic enzymes, dopamine-β-hydroxylase (DBH, EC 1.14.17.1) and phenylethanolamine N-methyl transferase (PNMT, EC 2.1.1.28) (Kvetňanský *et al.*, 1970a, b, 1971a; Thoenen, 1970, 1972). The long-term regulation of DBH by repeated stress and of PNMT by cold stress were also found to be mediated by elevated mRNA levels (Weisberg *et al.*, 1989; McMahon *et al.*, 1992; Sabban *et al.*, 1992). In contrast dopa decarboxylase was not affected by stress (Thoenen, 1972; Otten *et al.*, 1973).

In this chapter, we review the studies examining regulation of TH and other catecholamine biosynthetic enzymes by various types of stressors. The short and long-term stress-induced changes in activities and what is known of the molecular regulation of these enzymes in stress is discussed. Finally, we review the findings that have begun to delineate the mechanism of stress-induced induction of the catecholamine biosynthetic enzymes.

STRESS-INDUCED CHANGES IN TYROSINE HYDROXYLASE (TH)

Short-term regulation of tyrosine hydroxylase activity by stress

Several types of stressors were found to produce a very rapid increase in TH activity without changes in the number of enzyme molecules (Weiner *et al.*, 1978; Fluharty *et al.*, 1985a; Masserano *et al.*, 1981) whereas other stressors, which lead to long-term changes in the number of TH molecules, failed to elicit a rapid activation of TH. These differences indicate that the activation and induction of TH are two different processes.

Under the influence of some stressors, short-term increases in catecholamine release lead to a rapid activation of TH, which is often expressed as an increased affinity of TH for its pterin cofactor. This TH activation has been demonstrated by stressors such as decapitation (Masserano and Weiner, 1979), electroconvulsive shock (Masserano *et al.*, 1981), formalin-induced tissue damage (Masserano and Weiner, 1981), glucoprivation resulting from administration of insulin or

2-deoxy-D-glucose (Fluharty *et al.*, 1983, 1985a), hypotension (Fluharty *et al.*, 1983, 1985a). The short-term activation of TH after acute insulin administration is blocked by prior adrenal denervation (Fluharty *et al.*, 1985a). When an increase in maximal TH activity develops, the increased affinity of the enzyme for cofactor disappears (Fluharty *et al.*, 1985a).

It is known that the short-term regulation of TH activity involves covalent modification of TH via phosphorylation (Zigmond *et al.*, 1989). Several protein kinases have been shown to phosphorylate TH including cAMP-dependent, Ca^{2+}/calmodulin-dependent, Ca^{2+}/phospholipid-dependent and cGMP-dependent protein kinases, TH kinase and kinase N (in NGF-treated PC12 cells). There is a large body of research on the phosphorylation of TH by these protein kinases (see reviews Zigmond *et al.*, 1989; Masserano *et al.*, 1989), however little data exists on the mechanism of TH phosphorylation under stress conditions.

In contrast to the above mentioned stressors, cold, however, is a stressor which does not lead to the rapid activation of pre-existing TH molecules (Fluharty *et al.*, 1983, 1985a). The *in vivo* biosynthesis of adrenal catecholamines was also not changed during a short cold exposure, suggesting that cold does not increase input to the adrenal medulla until its exposure is prolonged (Fluharty *et al.*, 1985a). In spite of elevated TH mRNA levels within 1 h of cold exposure, TH activity does not increase until 12 h later (Baruchin *et al.*, 1990; Miner *et al.*, 1992). Increases in adrenomedullary biopterin levels during cold exposure may facilitate increase in the rate of *in vivo* tyrosine hydroxylation, compensating for the lack of short-term activation of TH (Baruchin *et al.*, 1990).

Long-term regulation of tyrosine hydroxylase activity by stress

During long-term regulation of TH activity, there is a gradual increase in maximal TH activity due to the induction of higher levels of TH protein (see reviews Costa *et al.*, 1974, 1977; Thoenen *et al.*, 1979; Zigmond, 1980; Masserano *et al.*, 1989). Various stressors have been shown to increase the maximal TH activity in numerous cellular locations.

Regulation of TH in adrenals under stress

Cold stress. Cold stress was found to increase TH activity in adrenal medulla (Thoenen, 1970; Kvetňanský *et al.*, 1971b; Thoenen, 1972; Hoeldtke *et al.*, 1974; Fluharty *et al.*, 1983, 1985a; Kiran and Ulus, 1992). Sprague-Dawley rats kept at 3 °C, showed little change in adrenal TH activity by 1 day, however when examined after 7 and 21 days TH activity was elevated about 50%, but had returned to near control values after 42 days (Kvetňanský *et al.*, 1971b). In some of the studies, rats were shaved and exposed to 5 °C for various lengths of time. In the shaved animals there was little change in the induction of TH activity during the first 2 days, but when left for 1 or 3 weeks adrenal TH activity was much higher relative to unshaved animals at 3 °C (about 3–4-fold control values) (Fluharty et al., 1985a).

Interestingly adrenal medullary biopterin increased after the onset of cold exposure and rose to about 2-fold that of unstressed controls after 1 day, and remained at that level with prolonged cold (Baruchin et al., 1990).

Hoeldtke et al. (1974) showed that there was an increase in immunoreactive TH protein in response to cold exposure and immobilization stress. The kinetic properties of the stress-induced enzyme were identical to the control enzyme. There was a two fold increase in TH protein by immunotitration, however the stress-induced enzyme had the same K_m for tetrahydrobiopterin and tyrosine and also the same K_i for norepinephrine as that from control animals.

Fukuhara et al. (1992) reported elevated adrenal TH activity in so called SART stress (specific alteration of rhythm in temperature) in which rats were alternatively transferred every hour from 24 °C to −3 °C for 5 days.

Immobilization stress. The first demonstration of physiologically (not pharmacologically) induced elevation of adrenal TH activity was reported in repeatedly immobilized rats by Kvetňanský et al. (1969; 1970b). Kenessey and Huszti (1976) and Kiran and Ulus (1992) also found an increase in adrenal TH activity with repeated immobilization stress. Kvetňanský et al. (1969; 1970b) showed that after a single 2.5 h immobilization, there was a small (about 25%) elevation in TH activity, which in the majority of experiments was at the border of significance. However, after seven such daily immobilizations, TH activity had risen to over 3 times the control levels (Kvetňanský et al., 1969, 1970b). No further increase in TH activity was observed in adrenals of rats that were immobilized daily for 6 weeks. There was no significant difference in TH activity immediately, or 6 h, after the last immobilization. Repeated immobilization stress of 2.5 h for 6 days was also shown to increase TH immunoreactive protein levels about 3-fold (Hoeldtke et al., 1974).

After cessation of the daily immobilization stress (after the 7th immobilization), TH activity decreased towards pre-immobilization levels with a half-life of about 3 days, which was related to a diminishing rate of TH synthesis (Kvetňanský et al., 1970b). Similar conclusions on the turnover rate of TH under stress were reported by Chuang and Costa (1974) and Chuang et al. (1975).

Adrenal TH activity under stress was also characterized by measuring kinetic parameters, such as K_m for the substrate and cofactor (Hoeldtke et al., 1974; Blazicek and Kvetňanský, 1989). A single immobilization of rats did not change either the K_m or the V_{max} as compared to the unstressed control group. Following repeated immobilization, the K_m remained unchanged whereas the V_{max} for both substrate and cofactor were significantly increased. These results confirmed the idea that repeated immobilization stress induced increased synthesis of TH protein, and this in turn resulted in elevated TH activity, while the affinity of the enzyme for both substrate and cofactor remained unchanged (Blazicek and Kvetňanský, 1989; Hoeldtke *et al.*, 1974).

Different increases in TH activity during repeated immobilization in various inbred strains of rats have been reported (Cooper and Stolk, 1979; Stolk *et al.*, 1980). The increase in adrenal TH and DBH activity with repeated immobilization was found to be more pronounced in spontaneously hypertensive rats than in normotensive rats (Kvetňanský *et al.*, 1976a; Nagatsu and Kato, 1980) suggesting that SHR rats are more susceptible to stress than WKY rats.

It is evident that immobilization stress is one of the most intensive stimuli for the induction of TH. It elevates TH activity and protein levels about 3-fold and these elevations are among the largest long-term changes reported for TH *in vivo*.

The increased maximal adrenal TH and DBH activity (Kvetňanský *et al.*, 1970b, 1971a, 1984), together with increased adrenal catecholamine content, urinary excretion and plasma secretion (Kvetňanský and Mikulaj, 1970; Kvetňanský *et al.*, 1978, 1984) suggest that the adrenal medulla responds to repeated immobilization stress by an increased capacity to synthesize catecholamines. The administration of labelled catecholamine precursors clearly demonstrated a substantially increased *in vivo* biosynthesis of adrenal catecholamines in repeatedly immobilized rats (Kvetňanský *et al.*, 1971d). Using differently labeled precursors (^{14}C-tyrosine and ^{3}H-DOPA) it was shown that the increased catecholamine biosynthesis is primarily dependent on the step of tyrosine conversion to L-DOPA, catalyzed by TH, but after repeated immobilization, when dopamine formation is markedly accelerated, the step catalyzed by DBH, conversion of dopamine to norepinephrine, may also become rate limiting (Kvetňanský *et al.*, 1971d). There was a very good correlation between elevated adrenal TH and DBH activity *in vitro* and the rate of catecholamine synthesis measured *in vivo* in control and immobilized rats (Kvetňanský *et al.*, 1971d).

Denervation of the adrenal gland by transection of the splanchnic nerve completely prevented the immobilization-induced elevation of TH activity (Kvetňanský *et al.*, 1970b; Kvetňanský, 1973). Hypophysectomy reduced adrenal TH activity but repeated immobilization of hypophysectomized rats significantly elevated activity of this enzyme, although not to the levels seen in sham-operated immobilized animals (Kvetňanský *et al.*, 1970b; Kvetňanský, 1973). Thus, both neural and endocrine factors appear to participate in the stress-induced rise in TH activity. The mechanism of regulation of TH will be discussed in more detail at the end of this chapter.

Other stressors. Intermittent repeated swimming stress of 5–7 min during a 2 h period, at a water temperature of 15 °C, also increased TH activity in the adrenal medulla and in the superior cervical and stellar ganglia (Otten *et al.*, 1973). The maximal increase in TH activity was about 2 days subsequent to the 2 h of intermittent swimming. A 1 h swimming stress was not sufficient to lead to the elevation in TH 2 days subsequently. Chlorisondamine, a preganglionic blocker, if injected within 1 h after the swimming stress inhibited the increase in TH activity. The increase in TH activity with swimming stress is likely to be transcriptional since it was inhibited by actinomycin D if injected immediately before or after the swimming stress (Otten *et al.*, 1973). Elevated activity of adrenal TH and other catecholamine biosynthetic enzymes after swimming or long-term physical load has also been reported by other authors (Roffman *et al.*, 1973; Bhagat and Horenstein, 1976; Mikulaj *et al.*, 1976; Parizkova and Kvetňanský, 1980).

Psychosocial stimulation (Axelrod *et al.*, 1970; Henry *et al.*, 1971, 1976; Pfeifer, 1976), isolation of animals (Maengwyn-Davies *et al.*, 1973; Thoa *et al.*, 1976) hypokinesia (Blazicek *et al.*, 1980), and chronic footshock (Stone *et al.*, 1978) also induced elevation of TH activity. Glucoprivation after administration of insulin or 2-deoxy-D-glucose increased TH activity to a very high extent, similar to immobilization stress (Kvetňanský *et al.*, 1971c; Fluharty *et al.*, 1983, 1985a; Kiran and Ulus, 1992).

The only stress situation in which a reduced adrenal TH activity has been reported is chronic exposure to heat at 34 °C (Petrovic and Janic-Sibalic, 1980).

Weightlessness as a possible stressor was studied in relation to changes in TH activity. Long-term space flight of rats (21 days) did not produce any significant elevation of adrenal TH or other catecholamine synthesizing enzymatic activities (Kvetňanský *et al.*, 1981a, 1981b). In some space experiments (Cosmos 782), in which the rats were euthanized 12–24 h after landing (due to weather conditions and/or other factors), TH activity showed a small but significant rise. Data of four other space experiments on rats, however, clearly documented that weightlessness, as one of the most important factors affecting living organisms during space flights, does not elevate the sympathetic-adrenal activity, including TH activity (Kvetňanský *et al.*, 1981a, b).

Effect of novel stressors. It has been found that when repeatedly immobilized, or cold acclimated rats, were exposed to a new acute stressor, the rise of plasma epinephrine and norepinephrine in response to this novel stressor was substantially higher than the response of naive animals to the same stressor (Kvetňanský *et al.*, 1984).

When a group of rats acclimated to high-altitude conditions (one year stay at an altitude of 1,350 m) was exposed to a single immobilization for 2 h and compared to a group of immobilized rats which had lived at 150 m above sea level, the adrenal response of the acclimated rats was more significant. These animals manifested a larger increase in adrenal TH activity and a larger reduction in adrenal catecholamine levels (Balaz *et al.*, 1980). Similar changes in adrenal medullary activity were seen after the long-term space flight of 18.5 days on board Cosmos 1129 (Kvetňanský *et al.*, 1981a). Although adrenal TH activity was unchanged in rats euthanized immediately after landing, compared to a control group, when these rats were subjected to an additional stress, repeated immobilization (2.5 h daily for 5 days), an increase in TH activity was observed which was much greater than that observed with control rats (not subjected to space flights) that underwent a parallel immobilization stress (Kvetňanský *et al.*, 1981a). Thus, although weightlessness and other factors associated with space flight do not change adrenal TH activity, they sensitize the adrenal medullary system to a greater response to a novel stressor on Earth.

Abercrombie *et al.* (1992), Adell *et al.* (1988) and Nisenboum *et al.* (1991) have supported this conclusion and extended it to several brain areas, where catecholamine biosynthesis and TH activity were not altered by chronic stress, but a greater elevation was observed when the chronically stressed rats were exposed to a novel stressor.

Regulation of TH in other regions under stress

Stress-induced changes in TH activity in the brain, ganglia, the heart and other organs have also been reported. A single 2.5 h immobilization stress did not change TH activity in the hypothalamus, however repeated immobilizations expressively increased TH activity (Kvetňanský *et al.*, 1976b). These changes in TH are in good agreement with changes in norepinephrine content in the hypothalamus, which is significantly reduced after the first immobilization and returns to control levels after repeated immobilizations, as a consequence of an increase in norepinephrine synthesis in the hypothalamus, which is significantly induced by repeated stress. In addition, an immobilization stress induced elevation of TH activity was also reported by Lamprecht *et al.* (1972) in the hypothalamus and by Palkovits *et al.* (1975) in individual hypothalamic nuclei especially in the arcuate nucleus.

Increased brain TH activity was found after repeated electroshock (Musacchio *et al.*, 1969; Modigh, 1976; Stone *et al.*, 1978; Iuvone and Dunn, 1986), repeated emotional stress (Weiss *et al.*, 1975), long-term cold exposure (Thoenen, 1970; Costa and Meek, 1974; Zigmond *et al.*, 1974; Richard *et al.*, 1988), and various stressors (Stone and McCarty, 1983; Nisenbaum *et al.*, 1991).

Immobilization and cold have also been shown to increase TH activity in sympathetic ganglia indicating that stress induces an increase in impulse flow from the CNS not only to the adrenal medulla but also to ganglia. This response, however, is selective for different stressors. Kiran and Ulus (1992) in an extensive study measured TH activity in 24 different sympathetic ganglia following exposure to different stressors. Immobilization (4 times, 6 h daily) increased TH activity primarily in the lumbar and sacral ganglia, slightly in the cervical ganglia but no changes in TH activity were found in the thoracic ganglia. Prolonged exposure to cold (64 h) increases TH activity mainly in the coeliac, L_2 and cervical inferior ganglia. Glucoprivation, by administration of 2-deoxy-D-glucose, which is known to elevate TH activity in the adrenal medulla, did not induce any change in TH activity in the sympathetic ganglia (Kiran and Ulus, 1992). Failure of this drug to alter TH in sympathetic ganglia indicated that 2-deoxy-D-glucose induced increases in impulse flow were highly selective to the adrenal medulla. Cold and immobilization induced increases in TH activity in superior cervical, coeliac and stellate ganglia were also reported by others (Thoenen, 1970, 1972; Otten et al., 1973; Ulus and Wurtman, 1979). Ganglionic TH activity also rose after electrical stimulation (Zigmond et al., 1980).

The results of these studies indicate that the increase in impulse flow from CNS to various zones of the peripheral sympathetic system and to the adrenal medulla is not uniform and shows a high degree of selectivity depending on the type of stressor.

Long-term stress has also been shown to increase TH activity in sympathetic nerve terminals. Repeated immobilization stress significantly increased both myocardial TH and DBH activities (Kvetňanský et al., 1984). Activation of cardiac TH activity was elicited by insulin-induced hypoglycemia (Fluharty et al., 1985b) and by short (1 h) or long-term (4 days) cold exposure at 5 °C (Fluharty et al., 1985a, b). Interestingly, the short-term cold exposure did not change TH activity in the adrenal medulla suggesting preferential activation of the peripheral sympathetic system by cold. On the other hand, rats exposed to intermittent −3 °C cold (SART stress) for 5 days did not show significant changes in TH activity in the heart, spleen or in the interscapular brown fat but had significantly elevated adrenal medullary TH activity (Fukuhara et al., 1992).

Molecular biological studies on the regulation of TH by stress

Regulation in adrenals

Cold stress. Several groups have found that cold stress is accompanied by increased TH mRNA levels (Stachowiak et al., 1985, 1986; Tank et al., 1985; Kaplan et al., 1987; Richard et al., 1988). Maximal induction of adrenal TH mRNA levels, of about 3-fold, was already observed within 3–6 h of exposure of shaved rats to cold stress at 5 °C. These levels remained constant during further exposure to cold. The increase in TH mRNA was eliminated by adrenal denervation (Stachowiak et al., 1986) and was greatly reduced by the ganglionic blocking agent, chlorisondamine (Stachowiak et al., 1988).

Surprisingly, the increase in TH protein levels lagged behind the rise in mRNA (Baruchin et al., 1990, Miner et al., 1992). As shown in Fig. 1, increases of

50–60% in TH protein were observed after 3 and 12 h, but the maximal increase of about 3- to 4-fold in TH protein was only obtained after 72 h of continuous cold. Despite the two fold rise in TH protein, only a modest increase in TH activity was detected following 1 day of cold stress. TH activity reached a maximal level several days subsequently (Hoeldtke *et al.*, 1974; Fluharty *et al.*, 1985a; Baruchin *et al.*, 1990; see Fig. 1). Thus in the case of cold stress, the initial rise in TH mRNA does not appear to be translated into immunoreactive protein until several days later, and there is a lag between the rise in TH protein and the elevation of TH activity. The mechanism of this lag remains to be determined. It is possible that the body temperature is lowered during the prolonged cold and this might effect the rate of translation.

The cold induced increase in TH appears to be regulated both at the pre- and post-translational level. The early rise in TH mRNA is likely to involve transcriptional activation. Indeed transcription of the TH gene has been shown to

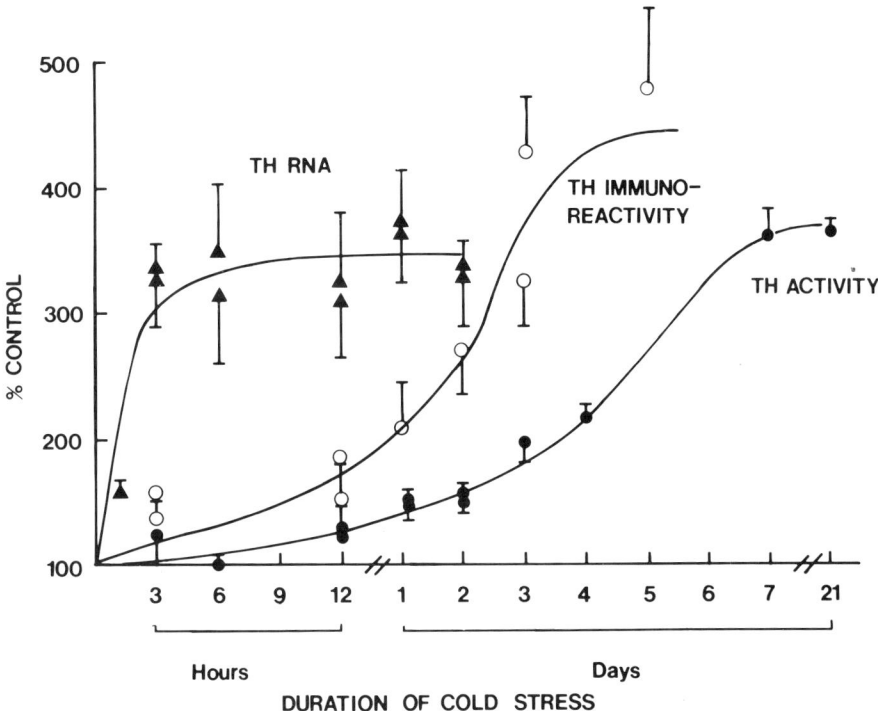

Figure 1. Time course of the effects of cold exposure on adrenal medullary levels of TH mRNA, protein, and enzyme activity. Animals were exposed to cold (5 °C) for the time periods specified, and TH immunoreactivity assayed by an immunoblot assay and TH mRNA by RNA dot-blot hybridization. Values are given as percent of non-stressed controls (±SEM). (From Miner *et al.*, 1992; with permission.)

be an important mode of long-term regulation. The 5' flanking region of the TH gene contains several consensus sequences which have been suggested to be involved in transcriptional activation of TH, including a cAMP regulatory element (CRE), AP1, AP2, POU/Oct, and SP1 sites (Cambi et al., 1989; D'Mello et al., 1989; Coker et al., 1988; Kim et al., 1991; Lewis et al., 1987; Carroll et al., 1991). The region containing the CRE was shown to be essential for regulation by cAMP (Lewis and Chikaraishi, 1987; Fader and Lewis, 1990) and appears to be also involved in regulation of TH by membrane depolarization (Kilbourne et al., 1992). The AP1 site has also been shown to be active in transcriptional regulation of TH. The AP1 sites have been shown to be TPA-inducible enhancer elements within the 5' region of phorbol ester inducible genes (Angel et al., 1987), which can interact with the immediate early genes, c-jun and c-fos, or members of this family. This region appears to mediate the increase in levels of TH mRNA with angiotensin (Stachowiak et al., 1990) and short-term NGF treatment of PC12 cells (Gizang-Ginsberg and Ziff, 1990).

The mechanism of the pre-translational regulation resulting in the rapid elevation in TH mRNA in response to cold stress is being investigated by Kaplan and coworkers by gel retardation assays using nuclear proteins from adrenal medulla of rats briefly exposed to cold stress. Miner et al. (1991; 1992) found that following 1 h of cold stress there was an increase in binding of the nuclear proteins, to the 5' upstream region of the TH gene containing the AP1 binding site. Thus, the elevation of TH mRNA in response to cold stress appears to proceed via increased binding to the AP1 site of the TH gene.

Immobilization stress. Immobilization stress of rats was found to lead to a large increase in TH mRNA levels in the adrenal (McMahon et al., 1992a; Sabban et al., 1992) and in the noradrenergic cell bodies of the *locus coeruleus* (Smith et al., 1991). A single 2 h immobilization was sufficient to cause about an eight fold rise in adrenal TH mRNA levels in Spraque Dawley rats (see Fig. 2). This is among the largest reported elevations in TH mRNA. Given the half life of TH mRNA (about 6–10 h, A. William Tank, personal communication), the very large rapid rise in TH mRNA following a single 2 h immobilization probably represents increased transcription. That this was so was confirmed using the transcriptional inhibitor, actinomycin D. Treatment of rats with actinomycin D (1 mg/kg) 30 min before and 45 min into a single 2 h immobilization stress, prevented the rise in TH mRNA.

The elevation in adrenal TH mRNA with a single 2 h immobilization was transient and 24 h later had returned to near control values. However, repeated daily immobilizations for as little as 2 days elicited a large rise in TH mRNA levels that was maintained even when examined 24 h later (Fig. 2). TH mRNA levels were similarly elevated following seven repeated daily immobilizations in

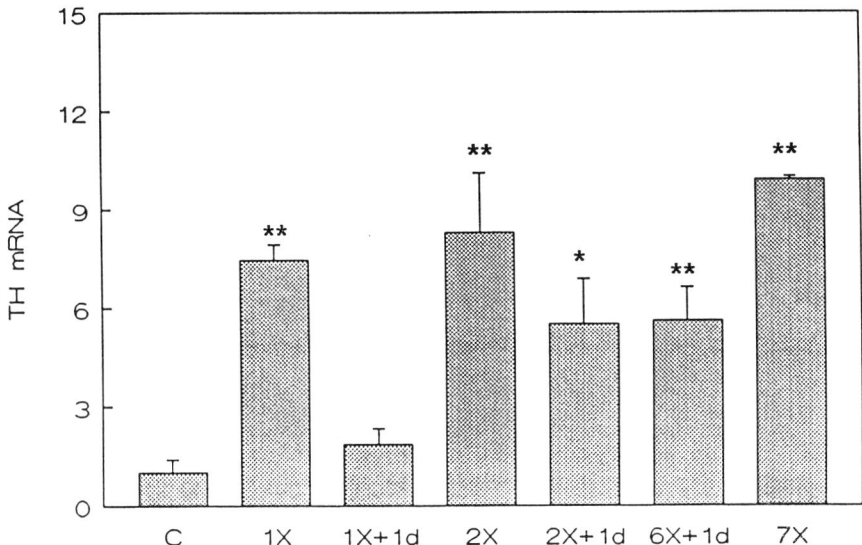

Figure 2. Effect of a varying number of immobilizations on TH mRNA levels. Levels of adrenal TH mRNA were determined in controls (C: $n = 4$), or immediately following 1 (1X: $n = 5$), 2 ($n = 6$) or 7 (7X: $n = 2$) daily 2 h immobilizations, or 1 day following 1 (1X + 1d: $n = 5$), 2 (2X + 1d: $n = 6$) or 6 (6X + 1d: $n = 4$) immobilizations. Values are expressed as mean ± SEM. $^{*}p \leqslant 0.01$ and $^{**}p \leqslant 0.001$ for t-test versus control group. (From McMahon *et al.*, 1992a; with permission.)

the adrenal (McMahon *et al.*, 1992a; see Fig. 2) as well as in *locus coeruleus*, as shown by *in situ* hybridization (Smith *et al.*, 1991).

Averaging over several experiments, adrenal TH mRNA levels were 4.1 ± 0.6 (mean ± SEM, $n = 9$) fold above levels in adrenals from control animals, 1 day following 6 daily immobilization stress (McMahon *et al.*, 1992a). This level is similar to the 4-fold rise in TH activity observed after seven repeated immobilizations (Kvetňanský *et al.*, 1970b). Therefore it is likely that the increase in activity is a direct consequence of the establishment of new higher steady state levels of TH mRNA. This is in the line with the finding that the increases in TH activity with immobilization stress resulted from an increase in V_{max} while other kinetic properties of the induced enzyme were unchanged (Hoeldtke *et al.*, 1974; Blazicek and Kvetňanský, 1989), indicating new steady state levels of TH protein.

The results of these studies also indicate that two immobilizations are sufficient to elicit the effect of repeated immobilizations. This would suggest that the occurrence of stress sufficient to lead to alterations in gene expression, may be more prevalent than previously considered.

Regulation in other tissues

An increase in TH activity has been reported in brainstem sections containing the _locus coeruleus_ following several days of cold stress (Stachowiak _et al._, 1986; Richard _et al._, 1988). The two-fold rise in TH mRNA in the _locus coeruleus_ following 4 days of stress was accompanied by a rise of 164% in TH protein, determined by immunoblots, and 140% in TH activity. Despite the increase in TH mRNA levels in rat adrenals and _locus coeruleus_, TH mRNA, protein and activity was unchanged in the substantia nigra with cold stress (Richard _et al._, 1988).

STRESS AND DOPA DECARBOXYLASE (DDC)

Cold stress, which elevated TH and DBH in rat adrenals, did not alter the activity of DDC. The activity of this enzyme was also unchanged in the superior cervical ganglion following 1–4 days of cold stress (Thoenen, 1972). Similarly, under conditions where TH activity was elevated following swimming stress, in superior cervical and stellate ganglia and in adrenals, DDC was not elevated in any of these tissues (Otten _et al._, 1973).

STRESS-INDUCED CHANGES IN DOPAMINE-β-HYDROXYLASE (DBH)

Short-term regulation of DBH activity by stress

Dopamine-β-hydroxylase does not appear to be subjected to short-term regulation during stress, as has been reported for TH.

Long-term regulation of DBH activity by stress

Regulation of DBH activity in adrenals

Cold stress. The effect of cold stress on dopamine-β-hydroxylase activity was examined. When rats were exposed to cold stress, adrenal DBH activity was elevated by 1 day of cold exposure and increased further after 7 and 21 days. The extent of the induction was about 50%. The magnitude of the elevation in DBH activity was similar to that observed for TH in the same study, except that DBH activity was already somewhat elevated following one day of cold, whereas the rise in TH activity was only evident at later times (Kvetňanský _et al._, 1971b). A rise in DBH activity in adrenals of rats exposed to 4 °C for 1–4 days, similar in magnitude and kinetics to that of TH, has also been reported (Thoenen _et al._, 1971; Thoenen, 1972). In this study there was a gradual rise in both TH and DBH activity to over twice control values.

Immobilization stress. Immobilization stress was found to elicit a large effect on adrenal DBH activity. Immediately following repeated immobilizations DBH activity was about 2-fold above that of unstressed control and rose to over 3-fold

6 h after the 7th or 42nd daily immobilization (Kvetňanský *et al.*, 1971a). Interestingly, DBH activity in the adrenal actually decreased from about 3-fold to 2-fold over levels in untreated control during the duration of the repeated immobilization, probably reflecting release of soluble DBH with norepinephrine during the immobilization. In line with this, serum DBH was found to be elevated with immobilization stress (Weinshilboum *et al.*, 1971).

Six hours after the immobilization stress, adrenal DBH activity was again about 3-fold above control levels. Even the 42nd immobilization further elevated DBH activity, when examined 6 h later. This further rise in DBH activity was inhibited by actinomycin D indicating that the rise in DBH activity is likely to be the result of elevated transcription. Treatment of rats, with the protein synthesis inhibitor, cycloheximide before the last of 7 daily repeated immobilizations, also prevented the post-stress rise in DBH activity. Denervation by transection of the splanchnic nerve markedly reduced the induction of DBH activity by immobilization stress, implicating transsynaptic mechanisms in the regulation of adrenal DBH by repeated immobilization (Kvetňanský *et al.*, 1971a).

Immobilization induced elevation of adrenal DBH activity has also been reported by others (Kenessey and Huszti, 1976; Blazicek *et al.*, 1980; Nagatsu and Kato, 1980; Stolk *et al.*, 1980).

Regulation of DBH in other locations and by other stressors

The effect of other stressors on adrenal DBH activity, as well as the effect of other stressors on DBH in non-adrenal tissues has not been studied as much as TH (Axelrod, 1972). Thoenen (1972) found that the superior cervical ganglia differed from the adrenal in its response to cold stress, in that while both TH and DBH activities were elevated in the adrenals, only TH activity was elevated in the superior cervical ganglion. Changes in adrenal DBH activity were found after glucoprivation (Viveros *et al.*, 1968) and physical exercise (Roffman *et al.*, 1973; Otten *et al.*, 1973; Bhagat and Horenstein, 1976).

Molecular biological studies on the regulation of DBH by stress

Cold and immobilization stress. The effect of cold stress on DBH mRNA levels has not previously been reported. The effect of immobilization stress on rat adrenal DBH mRNA levels was examined following a single and repeated immobilizations (McMahon *et al.*, 1992a; Sabban *et al.*, 1992). As shown in Fig. 3, a single 2 h immobilization, which greatly increased adrenal TH mRNA levels (Fig. 2), did not significantly alter adrenal DBH mRNA levels. However, 2 daily 2 h immobilizations led to a large increase in DBH mRNA levels. These levels were maintained when examined 1 day later and were similar to those observed one day subsequent to 6 daily immobilizations. Thus, the increase in

R. Kvetňanský and E. L. Sabban

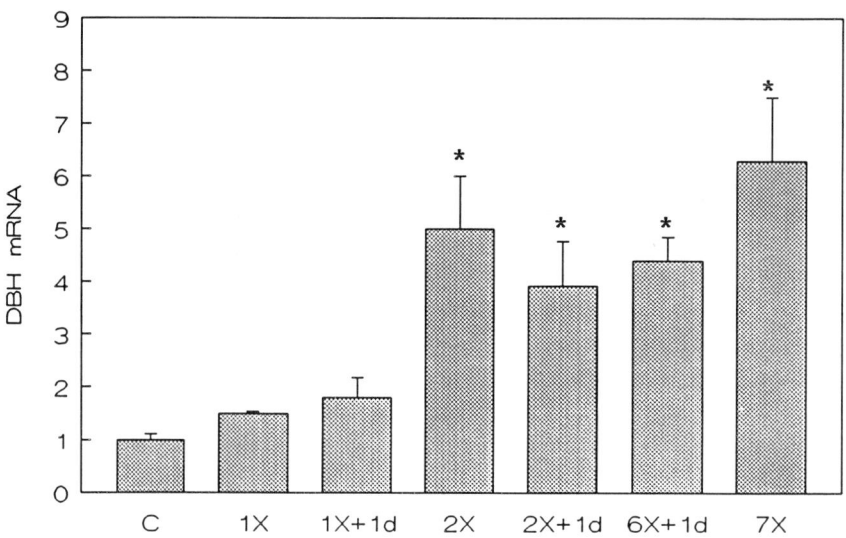

Figure 3. Effect of a varying number of immobilizations on DBH mRNA levels. Adrenal DBH mRNA levels were determined by Northern analysis, using a cDNA clone for rat DBH (McMahon *et al.*, 1990a). Levels of DBH mRNA were determined in controls (C: $n = 4$), or immediately following 1 (1X: $n = 5$), 2 (2X: $n = 6$), or 7 (7X: $n = 11$) daily 2 h immobilizations, or 1 day following 1 (1X + 1d: $n = 3$), 2 (2X + 1d : $n = 4$) or 6 (6X + 1d: $n = 2$) immobilizations. Values are expressed as mean ± SEM. *$p \leqslant 0.001$ for *t*-test versus control group. (Adapted from McMahon *et al.*, 1992a; with permission.)

DBH activity with repeated immobilization stress can be explained by attaining a new higher steady state level of DBH mRNA and presumably by increased synthesis of DBH protein.

The upstream region of the DBH gene has been found to contain several potential regulatory elements, including those involved in regulation by cAMP and glucocorticoids, but not an AP1 site (Kobayashi *et al.*, 1989; McMahon and Sabban, 1992b). It will be of interest to determine if these putative DNA regulatory elements are involved in the regulation of DBH mRNA levels by stress.

Especially interesting is how a single immobilization stress elevates TH mRNA without affecting DBH mRNA, and what is the mechanism whereby a second stress period can signal the elevation of both TH and DBH mRNA levels. Experiments in culture have been used to begin to study the common and divergent features of the regulation of TH and DBH mRNAs. Interestingly, short-term treatment of PC12 cells with membrane depolarizing agents, either elevated KCl or veratridine, specifically increased TH but not DBH mRNA levels. The maximum elevation occurred after 2–12 h and had returned to near control values by 24 h (Kilbourne and Sabban, 1990). A differential regulation of TH and DBH mRNA levels was also observed with increased PC12 cell density (Badoyannis *et al.*, 1991), in the *locus coeruleus* following chronic exposure to the antidepressant drug, imipramine (Nestler *et al.*, 1990). In contrast, several other treatments of PC12 cells, such as with cAMP analogs and glucocorticoids (McMahon *et al.*, 1990b) as well as with nicotine (Hiremagular, Nitahara, and Sabban, unpublished results) elevate both TH and DBH mRNA and may share common mechanisms with the repeated stress.

STRESS-INDUCED CHANGES IN PHENYLETHANOLAMINE N-METHYL-TRANSFERASE (PNMT)

Long-term regulation of PNMT by cold and immobilization stress

Cold stress led to a significant but small (about 20%) increase in PNMT activity. This was less than that observed for TH and DBH activities (Kvetňanský *et al.*, 1971b).

Repeated immobilization was shown to lead to a significant rise in adrenal PNMT activity (Kvetňanský *et al.*, 1970b). Immediately following a single immobilization, PNMT activity was not elevated. However 6 h after a single immobilization, a significant but small rise in PNMT activity was observed. Repeated daily immobilizations led to a significant elevation of PNMT activity. Thus, following seven repeated 2 h immobilizations adrenal medullary PNMT activity was about 50% higher than in unstressed controls and similar results were obtained following as many as 42 daily stresses. It should be noted that the magnitude of elevation of PNMT of about 50% by immobilization stress, is smaller than that observed for TH and DBH under identical conditions.

Molecular biological studies on the regulation of PNMT by stress

PNMT mRNA levels in the adrenal of rats exposed to cold stress (fur clipped and kept at 5 °C) displayed an elevation of about 50% above control values (Weisberg *et al.*, 1989). This rise was already apparent following 1 h of cold stress. The mRNA levels remained elevated at similar levels when examined after 7 days of chronic cold. PNMT protein levels, as examined by immunoreactivity, were also elevated 40–60%. Increased PNMT protein levels were observed within 12 h of the onset of cold stress and remained elevated for the 3 days examined (Weisberg *et al.*, 1989). In sum, it appears that PNMT is also regulated by chronic or repeated stress, but the magnitude of the regulation is considerably smaller than that of either TH or DBH.

MECHANISMS OF STRESS-INDUCED CHANGES IN CATECHOLAMINE BIOSYNTHETIC ENZYMES

Neural and humoral regulation in stress

It has been hypothesized, a long time ago, that an increase in the firing of sympathetic neurons elevates the rate of catecholamine biosynthesis in the adrenal medulla (Holland and Schumman, 1956) which may then increase the activity of the catecholamine biosynthetic enzymes.

Neural regulation. Neural regulation of adrenal catecholamine biosynthetic enzymes in stress was studied after one adrenal was denervated by severing the splanchnic nerve while the second intact adrenal served as a control. As shown in Fig. 4, denervation did not influence the levels of adrenal catecholamines in control animals, while in the stressed animal, there was no significant decrease of adrenal epinephrine in the denervated adrenal. The immobilization-induced depletion of adrenal epinephrine was also blocked by administration of hexamethonium (Kvetňanský *et al.*, 1971a). The increase in activity of TH and DBH by repeated immobilization in the intact adrenal, was completely blocked in the denervated adrenal, but PNMT activity was elevated in both intact and denervated adrenals following repeated immobilization stress (Kvetňanský *et al.*, 1970a, b, 1971a; Kvetňanský, 1973; see Fig. 4).

Many studies have consistently shown that the neuronal input to the adrenal is essential for the regulation of TH and DBH by a variety of other stressors, such as cold stress. This input is also required for the stress induced changes in TH mRNA levels (Stachowiak *et al.*, 1986). In this regard, the ganglionic blocking agent chlorisondamine, was able to inhibit most of the elevation of TH mRNA in response to cold stress (Stachowiak *et al.*, 1988).

Humoral regulation. The adrenal medulla was long held to be controlled only by nerve impulses carried to the medulla by the preganglionic sympathetic nerves. In 1966 Wurtman and Axelrod reported that glucocorticoids, which are secreted from the adrenal cortex, regulate the activity of PNMT in the adrenal medulla. The humoral regulation markedly influences stress-induced changes in adrenal catecholamines and in the catecholamine biosynthetic enzymes as summarized in Fig. 4. The activities of adrenal TH (Kvetňanský *et al.*, 1970a), DBH (Gewirtz *et al.*, 1971; Weinshilboum and Axelrod, 1970; Kvetňanský, 1973) and PNMT (Wurtman and Axelrod, 1966) decreased following hypophysectomy (hypox). In hypox rats, the TH and DBH activities were significantly elevated by repeated immobilization, but failed to attain the levels found in the adrenals of sham-operated rats. PNMT activity, however, was not elevated from the very low level by immobilization of hypox rats (Fig. 4).

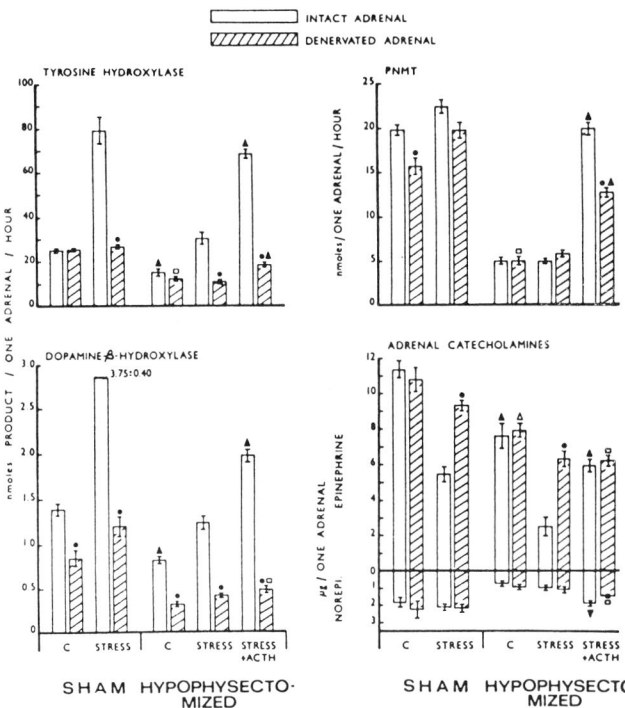

Figure 4. Effect of adrenal denervation (severing of nerve fibers from the superior mesenteric ganglia about 0.5 cm above the left adrenal; right adrenal served as intact control) on catecholamine-synthesizing enzyme activity in the adrenals of sham-operated and hypophysectomized rats exposed to stress. Animals were stressed by immobilization for 2.5 h daily and killed 6 h after the seventh immobilization (14 days after hypophysectomy and 9 days after denervation). ACTH (Acthar gel, 5 i.u. per rat, s.c.) was administered 60 min before each immobilization. Results are expressed as the mean values of 6–10 animals (±SEM). C = control. Statistical significance compared to intact adrenal: ⊗ = $p < 0.01$; ● = $p < 0.001$. Statistical significance compared to hypophysectomized + immobilized intact adrenal: △ = $p < 0.01$; ▲ = $p < 0.001$. Statistical significance compared to hypophysectomized + immobilized denervated adrenal: □ = NS; △ = $p < 0.05$; ⬭= $p < 0.001$. (From Kvetňanský, 1973; with permission.)

Adrenal denervation in hypox rats prevented the decrease of adrenal epinephrine, as well as the increase in adrenal TH and DBH usually seen after repeated immobilization (Kvetňanský *et al.*, 1970a; Gewirtz *et al.*, 1971). PNMT activity in the adrenals of immobilized hypox rats remained depressed in both intact and denervated adrenals. If ACTH (5 IU, s.c.) was administered to hypox rats before each immobilization, a marked increase in TH activity in the intact adrenal, and a small, though significant increase in TH was seen in the denervated adrenal (Kvetňanský *et al.*, 1970a; Kvetňanský, 1973; Fig. 4).

Thus, both neural and humoral factors appear to regulate TH and DBH activities in the adrenal medulla under stress, but the neural component appears to be the more important.

Treatment of hypox rats with ACTH restores PNMT in the intact gland to levels produced by repeated immobilization in normal rats, but denervation partially interferes with such a restoration of the enzyme (Kvetňanský et al., 1970a) (Fig. 4). Thus, humoral factors are more important than neural in determining the PNMT activity.

It was reported that the splanchnic nerve innervating the adrenal medulla activates or induces de novo synthesis of catecholamine synthesizing enzymes, while glucocorticoids predominantly inhibit degradation of PNMT and probably also DBH (Ciaranello, 1980). On the other hand glucocorticoids have also been shown to modulate the transsynaptic induction of TH in vivo during stress (Otten and Thoenen, 1975). The transsynaptic induction of TH appears to be mainly due to stimulation of acetylcholine nicotinic receptors. However VIP is thought to be as potent as acetylcholine in inducing the secretion of catecholamines from rat adrenal medulla (Malhotra et al., 1988). Costa et al. (1977) described the cascade mechanism involved in activation of adrenal TH by cold. Adrenal denervation substantially reduced all the steps involved in this process.

In summary, both neural and humoral components are involved in the regulation of adrenal catecholamine-synthesizing enzymes under stress. TH and DBH activities are affected by nerve activity and to a minor extent by the pituitary-adrenocortical axis. PNMT is predominantly regulated by glucocorticoids and to a small degree by nerve activity (reviewed in Kvetňanský, 1973; Axelrod and Reisine, 1984).

The mechanisms of regulation of TH, DBH and PNMT in the adrenal medulla of rats and mice under stress are also determined by genetic factors (Ciaranello et al., 1972; Stolk et al., 1980). The elevated activity of all three adrenal enzymes, induced by repeated immobilization stress, in the F344 strain of rats was completely dependent on neural regulation. However in another strain of rats (LEW) it was dependent on endocrine regulation (Stolk et al., 1980). By breeding these two strains of rats, hybrid rats were obtained which responded to repeated immobilization stress, in a similar way as did Sprague Dawley rats (used in our experiments), i.e. changes in TH and DBH were mainly dependent on neuronal input and PNMT changes on humoral regulation.

Central regulation in stress

Which areas of the CNS are responsible or participate in the stress-induced changes of catecholamine-synthesizing enzymes in the adrenal medulla? To address this question rats were spinalized at the thoracic level (Sourkes, 1985). For a few days after spinal transection adrenal TH activity changed little, but later on there was a gradual decrease in adrenal TH activity which could be blocked by splanchnictomy. Animals with one-side sectioned spinal cord (usually at the C_6–C_7) continued to show stress-induced increases in adrenal TH activity but only about a half of that obtained with sham operated animals (Gagner *et al.*, 1985). Concerning the neural regulation of the adrenal medulla, it is evident that the preganglionic sympathetic neurons that constitute the adrenal branch of the splanchnic nerve are located in the intermediolateral cell column of the spinal cord (Saper *et al.*, 1976; Gauthier *et al.*, 1979; Strack *et al.*, 1989). Strack *et al.* (1989) described five CNS areas that innervate sympathoadrenal preganglionic neurons: the caudal raphe nuclei, the ventromedial medulla, the rostral ventrolateral medulla, the A_5 noradrenergic cell group and the paraventricular hypothalamic nucleus. The anatomical organization of these different inputs is by itself complicated. A system of spinal interneurons also seems to be involved in this regulation (Strack *et al.*, 1989). The trajectory of descending pathways which terminate on the intermediolateral preganglionic cells is therefore still not clear.

Central regulation of catecholamine-synthesizing enzymes under stress is even less understood. Palkovits *et al.* (1984) examined the effect of hypothalamic deafferentation or transection of midbrain or medulla oblongata on immobilization induced elevation of catecholamine enzymes in the adrenal medulla. Deafferentation of the medial hypothalamus diminished immobilization-induced PNMT activity without affecting, adrenal TH or DBH activity. So hypothalamic cells which synthesize CRH (in the paraventricular nucleus), vasopressin (in the paraventricular and supraoptic nuclei), somatostatin (in the periventricular cells) or VIP may participate in the control of adrenal PNMT activity under stress via the pituitary-adrenocortical axis. Transection of the descending forbrain and limbic pathways at the midbrain-pontine junction lowered adrenal TH and DBH activities, but not PNMT activity, during repeated immobilization. Transection of the medulla oblongata influenced adrenal DBH activity but not that of TH and PNMT. These data suggest that an intact hypothalamo-pituitary connection is important for ACTH-induced elevated PNMT activity in repeated immobilization stress, while TH and DBH activities are mainly under spinal cord sympathetic control, which may be influenced by descending neuronal inputs (Palkovits *et al.*, 1984). Similar findings were reported following repeated immobilization in rats with septal or amygdaloid lesions (Kvetňanský, 1973). The pathway which initiates induction of adrenal TH by immobilization stress does not pass through a dopamine system (Sourkes, 1985).

Second messengers in regulation of catecholamine biosynthetic enzymes by stress

The transsynaptic induction of TH appears to be mainly due to stimulation of acetylcholine nicotinic receptors, since it is inhibited by ganglionic blockers.

However, treatment of rats with nicotine alone required high doses and long times to elevate TH and DBH activities (Slotkin and Seidler, 1976), although recently rapid and transient increases in TH transcription, mRNA levels and protein have been reported in nicotine treated rats (Fossom *et al.*, 1991). It appears that other components released from the splanchnic nerve, such as VIP or substance P, may also be important in the regulation of the adrenal catecholamine biosynthetic pathway in stress. Substance P has been found to modulate the nicotinic secretory response of bovine adrenal chromaffin cells both by inhibiting nicotinic activation of chromaffin cells and by protecting against desensitization of the nicotinic response (Livett *et al.*, 1979; Khalil *et al.*, 1988). VIP is thought to be as potent as acetylcholine in inducing the secretion of catecholamines from rat adrenal medulla (Malhotra *et al.*, 1988). Interestingly recent studies indicate that VIP and nicotine have a synergistic effect on the regulation of TH in adrenal chromaffin cells implying the potential importance of acetylcholine and its co-transmitter VIP for transsynaptic induction of TH (Olasmaa *et al.*, 1991).

The genes for TH, DBH and PNMT have been found to contain several regulatory genomic elements (Cambi *et al.*, 1989; D'Mello *et al.*, 1989; Kobayashi *et al.*, 1989; Ross *et al.*, 1990; McMahon and Sabban, 1992b). It will be interesting to determine which of these elements are involved in the regulation by stress.

Of particular interest is what constitutes the difference between a single and repeated immobilization. Clearly there is some kind of "memory"or change in the cells in response to a second repeated immobilization such that the response is different to that of a single immobilization, both in terms of sustained elevation of TH mRNA with repeated immobilizations as well as induction of DBH mRNA levels with repeated but not with a single stress (see Figs 2 and 3, and McMahon *et al.*, 1992a). Perhaps isolation of DNA binding protein in single versus repeated stress may enable a glimpse at the mechanism of adjustment to the longer treatment.

Acknowledgements

We gratefully acknowledge the support of grant NS28869 from the National Institutes of Health (to ELS). Dr Esther L. Sabban is the recipient of NIH Research Career Development Award, NS01121. We thank Dr Anne McMahon for help in preparation of the manuscript.

REFERENCES

Abercrombie, E. D., Nisenbaum, K. L. and Zigmond, M. J. (1992). Impact of acute and chronic stress on the release and synthesis of norepinephrine in brain: Microdialysis studies in behaving animals. In: *Stress: Neuroendocrine and Molecular Approaches*. R. Kvetňanský, R. McCarty and J. Axelrod (Eds). Gordon and Breach Sci. Publ., New York, pp. 29–42.

Adell, A., Garcia-Marquez, C., Armario, A. and Gelpi, D. (1988). Chronic stress increases sero-tonin and noradrenaline in rat brain and sensitizes their responses to a further acute stress. *J. Neurochem.* **50**, 1678–1681.

Angel, P., Imagawa, M., Chiu, R., Stein, B., Imbra, R. J., Rahmsdorf, H. J., Jonat, C., Herrlich, P. and Karin, M. (1987). Phorbol esther inducible-genes contain a common cis element recognised by a TPA-modulated trans-acting factor. *Cell* **49**, 729–739.

Axelrod, J. (1972). Dopamine-β-hydroxylase: Regulation of its synthesis and release from nerve terminals. *Pharmacol. Rev.* **24**, 233–243.

Axelrod, J., Mueller, R. A., Henry, J. P. and Stephens, P. M. (1970). Changes in enzymes involved in the biosynthesis and metabolism of noradrenaline and adrenaline after psychosocial stimulation. *Nature* **225**, 1059–1060.

Axelrod, J. and Reisine, T. D. (1984). Stress hormones: Their interaction and regulation. *Science* **224**, 452–459.

Badoyannis, H., Sharma, S. C. and Sabban, E. L. (1991). The differential effects of cell density and NGF on the expression of tyrosine hydroxylase and dopamine-β-hydroxylase in PC12 cells. *Molecular Brain Res.* **11**, 79–87.

Balaz, V., Balazova, E., Blazicek, P. and Kvetňanský, R. (1980). The effect of one-year acclimatiza-tion of rats to mountain conditions on plasma catecholamines and dopamine-β-hydroxylase ac-tivity. In: *Catecholamines and Stress: Recent Advances.* E. Usdin, R. Kvetňanský and I. J. Kopin (Eds). Elsevier North Holland, New York, pp. 259–264.

Baruchin, A. M., Weisberg, E. P., Miner, L. L., Ennis, D., Nisenbaum, L. K., Naylor, E., Strick-er, E. M., Zigmond, M. J. and Kaplan, B. B. (1990). Effects of cold exposure on rat adrenal tyrosine hydroxylase: An analysis of RNA, protein, enzyme activity and cofactor levels. *J. Neu-rochem.* **54**, 1769–1775.

Bhagat, B. D. and Horenstein, S. (1976). Modulation of adrenal medullary enzymes by stress. In: *Catecholamines and Stress.* E. Usdin, R. Kvetňanský and I. J. Kopin (Eds). Pergamon Press, Oxford, pp. 257–264.

Blazicek, P. and Kvetňanský, R. (1989). Kinetic parameters of rat adrenal TH and PNMT under acute and repeated stress. In: *Stress: Neurochemical and Humoral Mechanisms.* G. R. Van Loon, R. Kvetňanský, R. McCarty and J. Axelrod (Eds). Gordon and Breach Sci. Publ., New York, pp. 787–797.

Blazicek, P., Kvetňanský, R., Tigranjan, R. A., Langos, J. and Macho, L. (1980). Catecholamines and their synthesizing enzymes in rats exposed to prolonged hypokinesia. In: *Catecholamines and Stress Recent Advances.* E. Usdin, R. Kvetňanský and I. J. Kopin (Eds). Elsevier North Holland, New York, pp. 349–354.

Cambi, F., Fung, B. and Chikaraishi, D. M. (1989). 5' Flanking DNA sequences direct cell-specific expression of rat tyrosine hydroxylase. *J. Neurochem.* **53**, 1656–1659.

Carrol, J. M., Kim, W. S., Kim, K. T., Goodman, H. M. and Joh, T. H. (1991). Effects of sec-ond messenger system activation on functional expression of tyrosine hydroxylase fusion gene constructs in neuronal and non-neuronal cells. *J. Mol. Neurosci.* **3**, 65–74.

Chuang, D. and Costa, E. (1974). Biosynthesis of tyrosine hydroxylase in rat adrenal medulla after exposure in cold. *Proc. Natl. Acad. Sci. USA* **71**, 4570–4574.

Chuang, D., Zsilla, G. and Costa, E. (1975). Turnover rate of tyrosine hydroxylase during trans-synaptic induction. *Mol. Pharmacol.* **11**, 784–794.

Ciaranello, R. D. (1980). Regulation of adrenal catecholamine biosynthetic enzymes: Integration of neuronal and humoral stimuli in response to stress. In: *Catecholamines and Stress: Recent Advances.* E. Usdin, R. Kvetňanský and I. J. Kopin (Eds). Elsevier North Holland, New York, pp. 317–327.

Ciaranello, R. D., Dornbush, J. N. and Barchas, J. D. (1972). Regulation of adrenal phenylethanol-amine N-methyltransferase activity in three inbred mouse strains. *Mol. Pharmacol.* **8**, 511–520.

Coker III, G. T., Vinnedge, L. and O'Malley, K. (1988). Characterization of rat and human tyrosine hydroxylase genes: Functional expression of both promoters in neuronal and non-neuronal cell types. *Biochem. Biophys. Res. Commun.* **157**, 1341–1347.

Cooper, D. O. and Stolk, J. M. (1979). Differences between inbred rat strains in the alteration of adrenal catecholamine synthesizing enzyme activities after immobilization stress. *Neurosci.* **4**, 1163–1172.

Costa, E., Chuang, D. M., and Guidotti, A. (1977). Adrenal medulla: A model to study the trans-synaptic regulation of gene expression. In: *Structure and Function of Monoamine Enzymes.* E. Usdin, N. Weiner, M. B. H. Youdim (Eds). Marcel Dekker, New York, pp. 279–310.

Costa, E., Guidotti, A. and Hanbauer, I. (1974). Do cyclic nucleotides promote the transsynaptic induction of tyrosine hydroxylase? *Life Sci.* **14**, 1169–1188.

Costa, E. and Meek, J. L. (1974). Regulation of biosynthesis of catecholamines and serotonin in the CNS. *Ann. Rev. Pharmacol.* **14**, 491–511.

D'Mello, S. R., Turzai, L. M., Gioio, A. E. and Kaplan, B. B. (1989). Isolation and structural characterization of the bovine tyrosine hydroxylase gene. *J. Neurosci. Res.* **23**, 31–40.

Fader, D. and Lewis, E. J. (1990). Interaction of cyclic AMP and cell-cell contact in the control of tyrosine hydroxylase RNA. *Mol. Brain Res.* **8**, 25–29.

Fluharty, S. J., Snyder, G. L., Stricker, E. M. and Zigmond, M. J. (1983). Short- and long-term changes in adrenal tyrosine hydroxylase activity during insulin-induced hypoglycemia and cold stress. *Brain Res.* **267**, 384–387.

Fluharty, S. J., Snyder, G. L, Zigmond, M. J. and Stricker, E. M. (1985a). Tyrosine hydroxylase activity and catecholamine biosynthesis in the adrenal medulla of rats during stress. *J. Pharmacol. and Exp. Ther.* **233**, 32–38.

Fluharty, S. J., Rabow, L. E, Zigmond, M. J. and Stricker, E. M. (1985b). Tyrosine hydroxy-lase activity in the sympathoadrenal system under basal and stressful conditions: Effect of 6-hydroxydopamine. *J. Pharmacol. Exp. Ther.* **235**, 354–360.

Fossom, L. H., Carlson, C. D. and Tank, A. W. (1991). Stimulation of tyrosine hydroxylase gene transcription by nicotine in rat adrenal medulla. *Mol. Pharmacol.* **40**, 193–202.

Fukuhara, K., Kvetňanský, R., Weise, V. K., Ohara, H., Yoneda, R. and Kopin, I. J. (1992). Corre-lation of plasma catecholamine levels with tissue tyrosine hydroxylase activity in SART-stressed rats. In: *Stress : Neuroendocrine and Molecular Approaches.* R. Kvetňanský, R. McCarty and J. Axelrod (Eds). Gordon and Breach Sci. Publ., New York, pp. 881–889.

Gagner, J. P., Gauthier, S. and Sourkes, T. L. (1985). Descending spinal pathways mediating the response of adrenal tyrosine hydroxylase and catecholamines to insulin and 2-deoxyglucose. *Brain Res.* **325**, 187–197.

Gauthier, S., Gagner, J. P. and Sourkes, T. L. (1979). Role of descending spinal pathways in the regulation of adrenal tyrosine β-hydroxylase. *Exp. Neurol.* **66**, 42–54.

Gewirtz, G. P., Kvetňanský, R., Weise, V. K. and Kopin, I. J. (1971). Effect of hypophysectomy on adrenal dopamine-β-hydroxylase activity in the rat. *Mol. Pharmacol.* **7**, 163–168.

Gizang-Ginsberg, E. and Ziff, E. B. (1990). Nerve growth factor regulates tyrosine hydroxylase gene transcription through a nucleoprotein complex that contains c-fos. *Genes and Develop.* **4**, 477–491.

Gordon, R., Spector, S., Sjoerdsma, A. and Udenfriend, S. (1966). Increased synthesis of norepinephrine and epinephrine in the intact rat during exercise and exposure to cold. *J. Pharmacol. Exp. Ther.* **153**, 440–447.

Henry, J. P., Kross, M. E., Stephens, P. M. and Watson, F. M. C. (1976). Evidence that differing psychosocial stimuli lead to adrenal cortical stimulation by autonomic or endocrine pathways. In: *Catecholamines and Stress.* E. Usdin, R. Kvetňanský and I. J. Kopin (Eds). Pergamon Press, Oxford, pp. 457–466.

Henry, J. P., Stephens, P. M., Axelrod, J. and Mueller, R. A. (1971). Effect of psychosocial stimulation of the enzymes involved in the biosynthesis and metabolism of noradrenaline and adrenaline. *Psychosomatic Med.* **33**, 227–237.

Hoeldtke, R., Lloyd, T. and Kaufman, S. (1974). An immunochemical study of the induction of tyrosine hydroxylase in rat adrenal glands. *Biochem. Biophys. Res. Commun.* **57**, 1045–1053.

Holland, W. C. and Schumann, H. J. (1956). Formation of catecholamines during splanchnic stimulation of the adrenal gland of the cat. *Br. J. Pharmacol.* **11**, 449–453.

Iuvone, P. M. and Dunn, A. J. (1986). Tyrosine hydroxylase activation in mesocortical 3,4-dihydroxyphenylethylamine neurons following footshock. *J. Neurochem.* **47**, 837–844.

Kaplan, B. B., Stachowiak, M. K., Stricker, E. M. and Zigmond M. J. (1987). Regulation of adrenal tyrosine hydroxylase gene expression during cold stress. In: *Amino Acids in Health and Disease; New Perspectives.* S. Kaufman (Ed.). Alan R. Liss, New York, pp. 285–302.

Kenessey, A. and Huszti, Z. (1976). The effect of monoamine oxidase inhibitors on the synthesis and degradation of catecholamines in immobilized rats. In: *Catecholamines and Stress.* E. Usdin, R. Kvetňanský and I. J. Kopin (Eds). Pergamon Press, Oxford, pp. 331–340.

Khalil, Z., Marley, P. D. and Livett, B. G. (1988). Effect of substance P on nicotine-induced desensitization of cultured bovine adrenal chromaffin cells: possible receptor subtypes. *Brain Res.* **459**, 282–288.

Kilbourne, E. J., Nankova, B., Lewis, E., McMahon, A., Osaka, H., Sabban, D. B. and Sabban, E. L. (1992). Regulated expression of the tyrosine hydroxylase gene by membrane depolarization: Identification of the responsive element and possible second messengers. *J. Biol. Chem.* (submitted).

Kilbourne, E. and Sabban, E. L. (1990). Differential effect of membrane depolarization on levels of tyrosine hydroxylase and dopamine-β-hydroxylase mRNAs in PC12 pheochromocytoma cells. *Mol. Brain Res.* **8**, 121–27.

Kim, K. S., Carroll, J. M., Lee, M. K., Park, D. H. and Joh, T. H. (1991). Functional dissection and molecular analysis of the upstream sequence of tyrosine hydroxylase (TH) gene. *Soc. for Neurosci.* **17**, Abstract number 211.1.

Kiran, B. K. and Ulus, I. H. (1992). Selective response of rat peripheral sympathetic nervous system to various stress situations. In: *Stress: Neuroendocrine and Molecular Approaches.* R. Kvetňanský, R. McCarty and J. Axelrod (Eds). Gordon and Breach Sci. Publ., New York, pp. 561–568.

Kobayashi, K., Kurosawa, Y., Fujita, K. and Nagatsu, T. (1989). Human dopamine-β-hydroxylase gene: Two mRNA types having different 3′-terminal regions are produced through alternative polyadenylation. *Nucleic Acid Res.* **17**, 1089–1102.

Kvetňanský, R. (1973). Transsynaptic and humoral regulation of adrenal catecholamine synthesis in stress. In: *Frontiers in Catecholamine Research.* E. Usdin and S. Snyder (Eds). Pergamon Press, Oxford, pp. 223–229.

Kvetňanský, R., Albrecht, I., Torda, T., Saleh, N., Jahnova, E. and Mikulaj, L. (1976a). Effect of stress on catecholamine synthesizing and degrading enzymes in control and spontaneously hypertensive rats. In: *Catecholamines and Stress.* E. Usdin, R. Kvetňanský and I. J. Kopin (Eds). Pergamon Press, Oxford, pp. 237–247.

Kvetňanský, R., Gewirtz, G. P., Weise, V. K. and Kopin, I. J. (1970a). Effect of hypophysectomy on immobilization-induced elevation of tyrosine hydroxylase and phenylethanolamine N-methyltransferase in rat adrenal. *Endocrinology* **87**, 1323–1329.

Kvetňanský, R., Gewirtz, G. P., Weise, V. K. and Kopin, I. J. (1971a). Enhanced synthesis of adrenal dopamine-β-hydroxylase induced by repeated immobilization in rats. *Mol. Pharmacol.* **7**, 81–86.

Kvetňanský, R., Gewirtz, G. P., Weise, V. K. and Kopin, I. J. (1971b). Catecholamine-synthesizing enzymes in the rat adrenal gland during exposure to cold. *Am. J. Physiol.* **220**, 928–931.

Kvetňanský, R., McCarty, R. and Axelrod, J. (Eds) (1992). *Stress: Neuroendocrine and Molecular Approaches.* Gordon and Breach Sci. Publ., New York, pp. 1–1028.

Kvetňanský, R. and Mikulaj, L. (1970). Adrenal and urinary catecholamines in rats during adaptation to repeated immobilization stress. *Endocrinology* **87**, 738–743.

Kvetňanský, R., Mitro, A., Palkovits, M., Brownstein, M., Torda, T., Vigas, M. and Mikulaj, L. (1976b). Catecholamines in individual hypothalamic nuclei in stressed rats. In: *Catecholamines and Stress.* E. Usdin, R. Kvetňanský and I. J. Kopin (Eds). Pergamon Press, Oxford, pp. 39–50.

Kvetňanský, R., Nemeth, S., Vigas, M., Oprsalova, Z. and Jurcovicova, J. (1984). Plasma catecholamines in rats during adaptation to intermittent exposure to different stressors. In: *Stress: The Role of Catecholamines and Other Neurotransmitters,* **Vol. 1**. E. Usdin, R. Kvetňanský and J. Axelrod (Eds). Gordon and Breach Sci. Publ., New York, pp. 537–562.

Kvetňanský, R., Silbergeld, S., Weise, V. K. and Kopin, I. J. (1971c). Effects of restraint on rat adrenomedullary respones to 2-deoxy-D-glucose. *Psychopharmacologia (Berlin)* **20**, 22–31.

Kvetňanský, R., Sun, C. L., Lake, C. R., Thoa, N. B., Torda, T. and Kopin, I. J. (1978). Effect of handling and forced immobilization on rat plasma levels of epinephrine, norepinephrine and dopamine-β-hydroxylase. *Endocrinology* **103**, 1868–1874.

Kvetňanský, R., Torda, T., Macho L., Tigranjan, R. A., Serova, L. and Genin, A. M. (1981a). Effect of weightlessness on sympathetic-adrenomedullary activity of rats. *Acta Astronautica* **8**, 469–481.

Kvetňanský, R., Vigas, M., Tigranjan, R. A., Nemeth, S. and Macho, L. (1981b). Activity of the sympathetic-adrenomedullary system in rats after space flight on the Cosmos biosatellites. *Adv. Space Res.* **1**, 187–192.

Kvetňanský, R., Weise, V. K., Gewirtz, G. P. and Kopin, I. J. (1971d). Synthesis of adrenal catecholamines in rats during and after immobilization stress. *Endocrinology* **89**, 46–49.

Kvetňanský, R., Weise, V. K. and Kopin, I. J. (1969). Effect of repeated immobilization stress on rat adrenal tyrosine hydroxylase, dopamine-β-hydroxylase and phenylethanolamine N-methyltransferase. *Pharmacologist* **11**, 274.

Kvetňanský, R., Weise, V. K. and Kopin, I. J. (1970b). Elevation of adrenal tyrosine hydroxylase and phenylethanolamine-N-methyltransferase by repeated immobilization of rats. *Endocrinology* **87**, 744–749.

Lamprecht, F., Eichelman, B., Thoa, N. B., Williams, B. B and Kopin, I. J. (1972). Rat fighting behavior: Serum dopamine-β-hydroxylase and hypothalamic tyrosine hydroxylase. *Science* **177**, 1214–1215.

Levitt, M., Spector, S., Sjoerdsma, A. and Udenfriend, S. (1965). Elucidation of the rate-limiting step in norepinephrine biosynthesis in the perfused guinea-pig heart. *J.Pharmacol. Exp. Ther.* **148**, 1–8.

Lewis, E. J. and Chikaraishi, D. M. (1987). Regulated expression of the tyrosine hydroxylase gene by epidermal growth factor. *Mol. Cell. Biol.* **7**, 3332–3336.

Lewis, E., Harrington, C. A. and Chikaraishi, D. M. (1987). Transcriptional regulation of the tyrosine hydroxylase gene by glucocorticoid and cAMP. *Proc. Natl. Acad. Sci. USA* **84**, 3550–3554.

Livett, B. G., Kozousek, V., Mizobe, F. and Dean, D. M. (1979). Substance P inhibits nicotinic activation of chromaffin cells. *Nature* **278**, 256–257.

Maengwyn-Davies, G. D., Johnson, D. G., Thoa, N. B., Weise, V. K. and Kopin, I. J. (1973). Influence of isolation and of fighting on adrenal tyrosine hydroxylase and phenylethanolamine N-methyltransferase activities in three strains of mice. *Psychopharmacologia* **28**, 339–350.

Malhotra, R. K., Wakade, T. D. and Wakade, A. R. (1988). Vasoactive intestinal polypeptide and muscarine mobilises intracellular Ca^{2+} through breakdown of phosphoinositides to induce catecholamine secretion. Role of IP3 in exocytosis. *J.Biol. Chem.* **262**, 2123–2126.

Masserano, J. M., Takimoto, G. S. and Weiner, N. (1981). Electroconvulsive shock increases tyrosine hydroxylase activity in the brain and adrenal gland of the rat. *Science* **214**, 662–665.

Masserano, J. M., Vulliet, P. R., Tank, A. W. and Weiner, N. (1989). The role of tyrosine hydroxylase in the regulation of catecholamine synthesis. In: *Catecholamines II, Handbook of Experimental Pharmacology,* **Vol. 90/II,** *Trendelenburg.* N. Weiner (Ed.). Springer-Verlag, Berlin, pp.427–469.

Masserano, J. M. and Weiner, N. (1979). The rapid activation of adrenal tyrosine hydroxylase by decapitation and its relationship to a cyclic AMP-dependent phosphorylating mechanism. *Mol. Pharmacol.* **16**, 513–528.

Masserano, J. M. and Weiner, N. (1981). The rapid activation of tyrosine hydroxylase by the subcutaneous injection of formaldehyde. *Life Sci.* **29**, 2025–2029.

McCarty, R. and Stone, E. A. (1984). Chronic stress and regulation of the sympathetic nervous system. In: *Stress: The Role of Catecholamines and Other Neurotransmitters,* **Vol. 1.** E. Usdin, R. Kvetňanský and J. Axelrod (Eds). Gordon and Breach Sci. Publ., New York, pp. 563–576.

McMahon, A., Geertman, R. and Sabban, E. L. (1990a). Rat dopamine-β-hydroxylase: Molecular cloning and characterization of the cDNA and regulation of the mRNA by reserpine. *J. Neurosci. Res.* **25**, 395–404.

McMahon, A., Badoyannis, H., Sharma, S. C. and Sabban, E. L. (1990b). Parallel regulation of mRNAs for dopamine-β-hydroxylase and tyrosine hydroxylase by dexamethasone, cAMP and growth factors in PC12 cells. *Soc. for Neurosci.* **16**, Abstract number 397.4.

McMahon, A., Kvetňanský, R., Fukuhara, K., Weise, V. K., Kopin, I. J. and Sabban, E. L. (1992b). Regulation of tyrosine hydroxylase and dopamine-β-hydroxylase mRNA levels in rat adrenals by a single and repeated immobilization stress. *J. Neurochem.* **58**, 2124–2130.

McMahon, A. and Sabban, E. L. (1992b). Regulation of expression of dopamine-β-hydroxylase in PC12 cells by glucocorticoids and cAMP analogs. *J. Neurochem.* (submitted).

Mikulaj, L., Kvetňanský, R., Murgas, K., Parizkova, J. and Vencel, P. (1976). Catecholamines and corticosteroids in acute and repeated stress. In: *Catecholamines and Stress.* E. Usdin, R. Kvetňanský and I. J. Kopin (Eds). Pergamon Press, Oxford, pp. 445–454.

Miner, L. L., Baruchin, A. and Kaplan, B. B. (1992). Transsynaptic modulation of rat adrenal tyrosine hydroxylase gene expression during cold stress. In: *Stress: Neuroendocrine and Molecular Approaches.* R. Kvetňanský, R. McCarty and J. Axelrod (Eds). Gordon and Breach Sci. Publ., New York, pp. 313–324.

Miner, L. L., Pandalai, S. P. and Kaplan, B. B. (1991). Cold-induced alterations in the binding of adrenomedullary nuclear proteins to the promoter region of the tyrosine hydroxylase (TH) gene. *Soc. for Neurosci.* **17**, Abstract number 101.7.

Modigh, K. (1976). Long-term effects of electroconvulsive shock therapy on synthesis, turnover and uptake of brain monoamines. *Psychopharmacol.* **49**, 179–185.

Musacchio, J. M., Julou, L., Kety, S. S. and Glowinski, J. (1969). Increase in rat brain tyrosine hydroxylase activity produced by electroconvulsive shock. *Proc. Natl. Acad. Sci. USA* **63**, 1117–1119.

Nagatsu, T. and Kato, T. (1980). Catecholamine synthesizing enzymes of spontaneously hypertensive rats under stress. In: *Catecholamines and Stress: Recent Advances.* E. Usdin, R. Kvetňanský and I. J. Kopin (Eds). Elsevier North Holland, New York, pp. 339–348.

Nagatsu, T., Levitt, T. and Udenfriend, S. (1964). Tyrosine hydroxylase — The initial step in norepinephrine biosynthesis. *J. Biol. Chem.* **239**, 2910–2917.

Nestler, E. J., McMahon, A., Sabban, E. L., Tallman, J. F. and Duman, R. S. (1990). Chronic antidepressant administration decreases the expression of tyrosine hydroxylase in the rat *locus coeruleus. Proc. Natl. Acad. Sci. USA* **87**, 7522–7526.

Nisenbaum, L. K., Zigmond, M. J., Sved, A. F. and Abercrombie, E. D. (1991). Prior exposure to chronic stress results in enhanced synthesis and release of hippocampal norepinephrine in response to a novel stressor. *J. Neurosci.* **11**, 1478–1484.

Olasmaa, M., Guidotti, A., Grayson, D. and Costa, E. (1991). Regulation of medullary tyrosine hydroxylase (TH) expression by nicotinic and VIP receptor activation. *J. Neurochem.* **57**, S20B.

Otten, U., Paravicini, U., Oesch, F. and Thoenen, H. (1973). Time requirement for the single steps of transsynaptic induction of tyrosine hydroxylase in the peripheral sympathetic nervous system. *Naunyn-Schmiedeberg's Arch. Pharmacol.* **280**, 117–127.

Otten, U. and Thoenen, H. (1975). Circadian rhythm of tyrosine hydroxylase induction by short-term cold stress: Modulatory action of glucocorticoids in newborn and adult rats. *Proc. Natl. Acad. Sci. USA* **72**, 1415–1419.

Palkovits, M., Brownstein, M. J., Weise, V. K. and Kopin, I. J. (1984). Effect of central nervous system neurons on adrenal catecholamines under basal and stress conditions. In: *Stress: The Role of Catecholamines and Other Neurotransmitters, Vol. 2.* E. Usdin, R. Kvetňanský and J. Axelrod (Eds). Gordon and Breach Sci. Publishers, New York, pp. 609–616.

Palkovits, M., Kobayashi., R. M., Kizer, J. S., Jacobowitz, D. M. and Kopin, I. J. (1975). Effect of stress on catecholamines and tyrosine hydroxylase activity in individual hypothalamic nuclei. *Neuroendocrinol.* **18**, 144–153.

Parizkova, J. and Kvetňanský, R. (1980). Catecholamine metabolism and compositional growth in exercised and hypokinetic male rats. In: *Catecholamines and Stress: Recent Advances.* E. Usdin, R. Kvetňanský and I. J. Kopin (Eds). Elsevier North Holland, New York, pp. 355–358.

Petrovic, V. M. and Janic-Sibalic, V. (1980). Catecholamine synthesizing and degrading enzymes in the heat stressed or adapted rats. In: *Catecholamines and Stress: Recent Advances.* E. Usdin, R. Kvetňanský and I. J. Kopin (Eds). Elsevier North Holland, New York, pp. 365–370.

Pfeifer, W. D. (1976). Modification of adrenal tyrosine hydroxylase activity in rats following manipulations in infancy. In: *Catecholamines and Stress.* E. Usdin, R. Kvetňanský and I. J. Kopin (Eds). Pergamon Press, Oxford pp. 265–270.

Richard, F., Faucon-Biguet, N., Labatut, R., Rollet, D., Mallet, J. and Buda, M. (1988). Modulation of tyrosine hydroxylase gene expression in rat brain and adrenals by exposure to cold. *J. Neurosci. Res.* **20**, 32–37.

Roffman, M., Freedman, L. S. and Goldstein, M. (1973). The effect of acute and chronic swim stress on dopamine-β-hydroxylase activity. *Life Sci.* **12**, 369–376.

Ross, M. E., Evinger, M., Hyman, S. E., Carroll, J. M., Mucke, L., Comb, M., Reis, D. J., Joh, T. H. and Goodman, H. M. (1990). Identification of a functional glucocorticoid response element in the phenylethanolamine N-methyltransferase promoter using fusion genes introduced into chromaffin cells in primary culture. *J. Neurosci.* **10**, 520–530.

Sabban, E. L., Kvetňanský, R., McMahon, A., Fukuhara, K., Kilbourne, E. and Kopin, I. J. (1992). Stressors regulate mRNA levels of tyrosine hydroxylase and dopamine-β-hydroxylase in adrenals *in vivo* and in PC12 cells. In: *Stress: Neuroendocrine and Molecular Approaches.* R. Kvetňanský, R. McCarty and J. Axelrod (Eds). Gordon and Breach Sci. Publ., New York, pp. 325–335.

Saper, C. B., Loewy, A. D., Swanson, L. W. and Cowan, W. M. (1976). Direct hypothalamo-autonomic connections. *Brain Res.* **117**, 305–312.

Slotkin, T. A. and Seidler, F. J. (1976). Acute and chronic effects of nicotine on synthesis and storage of catecholamines in the rat adrenal medulla. *Life Sci.* **16**, 1613–1622.

Smith, M. A., Brady, L. S., Glowa, J., Gold, P. W. and Herkenham, M. (1991). Effects of stress and adrenalectomy on tyrosine hydroxylase mRNA levels in the *locus coeruleus* by *in situ* hybridization. *Brain Res.* **544**, 26–32.

Sourkes, T. L. (1985). Neurotransmitters and central regulation of adrenal functions. *Biol. Psychiatry* **20**, 182–191.

Stachowiak, M. K., Fluharty, S. J., Stricker, E. M., Zigmond, M. J. and Kaplan, B. B. (1986). Molecular adaptations in catecholamine biosynthesis induced by cold stress and sympathectomy. *J. Neurosci. Res.* **16**, 13–24.

Stachowiak, M. K., Jiang, H. K., Poisner, A. M., Tuominen, R. K. and Hong, J. S. (1990). Short and long-term regulation of catecholamine biosynthetic enzymes by angiotensin in cultured adrenal medullary cells. *J. Biol. Chem.* **265**, 4694–4702.

Stachowiak, M. K., Sebbane, R., Stricker, E. M., Zigmond, M. J. and Kaplan, B. B. (1985). Effect of chronic cold exposure on tyrosine hydroxylase mRNA in rat adrenal gland. *Brain Res.* **359**, 356–359.

Stachowiak, M. K., Stricker, E. M., Zigmond, M. J. and Kaplan, B. B. (1988). A cholinergic antagonist blocks cold stress-induced alterations in rat adrenal tyrosine hydroxylase mRNA. *Mol. Brain Res.* **3**, 193–196.

Stolk, J. M, Stolk, M. D., Hurst, J. H., Harris, P. Q. and Cooper, D. O. (1980). Genetic factors may determine the mechanisms responsible for altering rat adrenal gland catecholamine synthetic enzymes in response to stress. In: *Catecholamines and Stress: Recent Advances.* E. Usdin, R. Kvetňanský and I. J. Kopin (Eds). Elsevier North Holland, New York, pp. 329–338.

Stone, E. A., Freedman, L. S. and Morgano, L. E. (1978). Brain and adrenal tyrosine hydroxylase activity after chronic footshock stress. *Pharmacol. Biochem. Behav.* **9**, 551–553.

Stone, E. A. and McCarty, R. (1983). Adaptation to stress: Tyrosine hydroxylase activity and catecholamine release. *Neurosci. Biobehav. Rev.* **7**, 29–34.

Strack, A. V., Sawyer, W. B., Platt, K. B. and Loewy, A. D. (1989). CNS cell groups regulating the sympathetic outflow to adrenal gland as revealed by transneuronal cell body labeling with pseudorabies virus. *Brain Res.* **491**, 274–296.

Tank, A. W., Lewis, E. J., Chikaraishi, D. M. and Weiner, N. (1985). Elevation of RNA coding for tyrosine hydroxylase in rat adrenal gland by reserpine treatment and exposure to cold. *J. Neurochem.* **45**, 1030–1033.

Thoa, N. B., Tizabi, Y., Kopin, I. J. and Maengwyn-Davies, G. D. (1976). Alterations of mouse adrenal medullary catecholamines and enzymes in response to attack: Effect of pre- and post-treatment with phenobarbital. *Psychopharmacol.* **51**, 53–57.

Thoenen, H. (1970). Induction of tyrosine hydroxylase in peripheral and central adrenergic neurons by cold-exposure of rats. *Nature* **228**, 861–862.

Thoenen, H. (1972). Comparison between the effect of neuronal activity and nerve growth factor on the enzymes involved in the synthesis of norepinephrine. *Pharmacol. Rev.* **24**, 255–267.

Thoenen, H., Kettler, R., Burkard, W. and Saner, A. (1971). Neurally mediated control of enzymes involved in the synthesis of norepinephrine: are they regulated as an operational unit? *Naunyn-Schmiedeberg's Arch. Pharmacol.* **270**, 146–160.

Thoenen, H., Otten, U. and Schwab, M. (1979). Orthograde and retrograde signals for the regulation of neuronal gene expression: The peripheral sympathetic nervous system as a model. In: *The Neurosciences: Fourth Study Program.* F. O. Schmitt and F. G. Worden (Eds). MIT Press, Cambridge, MA, pp. 911–928.

Ulus, I. H. and Wurtman, R. J. (1979). Selective response of rat peripheral sympathetic nervous system to various stimuli. *J. Physiol. (London)* **293**, 513–523.

Usdin, E., Kvetňanský, R. and Axelrod, J. (Eds) (1984). *Stress: The Role of Catecholamines and Other Neurotransmitters, Vol. 1 and 2.* Gordon and Breach Sci. Publ., New York, pp. 1–1075.

Usdin, E., Kvetňanský, R. and Kopin, I. J. (Eds) (1976). *Catecholamines and Stress.* Pergamon Press, Oxford, pp. 1–631.

Usdin, E., Kvetňanský, R. and Kopin, I. J. (Eds) (1980). *Catecholamines and Stress: Recent Advances.* Elsevier North Holland, New York, pp. 1–618.

Van Loon, G. R., Kvetňanský, R., McCarty, R. and Axelrod, J. (Eds) (1989). *Stress: Neurochemical and Humoral Mechanisms, Vol. 1 and 2.* Gordon and Breach Sci. Publ., New York, pp. 1–1073.

Viveros, O. H., Arqueros, L. and Kirshner, N. (1968). Release of catecholamines and dopamine-β-hydroxylase from the adrenal medulla. *Life Sci.* **7**, 609–618.

Weiner, N., Lee, F. L., Dreyer, E. and Barnes, E. (1978). The activation of tyrosine hydroxylase in noradrenergic neurons during acute nerve stimulation. *Life Sci.* **22**, 1197–1215.

Weinshilboum, R. M. and Axelrod, J. (1970). Dopamine-β-hydroxylase activity in the rat after hypophysectomy. *Endocrinology* **87**, 894–900.

Weinshilboum, R., Kvetňanský, R., Axelrod, J. and Kopin, I. J. (1971). Elevation of serum dopamine-β-hydroxylase activity with forced immobilization. *Nature, New Biol.* **230**, 287–288.

Weisberg, E. P., Baruchin, A., Stachowiak, M. K., Stricker, E. M., Zigmond, M. J. and Kaplan, B. B. (1989). Isolation of a rat adrenal cDNA clone encoding phenylethanolamine N-methyltransferase and cold-induced alterations in adrenal PNMT mRNA and protein. *Molec. Brain Res.* **6**, 159–166.

Weiss, J. M., Glaser, H. I., Pohorecky, L. A., Brick, J. and Miller, N. E. (1975). Effect of chronic exposure to stressors on avoidance-escape behaviour and on brain norephinephrine. *Psychosomatic Med.* **37**, 522–534.

Wurtman, R. J. and Axelrod, J. (1966). Control of enzymatic synthesis of adrenaline in the adrenal medulla by adrenal cortical steroids. *J. Biol. Chem.* **241**, 2301–2305.

Zigmond, R. E. (1980). The long-term regulation of ganglionic tyrosine hydroxylase by preganglionic nerve activity. *Fed. Proc.* **39**, 3003–3008.

Zigmond, R. E., Chalazonitis, A. and Joh, T. (1980). Preganglionic nerve stimulation increases the amount of tyrosine hydroxylase in the rat superior cervical ganglion. *Neurosci. Letters* **20**, 61–65.

Zigmond, R. E., Schon, F. and Iversen, L. L. (1974). Increased tyrosine hydroxylase activity in the *locus coeruleus* of rat brain stem after reserpine treatment and cold stress. *Brain Res.* **70**, 547–552.

Zigmond, R. E., Schwarzchild, M. A. and Rittenhouse, A. R. (1989). Acute regulation of tyrosine hydroxylase by nerve activity and by neurotransmitters via phosphorylation. *Rev. Neurosci.* **12**, 415–461.

Tyrosine Hydroxylase, pp. 283–294
M. Naoi *et al.* (Eds)
© VSP 1993

TYROSINE HYDROXYLASE
IN NIGROSTRIATUM AND HYPOTHALAMUS:
LINKAGE TO THE CARDIOVASCULAR SYSTEM

NATALIE ALEXANDER

*University of Southern California School of Medicine, Department of Medicine,
Diabetes, Hypertension and Nutrition Division, Los Angeles, CA 90033, USA*

INTRODUCTION

It is my intention to present the major results of studies done over a period of years in collaboration with Dr Nagatsu and his colleagues. The studies all focused on a little understood connection between the cardiovascular system, represented centrally by arterial baroreceptor afferent pathways, and tyrosine hydroxylase (TH) activity in the major dopaminergic systems of the CNS.

The sinoaortic denervated (SAD) rat is a model of chronic arterial baroreceptor deficiency that shows elevated indices of catecholaminergic activity both peripherally (Alexander *et al.*, 1976; Alexander *et al.*, 1980) and centrally (Saavedra and Alexander, 1983; Nakata *et al.*, 1990). Cardiovascular characteristics of the SAD model is moderate hypertension and arterial pressure lability (Junqueira and Krieger, 1976; Alexander and Velasquez, 1980). The latter is movement related and typified by both dips and peaks in arterial pressure. Lesions of the A2 region of the nucleus of the tractus solitarius, a major region for termination of sinoaortic afferents, also induces arterial pressure lability (Snyder *et al.*, 1978).

We hypothesized that in the SAD model, pressure lability, in part, is due to the loss of integration between cortico-somatomotor and baroreceptor afferents. We tested this possibility by measuring TH activity, DA content and DA release in the nigrostriatal system of SAD rats (Alexander *et al.*, 1984a; Alexander *et al.*, 1987a, 1988). In addition, we tested for functional reciprocity between nigrostriatum and baroreceptor input by lesioning or stimulating the substantia nigra, pars compacta, of intact and SAD rats, respectively, and studying the effect on the cardiovascular system (Alexander *et al.*, 1986).

SAD rats show time-related, regionally specific changes of TH in the hypothalamus (Alexander and Morris, 1988). Recently, we examined the possibility that sinoaortic nerves affect TH activity differently on the 2 sides of the brain (Alexander *et al.*, 1990). The idea arose from the fact that, although carotid sinus nerves are alike bilaterally, the left aortic nerve is considerably larger than the right one. This anatomical difference in aortic nerves suggested that in SAD animals, effects produced by denervation of the left side would be greater than those from the right causing asymmetrical central changes. We studied this problem in 7-day left- and right-SAD rats by measuring TH on right and left sides of the brain in 4 hypothalamic regions; the assumption was that intact rats would have equal TH activity on the 2 sides of the brain.

METHODS

Tyrosine hydroxylase (TH)

The *in vivo* activity of TH was determined by measuring the rate of accumulation of L-3,4,dihydroxyphenylalanine (DOPA) after inhibition of DOPA decarboxylase by NSD 1015, 100 mg/kg, injected intraperitonal injection (i.p.) 30 min prior to decapitation. DOPA in micropunched brain samples was assayed by radioenzymatic assay as described elsewhere (Alexander *et al.*, 1990).

The *in vitro* activity of TH was determined by radiometric microassay of micropunched brain samples as described elsewhere (Alexander and Morris, 1988). The latter was modified into a micro-micro assay which increased sensitivity for small unilateral brain samples. TH protein was measured by a sensitive immunoassay (Makio *et al.*, 1984).

Catecholamines

Concentrations of catecholamines in micropunched brain samples were measured by one of 2 methods. DA and its metabolites, dihydroxyphenylacetic acid (DOPAC) and homovanillic acid (HVA) were measured by HPLC-EC (Alexander *et al.*, 1988). NE and E and DA, for hypothalamic studies, were measured by radioenzymatic assay (Alexander *et al.*, 1984b).

Microdialysis

A removable microdialysis probe with a 3 mm long dialysis bag at the tip was constructed (Ozaki *et al.*, 1987). One week prior to a study, rats were anesthetized with sodium pentobarbital and a guideshaft with occluder was implanted stereotaxically. On the study day, the occluder was removed and a probe inserted through the guide into the targeted striatal site (Alexander *et al.*, 1988).

Substantia nigra (SN) and striatal lesions

Bilateral radio frequency lesions (0.8 mA, 30 s duration) in anesthetized rats (Brevital, 50 mg/kg) were targeted for the SN ($n = 10$ acute studies; $n = 7$ chronic studies) or the adjacent ventral tegmentum (VTA) ($n = 2$ acute studies). The latter served as controls along with another control group ($n = 6$ acute studies;

n = 6 chronic studies) in which electrodes were lowered into the SN only. Another group (n = 4 chronic studies) received bilateral 6-OHDA injections (4 μg) into caudate nucleus (CN) of the striatum; controls received vehicle (n = 2) (Alexander *et al.*, 1986).

Sinoaortic denervation

Male Wistar rats, body weight 280–300 g, were used. The Krieger method of sinoaortic denervation was performed with the rats under Innovar anesthesia (0.03 ml/rat intramuscular injection (i.m.)) (Krieger, 1964). Sham-operated (SO) rats were anesthetized and neck tissues dissected.

RESULTS

Nigrostriatal studies

Study 1. Objective. To determine if SAD altered TH activity and DA in the nigrostriatum of SAD and SO rats sacrificed 3 days post-surgery. *In vitro* TH activity was measured in pooled micropunched samples of CN and SN, and DA in CN.

Results. The data in Fig. 1 show that TH activity is significantly reduced in caudate nucleus (CN) and SN of 3-day SAD rats. Furthermore, the concentration of DA is significantly reduced in CN of the SAD rats.

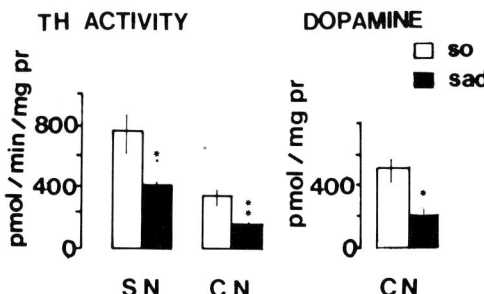

Figure 1. Tyrosine hydroxylase (TH) activity in caudate nucleus (CN) and in substantia nigra (SN) and dopamine (DA) concentration in CN of 3-day SAD and SO rats.

Study 2. Objective. To determine if the effect of SAD on striatal DA was regional and/or localized to specific intra-regional sites in 3-day SAD and SO rats.

Consecutive micropunched brain samples were obtained throughout the rostral-caudal (A–P) length of the dorsal, lateral and ventral regions of the CN. Two consecutive samples were pooled and the DA content was measured.

Results. The cumulative DA concentration for each region was significantly lower in SAD than SO rats as noted in Table 1. Figure 2 shows the A–P intra-regional concentrations of DA in the dorsal, lateral and ventral regions of CN. Asterisks indicate which local sites were significantly different between 3-day SAD and SO rats.

Table 1.

Dopamine in three areas of caudate nucleus in 3-day SAD and SO rats

	Dopamine (ng/mgm prot.)		
	Dorsal	Lateral	Ventral
SO (6)	74 ± 8	59 ± 6	52 ± 5
SAD (6)	47 ± 4	36 ± 3	29 ± 3
P	< 0.001	< 0.001	< 0.001

Combined values of 7 or 8 micropunched samples from consecutive slices of the CN.

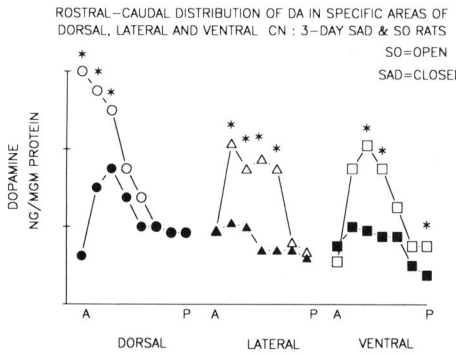

Figure 2.   Rostral-caudal (A–P) distribution of DA in specific sites within the dorsal, lateral and ventral regions of the CN: 3-day SAD (closed symbols) and SO (opens symbols) rats.

Study 3. Objective. To determine if the dynamic process of DA release is attenuated by SAD. For this, striatal extracellular DA was measured by microdialysis in SAD and SO freely moving rats. One to 7 days, average 4 days, after SAD or SO, a probe was inserted into the striatum and perfused with artificial CSF at

2 μl/min. Control dialysate samples were collected at 20 min intervals then rats received an injection of pargyline (75 mg/kg, i.p.) to block monoamine oxidase and thereby inhibit oxidation of DA. Samples were collected for the next 2 h.

Results. Figure 3 shows that striatal extracellular DA as well as its major metabolites DOPAC and HVA are significantly lower in 3-day SAD than SO rats. After pargyline treatment, striatal DA increased and reached a peak value at the end of 60 min in SO rats whereas in pargyline-treated SAD rats, DA reached a lower peak, sooner than in the SO rats (data not shown). The difference between rates of DOPA accumulation in pargyline-treated SAD and SO rats did not achieve statistical significance, but it was about 3 times faster in SO rats. (Regression analysis: 0.29 vs 0.087 pmols/h, SO vs. SAD.) Both DA metabolites disappeared from the dialysate within 80–100 min showing that pargyline had effectively inhibited monoamine oxidase function.

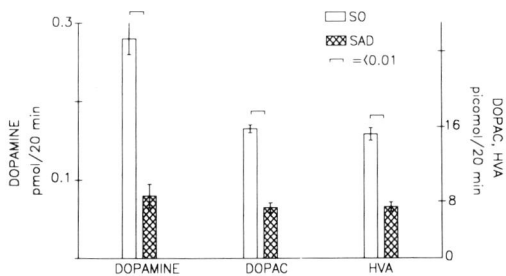

Figure 3. Dialysate concentrations of DA and dopamine metabolites, DOPAC and HVA, obtained by microdialysis of striatum in freely moving, 3-day SAD and SO rats.

Study 4. Objective. To determine if the baroreceptor-nigrostriatal pathways had functionally reciprocal activity. Two types of studies were done in which MAP, HR and motor behavior were monitored continuously from freely moving rats: 1) after production of bilateral SN lesions or 6-OHDA striatal lesions; 2–3 days (acute) or 10–14 days (chronic) and 2) before and after microinjection of apomorphine (APO), a DA agonist, into SN of 3-day SAD and SO rats.

The cardiovascular data were processed on line by computer with sampling parameters of 80 Hz, 0.1 s dur and 40 /min.

Results. Freely moving rats with radio-frequency lesions of the SN showed increased lability of MAP, as measured by the standard deviation (SD) of continuously recorded MAP, see Fig. 4. The increases ranged from 56 to 96% in

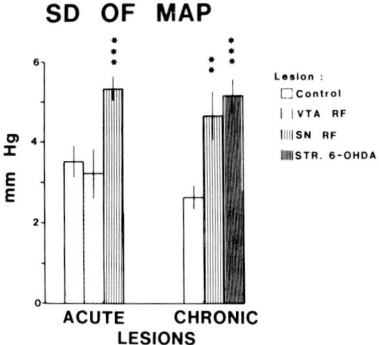

Figure 4. Standard deviation (SD) of MAP in freely moving rats with radiofrequency lesions of the substantia nigra (SN), or ventral tegmental area (VTA), or 6-OHDA lesions of striatum (STR); controls had electrodes lowered or vehicle injection only. (See Methods.)

rats with both acute and chronic lesions and were independent of nigrostriatal DA levels (data not shown). HR lability was not affected by SN lesions. Mean scores of motor activity were not significantly different for most activities, rearing, walking, etc. between SN lesioned rats and controls. Only frequency of head movement was higher in SN lesioned rats (data not shown).

Apomorphine, 20 μg microinjected into SN, produced stereotypical activity in both SO and SAD rats but only SAD rats showed a cardiovascular response, namely, a 30 min long reduction of arterial pressure and HR, see Fig. 5.

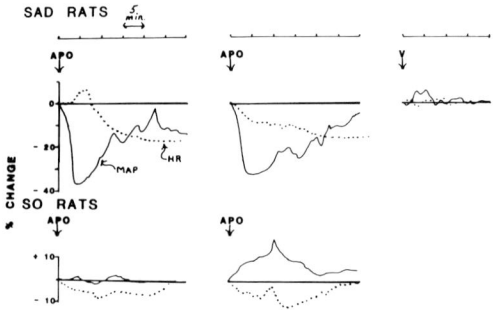

Figure 5. Effect of microinjections of apomorphine (20 μg) into SN on MAP (solid line) and HR (dotted line) of 2 SAD rats and one SO rat.

Hypothalamic studies

Objective. To determine if resection of the anatomically larger left sinoaortic nerves would have a greater influence than resection of right sinoaortic nerves on hypothalamic TH activity. In order to assess the effects of left or right SAD, we measured left and right TH activity in specific regions of the hypothalamus, assuming that TH activity would be the same on both sides of the brain in intact rats. Micropunch technique was used to collect right and left arcuate nucleus (ARC), median eminence (ME), paraventricular nucleus (PVN) and supraoptic nucleus (SON); regions that, in previous studies, showed a change in catecholamine metabolism induced by SAD (Alexander and Morris, 1988).

Results. Unexpectedly, we discovered that normal rats have lower TH activity on the left than right side of the ME and ARC, as measured by both *in vitro* and *in vivo* methods, whereas TH activity is the same on both sides of the PVN. See Tables 2 and 3. In an additional study of *in vivo* TH activity, asymmetry index (left/right ratio) for ME and SON were 0.62 ($p < 0.005$) and 0.98 (ns), respectively. Another unexpected finding was that the ARC of normal rats also had lower left than right side concentrations of norepinephrine (NE) and epinephrine (E) as well as DA; also left ME DA was lower than right, and no catecholamine asymmetry was found in PVN, see Fig. 6.

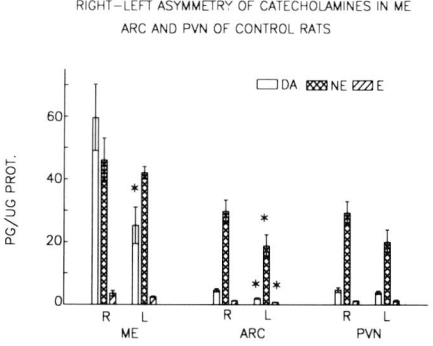

Figure 6. Right-left asymmetry of catecholamines in ME, ARC and PVN of control rats.

Table 3 shows the effects of unilateral SAD on hypothalamic bilateral *in vivo* TH activity. Right SAD (R-SAD) reduced contralateral TH activity in ARC and ME whereas left SAD (L-SAD) reduced TH on both sides of the brain and also raised TH activity in PVN; bilateral SAD had the same effect as L-SAD (data not shown).

Table 2.
in vitro TH activity* of right and left sides of the brain of male wistar
rats (pmols/min/mgm tissue prot.)

	Right Side	Left Side	Asymmetry index L/R × 100
ME	32 ± 4	16 ± 3	50
ARC	6 ± 0.7	3 ± 1	55
PVN	11 ± 3	10 ± 2	91

Means ± SEM (n = 6 − 7) *micro-micro assay

Table 3.
in vivo TH activity in SO, L-SAD* and R-SAD* RATS: dopa concentrations
on right and left sides of the brain

Region	Grp	DOPA (ng/mg prot.) Right Side	Left Side	Asymmetry index L/R × 100
ME	SO	50 ± 5	38 ± 4	, 0.77 ± 0.06[a]
	L-SAD	33 ± 4	13 ± 3[b]	0.47 ± 0.12[a]
	R-SAD	44 ± 5	18 ± 2[b]	0.45 ± 0.07[ab]
ARC	SO	16 ± 1	14 ± 0.6	0.87 ± 0.07
	L-SAD	10 ± 1[b]	8 ± 0.5[b]	0.88 ± 0.08
	R-SAD	13 ± 1	9 ± 0.6	0.72 ± 0.05[a]
PVN	SO	10 ± 1	12 ± 1	1.17 ± 0.07
	L-SAD	12 ± 1[b]	10 ± 1	0.84 ± 0.06[a]
	R-SAD	16 ± 2	11 ± 1	0.81 ± 0.17

Means ± SEM; (n = 6, 7); [a]paired *t*-test, P < 0.005 − 0.0005; [b]ANOVA vs SO,
P < 0.05 − 0.01; *L-SAD = left SAD, R-SAD = right SAD.

The bilateral asymmetry of TH activity both in control and SAD rats could
result from asymmetry of the amount of enzyme protein or enzyme configuration.
Table 4 shows that TH protein was lower on the left than right side of ME and
ARC in SO and R-SAD rats.

DISCUSSION

These studies showed that interruption of arterial baroreceptor afferents lowers
TH activity in both caudate nucleus and substantia nigra (pars compacta), (see
Fig. 1 and Alexander *et al.*, 1984). The accompanying reduction of DA concentration in the caudate nucleus we interpret as resulting from inhibition of the
activity of TH.

Table 4.

TH protein (ng/mg tissue prot.) on right and left sides of the brain of SO and R-SAD* rats

Region	Grp	Right Side	Left Side
ME	SO	0.59 ± 0.08	0.34 ± 0.03[b]
	R-SAD	1.13 ± 0.22	0.48 ± 0.12[a]
ARC	SO	0.36 ± 0.04	0.26 ± 0.04[c]
	R-SAD	0.55 ± 0.09	0.38 ± 0.05
PVN	SO	1.90 ± 0.48	2.05 ± 0.27
	R-SAD	2.25 ± 0.25	1.84 ± 0.13

Means ± SEM; [a]$P < 0.05$;[b]$P < 0.025$;[c]$P < 0.0025$ by paired t-test; *R-SAD = right SAD.

Tassin *et al*. (1976) described an A-P concentration gradient of striatal DA and our studies confirmed this and additionally showed that it was present in 3 different regions of the CN (see Fig. 2 and Alexander *et al*., 1987a). A 3-day deficit of arterial baroreceptor input reduced the overall DA concentration in each region of the CN about 40% below that of SO controls, Table 1. Only 3 or 4 intra-regional sites out of the 7 or 8 studied within each area achieved statistical significance between SAD and SO rats. It is curious that the most rostral region of the dorsal area, the mid-region of the lateral and the more caudal regions of the ventral caudate were statistically significantly lower in SAD rats. It suggests a spiral projection of axons passing from SN into ventral caudate, running laterally and finally terminating in the most rostral regions of the dorsal area. Thus, perhaps a specific subgroup of nigrostriatal neurons are mainly influenced by baroreflex activity.

All together these data suggest that under normal conditions, input from arterial baroreceptors through second order, as yet undefined, neuronal pathways, contributes to maintenance of DA stores in the nigrostriatal system, possibly through influences on TH activity.

The latter was supported by results obtained from the striatal microdialysis studies (Alexander *et al*., 1988a). The DA concentration in dialysate collected from the caudate of freely moving SAD and SO rats is about 50% lower in SAD than SO rats. That means that in SAD rats the rate of DA release and turnover is low, the latter being indicated by the reduced concentrations of DA metabolites, see Fig. 3. When monoamine oxidase was blocked by pargyline treatment, dialysate DA accumulated in SAD rats at about one-third the rate of SO rats. This probably reflects the reduced stores resulting from reduced DA synthesis but, we did not rule out the possibility that SAD rats had enhanced catechol-O-methyltransferase activity to account for low extracellular DA.

These changes in the nigrostriatal system of the SAD rat could have relevance to their characteristic lability of MAP. It was reported in the SAD that pressure lability is mainly a function of the animal's activity or behavior (Junqueira and Krieger, 1976). Thus limb movements or asynchronous sleep (REM) causes sharp, brief dips or more prolonged blood pressure lowering. On the other hand, certain specific stereotype behaviors and synchronous sleep, is associated with rising arterial pressure. Persons with Parkinson's disease often display postural hypotension (Senard et al., 1990), a condition that could be classified as lability of pressure.

The increase of over 50% in SD of MAP in lesioned rats, Fig. 4, represents a large relative increase since it occurred in rats with *intact* arterial baroreceptor afferent nerves which, of course, provide continuous feedback for stabilization of pressure. The specificity of SN lesions on pressure lability was supported by the absence of an effect in rats with lesions of the VTA. In some SN lesioned rats, motor activity may have contributed to lability of MAP but the occurrence of increased lability without increased motor behavior indicates it is not the only factor.

Lin et al. (1982) showed, in anesthetized rats, that lesions and electrical stimulation of SN enhanced and reduced, respectively, the reflex bradycardia induced by systemic epinephrine injections. This study along with ours suggest reciprocal neuronal projections between the nigrostriatum and CNS regions known to directly influence cardiovascular activity. A direct anatomical link between baroreceptor nerve terminals and SN or CN has not been reported. However, tracer studies (including our own unpublished observations) identified projections from amygdala to SN. The link with baroreceptors may be a 2 or 3 neuronal pathway because the nuc. tractus solitarius projects directly to amygdala and it also projects to loc. coeruleus which in turn projects to amygdala (Emson, 1983).

Further evidence in support of a functional interaction between the nigrostriatum and baroreflex pathways, was obtained from the study with apomorphine, the DA agonist. Even though both SAD and SO rats showed stereotype behaviors after apomorphine injection, only SAD rats showed cardiovascular responses. This suggests a nigrostriatal source of inhibition to sympathetic efferent nerve activity which, in the absence of baroreflex feedback, is manifest by a sustained reduction in arterial pressure.

The study of hypothalamic right-left TH activity unexpectedly revealed that the TIDA neuronal system in normal rats had lower TH activity on the left than right side of the brain and that this could be accounted for by lower TH immunospecific protein on the left side (Alexander et al., 1990). The reduced DA concentrations on the left sides of both ARC and ME supported the TH findings and indicated that DA synthesis was attenuated in the left TIDA system of normal male rats. Additionally, the NE and E concentrations of the left ARC were significantly lower than on the right side suggesting a role for brain stem afferents in the control of bilateral asymmetry of TH in the TIDA neurons.

The functional significance of these findings in TIDA neurons of normal rats is purely speculative at this time. The best known function of TIDA DA is its inhibitory effect on prolactin secretion; DA is released from the ME into the portal capillaries that carry blood to the pituitary. An earlier study found that serum prolactin is suppressed in 3-day SAD rats (Alexander *et al.*, 1987a). The above study of 7-day, unilateral SAD rats indicates that arterial baroreceptor input functions to minimize the bilateral asymmetry of TH in ME and ARC regions. Evidence for this was that 7-day right SAD rats had a significant further suppression of TH activity and TH protein on the left sides of the TIDA system. Left SAD suppresses TH activity on both sides of the brain. These studies supported our thesis that removal of the anatomically larger left sinoaortic input would have a more extensive effect in the CNS than right SAD. However, right SAD enhanced the degree of asymmetry in ARC and ME because it mainly affected one side of the brain.

In conclusion, these studies provide evidence linking the cardiovascular and motor systems and the cardiovascular and hypothalamic-hypophyseal systems. The link is between arterial baroreceptor secondary neuronal pathways and dopaminergic neurons of the nigrostriatum and hypothalamus. In both regions, TH activity and consequently DA concentrations and release were attenuated by interruption of arterial baroreflex afferents. This indicates that normally the tonic sinoaortic neural input contributes to control of TH activity and dopaminergic processes in these regions and thereby to integration of their activity with the cardiovascular system.

Acknowledgments

The author expresses her sincere appreciation to Professor Nagatsu and his co-workers for their collegiality, scientific support and friendship. This work was done, in part, with funding from the NSF Japan Programs No. 86–1328 and NIH Grant HL32532.

REFERENCES

Alexander, N., McClasky, J., Fung, K. and Maronde, R. (1976). Elevated dopamine beta-hydroxylase activity in rats with neurogenic hypertension. *Life Sci.* **18**, 652–665.

Alexander, N. and Velasquez, M. (1980). Blood pressure and plasma catecholamines in arterial baroreceptor denervated rats. In: *Arterial Baroreceptors and Hypertension.* P. Sleight (Ed.). Oxford University Press, New York, pp. 418–423.

Alexander, N., Velasquez, M., DeCuir, M. and Maronde, R. (1980). Indices of sympathetic activity in the sinoaortic denervated rat. *Am. J. Physiol.* **238**, H521–H526.

Alexander, N., Hirata, Y. and Nagatsu, T. (1984a). Reduced tyrosine hydroxylase activity in nigrostriatal system of sinoaortic-denervated rats. *Brain Res.* **299**, 380–382.

Alexander, N., Yoneda, S., Vlachakis, N. V. and Maronde, R. F. (1984b). Role of conjugation and red blood cells for inactivation of circulating catecholamines. *Am. J. Physiol.* **247**, R203–R207.

Alexander, N., Ross-Cisneros, F., Kogosov, E., DeCuir, M., Haun, C. K. and Nagatsu, T. (1986). Nigrostriatal and cardiovascular systems interactions. In: *Brain and Blood Pressure Control.* K. Nakamura (Ed.). Elsevier Science Pub., New York, pp. 61–66.

Alexander, N., Kogosov, E. and Haun, C. K. (1987a). Sinoaortic denervation changes catecholamines in caudate-putamen of rats. *Brain Res.* **444**, 320–324.

Alexander, N., Melmed, S. and Morris, M. (1987b). Suppressed serum prolactin in sinoaortic-denervated rats. *Am. J. Physiol.* **252**, R290–R293.

Alexander, N. and Morris, M. (1988). Effects of chronic sinoaortic denervation on central vasopressin and catecholamine systems. *Am. J. Physiol.* **255**, R768–R773.

Alexander, N., Nakahara, D., Ozaki, N., Kaneda, N., Sasaoka, T., Iwata, N. and Nagatsu, T. (1988). Striatal dopamine release and metabolism in sinoaortic-denervated rats by *in vivo* microdialysis. *Am. J. Physiol.* **254**, R396–R399.

Alexander, N., Kaneda, N., Ishii, A., Mogi, M., Harada, M. and Nagatsu, T. (1990). Right-left asymmetry of tyrosine hydroxylase in rat median eminence: influence of arterial baroreflex nerves. *Brain Res.* **23**, 195–198.

Emson, P. C. (1983). *Chemical Neuroanatomy.* Raven Press, New York.

Juniqueira, L. F. and Krieger, E. M. (1976). Blood pressure and sleep in the rat in normotension and in neurogenic hypertension. *J. Physiol. (London)* **259**, 725–735.

Krieger, E. M. (1964). Neurogenic hypertension in the rat. *Circ. Res.* **15**, 511–521.

Lin, M., Tsay, B. and Chen, F. (1982). Activation of dopaminergic receptors within the caudate-putamen complex facilitates reflex bradycardia in the rat. *Jpn. J. Physiol.* **32**, 431–442.

Makio, M., Kojima, K. and Nagatsu, T. (1984). Detection of inactive or less active forms of tyrosine hydroxylase in human adrenals by a sandwich enzyme immunoassay. *Anal. Biochem.* **138**, 265–275.

Nakata, T., Berard, W., Kogosov, E. and Alexander, N. (1991). Hypothalamic NE release and cardiovascular response to NaCl in sinoaortic-denervated rats. *Am. J. Physiol.* **260**, R733–R738.

Ozaki, N., Nakahara, D., Kaneda, N., Kiuchi, K., Okada, T., Kasahara, Y. and Nagatsu, T. (1987). *J. Neural Transm.* **70**, 241–250.

Saavedra, J. and Alexander, N. (1983). Catecholamines and phenylethanolamine-N-methyltransferase in selected brain nuclei and in the pineal gland of neurogenic hypertensive rats. *Brain Res.* **274**, 388–392.

Senard, J. M., Valet, P., Durrieu, G., Berlan, M., Tran, M.A., Montastruc, J. L., Rascol, A. and Montastruc, P. (1990). Adrenergic supersensitivity in Parkinsonians with orthostatic hypotension. *Eur. J. Clin. Invest.* **20**, 613–619.

Snyder, D., Nathan, N. and Reis, D. J. (1978). Chronic lability of arterial pressure produced by selective destruction of the catecholamine innervation of the nucleus tractus solitarii in the rat. *Circ. Res.* **43**, 662–671.

Tassin, J. P., Cheramy, A., Blanc, G., Thierry, A. M. and Glowinski, J. (1976). Topographical distribution of dopaminergic innervation and of dopaminergic receptors in the rat striatum. I. Microestimation of [^{3}H]dopamine uptake and dopamine content in microdiscs. *Brain Res.* **107**, 291–301.

Tyrosine Hydroxylase, pp. 295–311
M. Naoi *et al.* (Eds)
© VSP 1993

HYPERTENSION IN RATS FED A LOW CALCIUM DIET: ROLES OF CATECHOLAMINE SYNTHESIS AND SECRETION

AKIFUMI TOGARI

Department of Pharmacology, School of Dentistry, Aichi-Gakuin University, 1-100 Kusumoto-cho, Chikusa-ku, Nagoya 464, Japan

Abstract—Several properties of the hypertension evoked by a low calcium diet are described, and possible mechanisms for this hypertension are discussed. Elevation of blood pressure were observed in male Wistar rats (6 week-old) fed a low calcium diet for 3 weeks. The elevation was associated with decreased calcium level and increased parathyroid hormone (PTH) and catecholamine levels in plasma. In parathyroidectomized rats receiving a normal calcium diet, blood pressure did not rise, though the plasma calcium level decreased to a similar extent as that in rats fed a low calcium diet. The synthesis and secretion of adrenal catecholamines were increased in rats fed a low calcium diet but not in parathyroidectomized rats. These findings seem to indicate that hyperparathyroidism, not hypocalcemia, is involved in the elevation of blood pressure in rats fed a low calcium diet. The elevated blood pressure was reduced by a calcium antagonist, nifedipine, and calcium supplementation but not by an inhibitor of angiotensin-converting enzyme, captopril. Acute administration of synthetic PTH, which showed vasodilatory action in several hypertensive rats, had no hypotensive effect in rats fed a low calcium diet. From these evidences, it may be suggest that the hypertension is based on enhanced calcium uptake in vascular smooth muscle by chronic hyperparathyroidism and not on enhanced renin-angiotensin system. The relationship between hyperparathyroidism and enhanced catecholamine metabolism in regulating the blood pressure were discussed.

Key words: hypertension; calcium; parathyroid hormone; catecholamine; nifedipine; captopril.

INTRODUCTION

Alterations in calcium metabolism have been described in both human and experimental hypertension (McCarron *et al.*, 1981a; Wright and Rankin, 1982; McCarron and Morris, 1984; Massry *et al.*, 1986). Hypercalciuria, decreased plasma calcium concentrations, and increased parathyroid hormone (PTH) concentrations have been observed in the spontaneously hypertensive rat (SHR) (McCarron *et al.*, 1981a; Lau *et al.*, 1984b; Resnick, 1987a), the deoxycorticosterone (DOCA)-salt hypertensive rat (Kurtz and Morris, 1985), the Dahl salt hypertensive rat (Goulding and Gold, 1986; Umemura *et al.*, 1986), and in some patients with essential hypertension (McCarron, 1982a; Resnick *et al.*, 1983; Grobbee *et al.*, 1986; Hvarfner *et al.*, 1987).

As an important link between abnormal systemic calcium metabolism and hypertension, it has been observed that dietary calcium supplementation lowers blood pressure in SHR (Ayachi, 1979; McCarron *et al.*, 1981a; Lau *et al.*, 1984a; Kageyama *et al.*, 1986; Tenner *et al.*, 1989), in other experimental hypertensive rats (Kageyama *et al.*, 1987; DiPette *et al.*, 1990), and in human essential hypertension (McCarron and Morris, 1985; Luft *et al.*, 1986). This suggests that the level of dietary calcium significantly influences the development and mainte-nance of increased arterial pressure. Reductions in blood pressure due to a high calcium diet have been reported in normotensive rats as well as in hypertensive rats (McCarron, 1982b; Jones *et al.*, 1986). Similarly, blood pressure in nor-motensive young or pregnant rats was increased by calcium restriction (Itokawa *et al.*, 1974; Belizan *et al.*, 1981). These findings suggest that calcium balance is important in blood pressure regulation in both normal and hypertensive rats.

The increase in blood pressure in normal Wistar rats caused by calcium de-privation was first reported by Itokawa *et al.* in 1974. However, there is genetic variability in the blood pressure response to alterations in dietary calcium in rodents (Huie *et al.*, 1987). Recently, Itokawa *et al.*'s observation of increased blood pressure in calcium deficient normal Wistar rats was confirmed (Togari *et al.*, 1989) using a low calcium diet to evoke hypocalcemia and nutritional hyperparathyroidism (Table 1).

Table 1.

Composition of diet

Ingredients	%
Cerelose (Glucose H_2O)	64.5
Casein	18.0
Roughage celluflour	3.0
Cottonseed oil	10.5
Cystine	0.2
Choline chloride	0.2
Mineral mixture[a]	4.0
Vitamin mixture[b]	0.1

a) In the normal diet (milligrams per 100 g diet): $CaCO_3$ 750; K_2HPO_4 707; KH_2PO_4 553; KCl 1144; NaCl 418; $MgSO_4$ 318; $FeSO_4$ 35; $ZnSO_4$ 4.9; NaF 2.3; $CuSO_4$ 1.0; $MnSO_4$ 0.7; $(NH_4)_6Mo_7O_{24}$ 0.09; $CoCl_2$ 0.06. In the low calcium diet the minerals were the same except for the absence of $CaCO_3$.

b) Vitamin mixture (milligrams per 100 g diet): inositol 20.0; Ca-pantothenate 2.8; nicotinamide 2.0; thiamine 0.5; riboflavin 0.5; pyridoxine 0.5; folic acid 0.02; biotin 0.01; cyanocobalamin 0.002. Fat-soluble vitamins (E, K, A and D) were already dissolved in the cottonseed oil.

In the following review, several properties of the hypertension evoked by a low calcium diet are described, and possible mechanisms for this hypertension are discussed.

HIGH BLOOD PRESSURE IN RATS FED A LOW CALCIUM DIET

In young Wistar rats, a decrease in dietary calcium significantly increased the blood pressure, although blood pressure returned to normal after the dietary calcium was restored (Fig. 1).

This supports the experimental observations in normotensive and hypertensive rats that calcium balance may be important in blood pressure regulation. However, the mechanisms of both the hypertensive effect of calcium deprivation and the antihypertensive effect of calcium supplementation are still not well understood.

Does alteration in plasma calcium play a major role in the initial development and maintenance of high blood pressure in rats fed a low calcium diet? It has

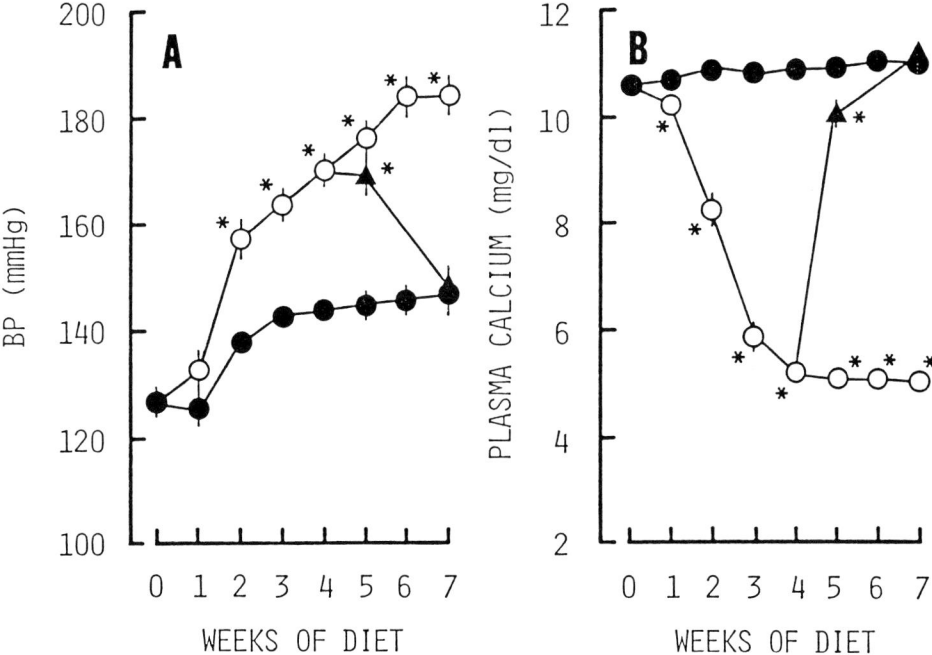

Figure 1. Changes in blood pressure (A) and plasma calcium (B) in rats. Rats were fed a normal calcium diet (●) or a low calcium diet (○) for 7 weeks. Some of the low calcium diet rats were changed to a normal calcium diet after 4 weeks (▲). Plasma calcium levels were measured in blood samples obtained from a tail vein with a heparinized capillary tube. Each point represents a mean value, and the bar indicates the corresponding SEM ($n = 6$–12). The rats were 3 weeks old at the start of the experiment. *$P < 0.01$, Statistically significant difference from the normal calcium diet group at the corresponding time, by Student's t-test or Cochran-Cox test.

been reported that parathyroidectomy reduces the hypertensive effect of mineralocorticoid treatment (Berthelot and Gairard, 1980) and impedes blood pressure increase in SHR (Schleiffer *et al.*, 1981; Mann *et al.*, 1984). Therefore, PTH seems to be one of the factors which regulate blood pressure in experimental hypertension. The hypertension in rats fed a low calcium diet was associated with hypocalcemia and hyperparathyroidism, but changes in plasma calcium concentration in the absence of PTH did not modify blood pressure in rats.

Table 2.

Influence of chronically affected parathyroid function on blood pressure, total plasma calcium, ionized plasma calcium, plasma phosphate, plasma alkaline and acid phosphatase, and plasma parathyroid hormone

	Intact rats		PTX rats[c]
	NCaD group[a]	LCaD group[b]	NCaD group
Blood pressure (mmHg)	149.9 ± 2.4	181.5 ± 3.7*	151.0 ± 4.0
Heart rate (beats/min)	383.0 ± 9.0	389.5 ± 4.7	374.8 ± 9.2
Total calcium (mg/dl)	10.3 ± 0.1	4.7 ± 0.1*	5.6 ± 0.3*
Ionized calcium (mg/dl)	5.3 ± 0.1	2.3 ± 0.1*	2.6 ± 0.2*
Phosphate (mg/dl)	6.2 ± 0.1	6.4 ± 0.2	13.3 ± 0.5*
Alkaline phosphatase (nmol/min/ml)	188.2 ± 8.0	559.2 ± 21.5*	206.0 ± 6.6
Acid phosphatase (nmol/min/ml)	14.9 ± 0.7	14.6 ± 1.5	13.5 ± 1.3
Parathyroid hormone (pg/ml)	32.7 ± 4.6	550.5 ± 53.0*	n.d.

a) Fed NCaD (Ca 0.3%; P 0.42%) for 7 weeks.
b) Fed LCaD (Ca 0.01%; P 0.42%) for 7 weeks.
c) Fed NCaD for 4 weeks, then parathyroidectomized, then fed NCaD for 3 weeks.
Each value is a mean ± SEM (n = 5–12). Age of rats used was 10 weeks. n.d.: not determined. *$P < 0.01$, Statistically significant difference from the normal calcium diet group, by Student's t-test or Cochran-Cox test.

As shown in Table 2, in parathyroidectomized rats, a decrease in plasma calcium did not increase the blood pressure. This suggests that it is not the decrease in plasma calcium but rather the increase in parathyroid function which causes hypertension in rats fed a low calcium diet.

Several possible mechanisms of the antihypertensive effect of calcium have been proposed. Calcium has been shown to decrease sympathetic nerve activity as judged by the systemic response to stress (Hatton *et al.*, 1987). Some investigators have reported that dietary calcium promotes natriuresis (Ayachi, 1979), although others reported that it does not (Lau *et al.*, 1984a; Kageyama *et al.*, 1986). Calcium may also lower blood pressure by virtue of its effect on PTH, vitamin D metabolites, or other hormones. Finally, dietary calcium has also been said to act via phosphate depletion (Lau *et al.*, 1984a). However, oral phosphate loading causes a decrease in blood pressure, not an increase as would be predicted by the phosphate depletion hypothesis (Jones *et al.*, 1986). In fact, Table 2 shows that the high blood pressure in young rats fed a low calcium diet returned to normal after they were fed a normal calcium diet, and phosphate was not involved in these changes.

EFFECT OF CAPTOPRIL ON HIGH BLOOD PRESSURE

What is the role of PTH in this kind of hypertension? One possible etiology of this hypertension is an alteration in the renin-angiotensin system, because calcium and PTH stimulate renin release *in vitro* (Chen and Poisner, 1976), and synthetic bovine PTH(1-34) increases plasma renin activity *in vivo* in saline-loaded dogs (Powell *et al.*, 1978). This possibility was examined pharmacologically using captopril, an inhibitor of angioten-sinconverting enzyme.

As shown in Fig. 2A, captopril did not affect blood pressure (5 and 10 mg/kg, p.o.). At 10 mg/kg p.o. captopril had its maximum antihypertensive effect in the two-kidney Goldbatt hypertensive rat (Laffan *et al.*, 1978). This evidence suggests that hypertension associated with hyperparathyroidism is not renin dependent, which is consistent with clinical observations that elevated blood pressure is not reduced by saralasin in patients with hyperparathyroidism (Zawada *et al.*, 1980). Resnick (1987b) emphasized the biochemical and clinical heterogeneity of human hypertension, and said that calcium supplementation in human hypertension could decrease blood pressure in low-renin essential hypertension, and increase it in high-renin hypertension. Hypertension in rats fed a low calcium diet, which is reversed by calcium supplementation but not by captopril, may therefore be similar to low-renin essential hypertension.

EFFECT OF NIFEDIPINE ON HIGH BLOOD PRESSURE

Treatment with a calcium antagonist as well as calcium supplementation were beneficial in hypertension associated with nutritional hyperparathyroidism (Figs 1 and 2). These data are consistent with recent clinical observations that essential hypertensive patients with low renin levels, low ionized calcium levels and elevated PTH levels respond to the calcium antagonist nifedipine, and to oral calcium supplementation (Resnick *et al.*, 1986). Nifedipine reduces vascular tone by inhibiting the influx of extracellular calcium via calcium channels into smooth muscle cells (Fleckenstein *et al.*, 1972; Nabata, 1977; Godfraind *et al.*, 1982). The hypotensive effect of nifedipine was more pronounced in rats fed a low calcium diet than in those fed a normal diet (Fig. 2B), which supports the idea that the basic change which occurs in vascular smooth muscle as hypertension develops is an increase in cell membrane permeability to calcium.

Several studies showed that PTH increases intracellular ionized calcium in a variety of tissues including the kidney, bone and vascular smooth muscle cells (Hruska *et al.*, 1986; Reid *et al.*, 1987; Yamaguchi *et al.*, 1987; Kawashima,

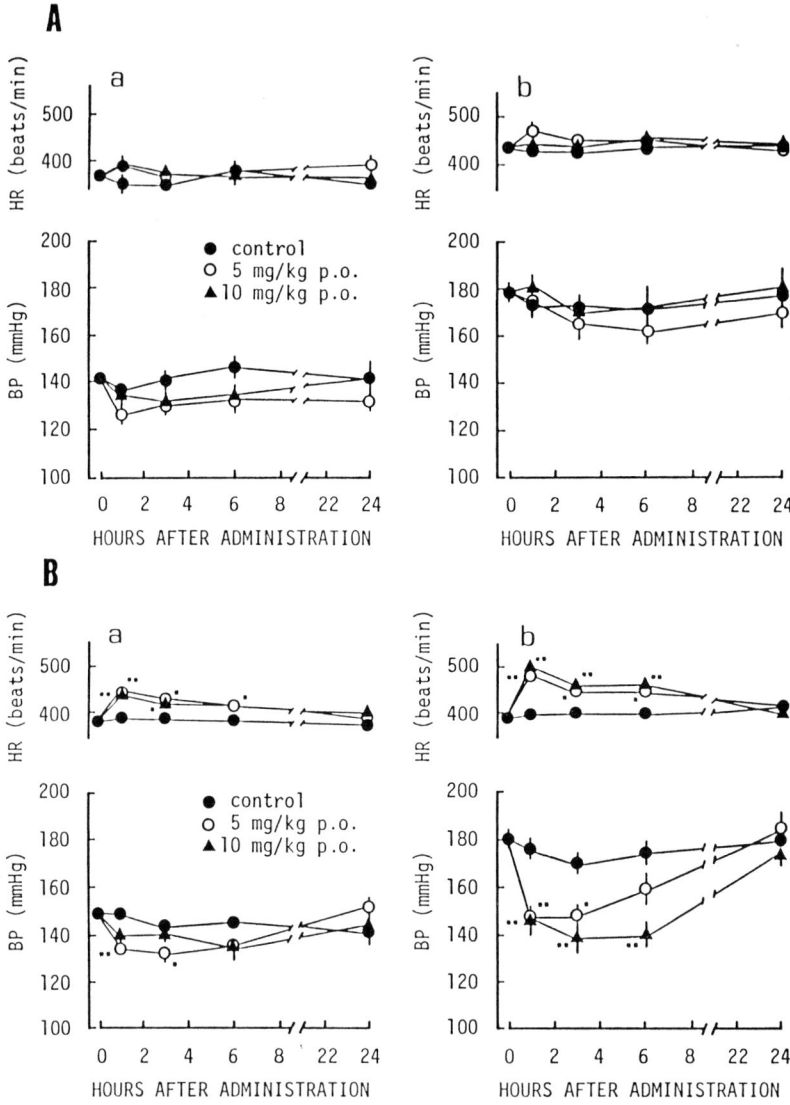

Figure 2. Effects of captopril (A) and nifedipine (B) on blood pressure (BP, lower panel) and heart rate (HR, upper panel) in rats fed a normal calcium diet (a) or a low calcium diet (b) for 5 weeks. Drug was given orally at time 0. Control rats were given the vehicle p.o. through a gastric tube. Each point represents a mean value. Bars indicate SEM ($n = 6$). The age of the rats used was 8 weeks. $^*P < 0.05$, $^{**}P < 0.01$, Statistically significant difference at the indicated time, determined by Dunnett's method for comparing multiple treatments with a control, using analysis of variance.

1990), suggesting that chronic hyperparathyroidism in rats may increase calcium influx and thereby increase vascular tone, leading to high blood pressure. In contrast to these results, Pang *et al.* (1988), demonstrated that PTH inhibited calcium uptake in rat tail artery. They also observed direct inhibition of L-type calcium channels in neuroblastoma and vascular smooth muscle cells (Pang *et al.*, 1990a).

There is considerable information available concerning the effects of calcium antagonists on plasma and tissue catecholamine levels in SHR. According to De Leeuw *et al.* (1985), the hypotensive effect of verapamil in SHR is accompanied by an insignificant rise in plasma noradrenaline and a significant fall in plasma adrenaline. Other studies have also found that the hypotensive effect of verapamil in SHR is accompanied by a decrease in plasma and adrenal corticosterone content, and a fall in catecholamine concentration in adrenal glands and myocardium (Nicolov *et al.*, 1988).

Thus, calcium antagonists probably lower blood pressure by direct action on hemodynamic pathways and also by their effects on neurohumoral and hormonal factors.

HYPERPARATHYROIDISM AND HYPERTENSION

Although hyperparathyroidism may cause blood pressure to be chronically high, there is conflicting evidence. First, infusion of physiologic levels of PTH partially inhibits the vasoconstrictor response to angiotensin II in experimental models, which is incompatible with a vasoconstrictor effect of PTH (Ellison and Mc-Carron, 1982); second, short-term suppression of endogenous PTH by calcium infusion causes increase of blood pressure (McCarron *et al.*, 1981b); and third, both systemic and regional acute infusion of PTH causes vasodilation (Schleiffer *et al.*, 1979; Crass and Pang, 1980; Pang *et al.*, 1980; Nakamura *et al.*, 1981; Ellison and McCarron, 1984; McCarron *et al.*, 1984). This evidence may indicate that there is a difference between the acute and chronic effects of PTH on the vasculature, or that another factor is involved under chronic hyperparathyoidism (possibly a catecholamine). It may also mean that a hypertensive substance, as yet not identified, is released from the hypertrophic parathyroid.

The hypotensive effect of PTH was observed in hypertensive rats as well as in normotensive rats. Studies in spontaneously hypertensive rats, animals with the two-kidney-one-clip model of hypertension, and those with DOCA and salt hypertension (Nakamura *et al.*, 1981) showed that acute administration of bovine or human PTH(1-34) reduces blood pressure in a dose-dependent fashion, and that the hypotensive responses of these animals to exogenous PTH are different. These results could mean that the vasodilatory action of PTH is influenced by the pathogenesis of the hypertension. It has been demonstrated that the effects of PTH are critically dependent on the calcium status of the rat (Anderson et al., 1983; McCarron and Morris, 1984). Recently, it was shown that dietary calcium deprivation, which causes hypocalcemia and nutritional hyperparathyroidism, also lowers sensitivity to exogenous human PTH(1-34) (Togari et al., 1990).

As shown in Fig. 3, subcutaneous administration of human PTH(1-34) (50 mg/kg) caused a significant decrease in blood pressure in rats fed a normal calcium diet. This did not occur in rats fed a low calcium diet. This difference in the cardiovascular actions of PTH between rats fed normal and low calcium diets was also observed by direct measurement of arterial blood pressure during anesthesia (Fig. 4).

The reduction in blood pressure, which reached its maximum within 1 to 2 min and lasted for 2 to 5 min, was more pronounced in rats fed a normal calcium diet than in rats fed a low calcium diet. These data suggest that the acute hypotensive action of PTH was attenuated in vasculature which was chronically exposed to PTH. They also indicate that the hypotensive action of human PTH(1-34) is pharmacological rather than physiological (Togari et al., 1990).

Experiments in which hypertension was prevented or reduced by parathyroidectomy have been used to argue for a hypertensive role for PTH. How-

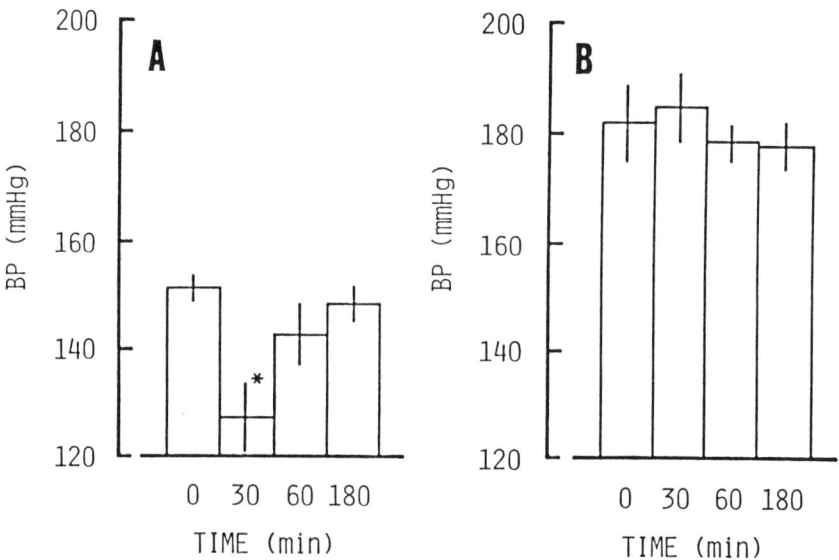

Figure 3. Effects of human PTH(1-34) on tail-cuff systolic blood pressure (BP) in rats fed a normal calcium diet (A) or a low calcium diet (B) for 7 weeks. Age of rats used was 10 weeks. After the measurement of blood pressure at 0 time, the drug was given subcutaneously. Date are plotted as the means, and bars illustrate the SEM ($n = 6$). $^{*}P < 0.01$, Statistically significant difference from 0 time value, by Student's t-test or Cochran-Cox test.

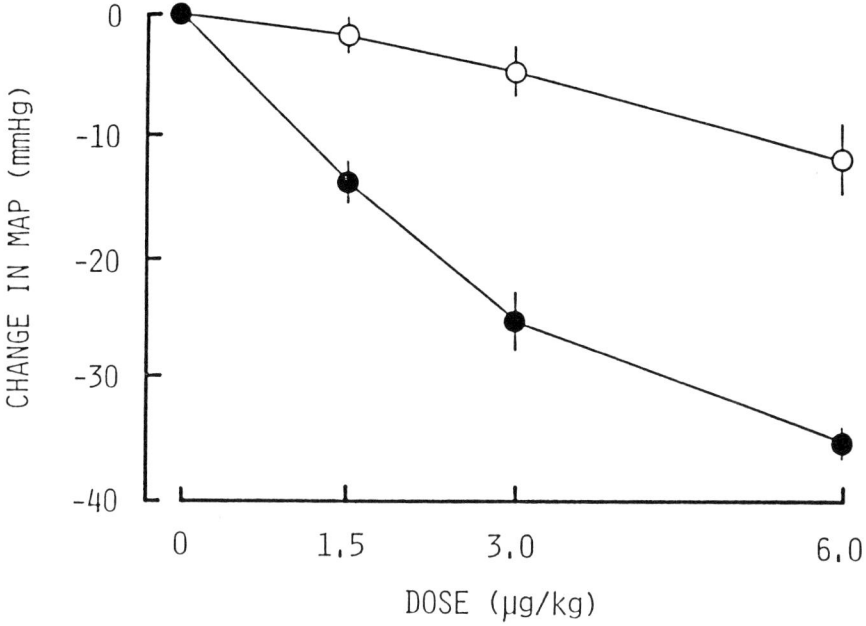

Figure 4. The hypotensive effects of human PTH(1-34) in rats fed a normal calcium diet (●) or a low calcium diet (○) for 8 weeks. Age of rats used was 11 weeks. The drug was given intravenously to rat anesthetized by intraperitoneal administration of ketamine-HCl (100 mg/kg). Change in mean arterial blood pressure (MAP) was defined as the difference between the MAP before injection and the lowest MAP after injection.

ever, these results only suggest the presence of a hypertensive substance in the parathyroid gland, and this substance need not be PTH. Thus far, PTH is the only hormone known to be secreted from the parathyroid gland. Recently, a new hypertensive factor, distinct from PTH, was found in plasma of SHR. In normotensive rats it elevates blood pressure after a certain delay, and it can increase calcium uptake in vascular smooth muscle (Pang and Lewanczuk, 1989; Pang *et al.*, 1990b). Histological examination showed that it is produced by a specific cell type in the parathyroid glands (Kaneko *et al.*, 1989). Therefore, it has been proposed that it be called "Parathyroid Hypertensive Factor (PHF)." The relationship between enhanced parathyroid function and hypertension may be explained by the presence of PHF.

INCREASES IN CATECHOLAMINE SYNTHESIS AND RELEASE UNDER HYPERPARATHYROIDISM

The relationships between calcium, PTH, catecholamines and blood pressure are complex and remain unexplained. There is evidence that calcium plays a central role in excitation-contraction coupling of vascular smooth muscle cells in response to norepinephrine. Calcium also appears to play a key role in the synthesis and release of catecholamines in the sympathetic nerve terminal and in the adrenal medulla. In 1984, Baksi et al. reported that a diet deficient in calcium and vitamin D alters brain and adrenal catecholamines (Baksi and Hughes, 1982a, b, 1984).

In rats fed a low calcium diet, Hagihara et al. (1990) measured plasma catecholamines, and also adrenal catecholamines, tyrosine hydroxylase (TH) and dopamine β-hydroxylase (DBH).

As shown in Fig. 5 the results indicated that after 7 weeks of calcium deprivation plasma norepinephrine and epinephrine concentrations were significantly higher than in the animals on a normal calcium diet. On the other hand, in parathyroidectomized rats plasma norepinephrine concentrations were low, but plasma epinephrine concentrations were normal. The rats on the low calcium diet had significantly less norepinephrine and epinephrine in their adrenal glands than those on the normal diet.

The activities of adrenal TH and DBH in all three groups are shown in Fig. 6, TH and DBH activities in rats on a low calcium diet were significantly higher than in those on a normal calcium diet and were also higher than in parathyroidectomized rats. The enzymatic activity of TH and DBH was about 1.5 times higher in rats on a low calcium diet. The increase in TH activity was evaluated by the V_{max} value measured under saturating concentrations of tyrosine and puterin cofactor, which suggests that there was an increase in the amount of the enzyme in the adrenal gland. Thus, these data show that plasma catecholamines and TH and DBH activities in the adrenal glands were higher in rats on a low calcium diet but not in parathyroidectomized rats, and suggest that hyperparathyroidism but not hypocalcemia is related to increased catecholamines. This also agrees with the earlier observations of Baksi and Hughes (1984) that adrenal catecholamine levels decrease in rats following calcium deprivation.

The mechanism underlying the relationship between hyperparathyroidism and enhanced catecholamine metabolism remains to be elucidated, and the functional role of changes in plasma catecholamines is still only partially understood. In SHR, the results of several studies indicate a relationship between catecholamines and hypertension. First, Ozaki et al. (1968) demonstrated that norepinephrine level in the adrenal gland significantly increased at 6 weeks of age and the increase lasted for one year. Second, enzymatic analyses have indicated that the activities of catecholamines synthesizing enzymes, including TH, dopa decarboxylase, DBH and phenylethanolamine N-methyltransferase, in the adrenal gland increase as hypertension develops (Nagatsu et al., 1971, 1973). Third, TH and DBH inhibitors including oudenone and fusaric acid have significant hypotensive effects (Hidaka et al., 1969; Umezawa et al., 1970). Fourth, the hypertension was not reduced by adrenalectomy (Aoki et al., 1973). Finally, the increase in adrenal TH and DBH under repeated stress was found to be

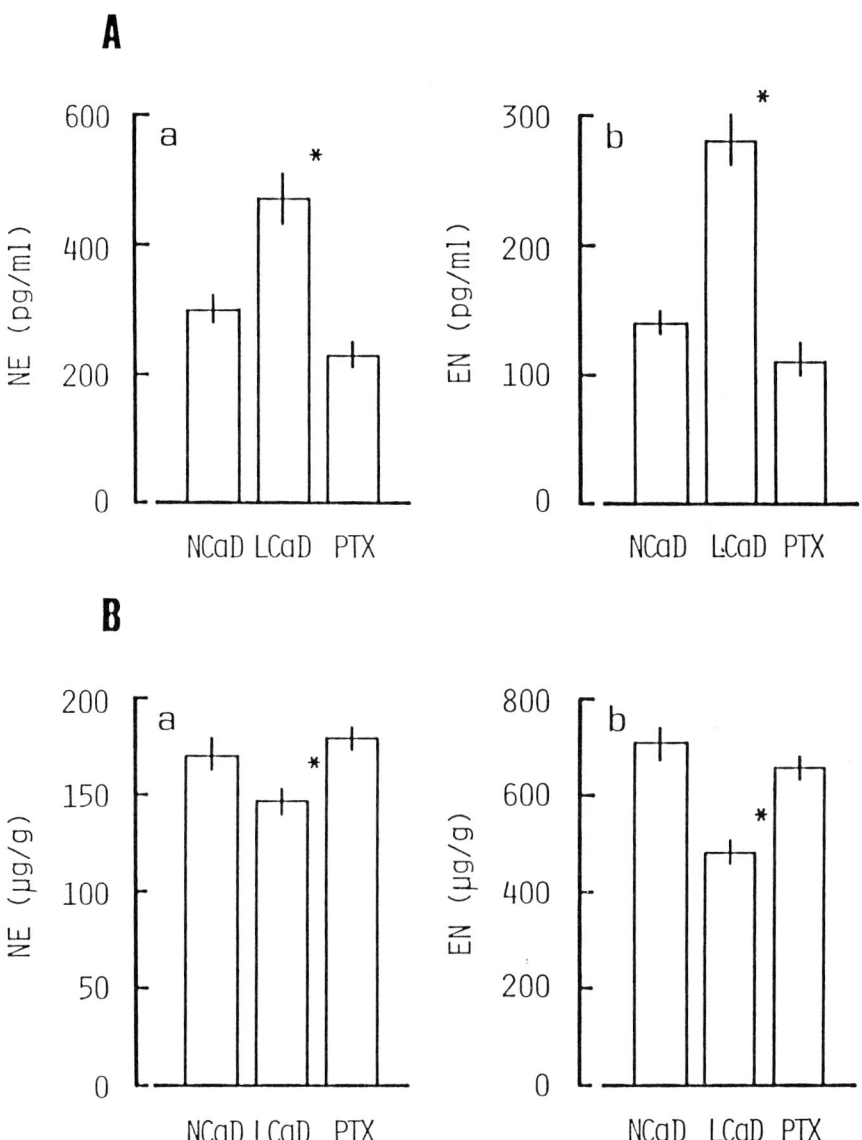

Figure 5. Catecholamine levels in plasma (A) and adrenal glands (B) of rats. Rats were fed a normal calcium diet (NCaD) or a low calcium diet (LCaD) for 7 weeks. Some of the normal calcium diet rats were parathyroidectomized (PTX) after 4 weeks and then placed on a normal calcium diet for 3 weeks. Age of rats used was 10 weeks. Epinephrine (EN) and norepinephrine (NE) levels were assayed using high performance liquid chromatography (HPLC) with electrochemical detection (ECD). Data are plotted as the means, and bars illustrate the SEM ($n = 6$–14). $^*P < 0.01$, Statistically significant difference from the NCaD group.

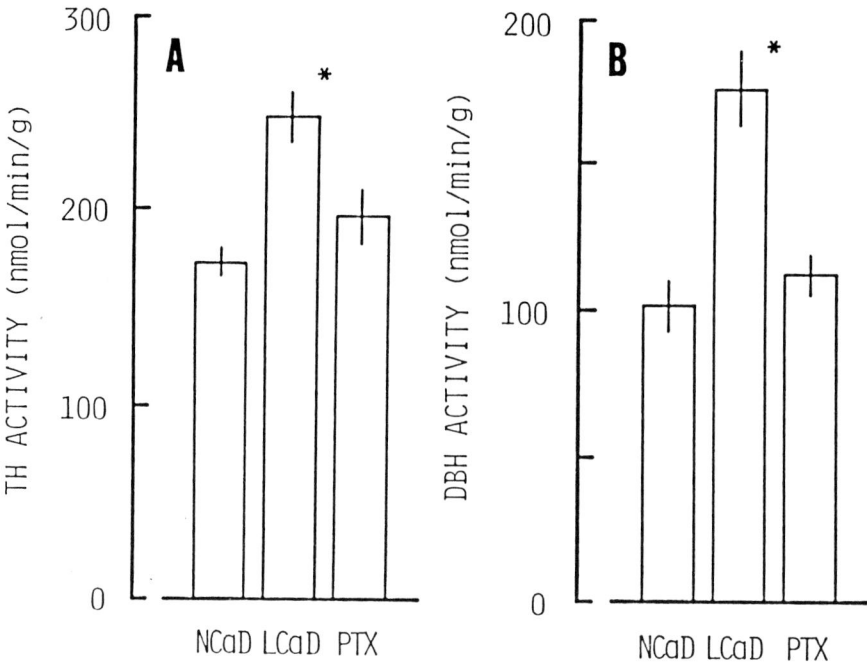

Figure 6. Activities of tyrosine hydroxylase (A) and dopamine β-hydroxylase (B) in adrenal glands of rats. Rats were fed a NCaD or a LCaD for 7 weeks. Some of NCaD rats were parathy-roidectomized (PTX) after 4 weeks and then placed on a NCaD for 3 weeks. Age of rats used was 10 weeks. Both enzyme activities were assayed using HPLC-ECD. Data are plotted as the means, and bars illustrate the SEM (n = 6-14). *P < 0.01, Statistically significant difference from the NCaD group.

more pronounced in SHR than in normotensive control Wistar rats (Kvetňanský et al., 1976). Thus, marked elevation in adrenal TH and DBH in SHR might not cause or maintain hypertension, but instead they may be secondary effects of the high susceptibility of SHR to stress during the development of hypertension. Similarly, the hypertension due to a low calcium diet may be based upon emotional and physical stress, which provoke the release of catecholamines from the sympatho-adrenal medullary system, because dietary calcium seems to be one of the important factors which influences the reactivity to stressful stimulation.

Increased sympatho-adrenal medullary activity might influence parathyroid function. The synthesis and secretion of parathyroid hormone is primarily regulated by the extracellular calcium concentration. Several *in vitro* and *in vivo* studies have also demonstrated that the beta-adrenergic nervous system has a role in the modulation of PTH release (Kukreja et al., 1975; Brown et al., 1977; Fischer and Blum, 1980). Under stress, this role may be physiologically important

(Ljunhgall *et al.*, 1984). Thus, increased catecholamines in rats on a low calcium diet may increase the blood pressure by directly affecting vascular muscle as well as by mediating the secretion of PTH.

These observations suggest that rather than causing hypertension, the increase in catecholamine synthesis and secretion may greatly increase hypertension in rats on a low calcium diet.

CONCLUSION

A decrease in dietary calcium significantly increased blood pressure in young Wistar rats, and this effect was reversed when dietary calcium was restored. This supports the experimental observations in normotensive rats and SHR that calcium balance may be important in blood pressure regulation. There is also some evidence that the increase in parathyroid function may play a major role in the development and maintenance of high blood pressure in rats fed a low calcium diet.

The hypertension associated with hyperparathyroidism was not influenced by captopril, suggesting that it is not renin-dependent, and treatment with calcium antagonists as well as calcium supplementation were beneficial. These phenomena are similar to those seen in human low-renin essential hypertension.

In the hypertensive rat, there are abnormally high levels of plasma catecholamines, and of TH and DBH activity in the adrenal glands. This increased catecholamine synthesis and secretion may be caused by hyperparathyroidism in rats fed a low calcium diet and may elevate the blood pressure by directly affecting vascular muscle. These increases enhance the secretion of PTH, leading to greater elevation of blood pressure.

In summary, it is likely that a low calcium diet stimulates the parathyroid, and thereby causes hypertension. This may involve increased secretion of PTH or PHF, or both, as well as increased catecholamine synthesis and secretion. High blood pressure would result from chronic exposure to these hormones and to other factors.

REFERENCES

Anderson, S., Grady, J. R., Ellison, D. H. and McCarron, D. A. (1983). Calcium balance and parathyroid hormone mediated vasodilation in the spontaneously hypertensive rat. *Hypertension* 5 (Suppl. I), I59–I63.

Aoki, K., Takikawa, K. and Hotta, K. (1973). Role of adrenal cortex and medulla in hypertension. *Nature* 241, 122–123.

Ayachi, S. (1979). Increased dietary calcium lowers blood pressure in the spontaneously hypertensive rat. *Metabolism* 28, 1234–1238.

Baksi, S. N. and Hughes, M. J. (1982a). Regional alterations of brain catecholamines by lead ingestion in adult rats: influence of dietary calcium. *Arch. Toxicol.* 50, 11–18.

Baksi, S. N. and Hughes, M. J. (1982b). Chronic vitamin D deficiency in the weaning rat alters catecholamine metabolism in the cortex. *Brain Res.* **242**, 387–390.

Baksi, S. N. and Hughes, M. J. (1984). Alteration of adrenal catecholamine levels in the rat after dietary calcium and vitamin D deficiencies. *J. Auton. Nerv. Syst.* **11**, 393–396.

Belizan, J. M., Pineda, O., Sainz, E., Menendez, L. A. and Villar, J. (1981). Rise of blood pressure in calcium-deprivated pregnant rats. *Am. J. Obstet. Gynecol.* **141**, 163–169.

Berthelot, A. and Gairard, A. (1980). Parathyroid hormone- and deoxycorticosterone acetate-induced hypertension in the rat. *Clin. Sci.* **58**, 365–371.

Brown, E. M., Hurwitz, S. and Aurbach, G. D. (1977). Beta-adrenergic stimulation of cyclic AMP content and parathyroid hormone release from isolated bovine parathyroid cells. *Endocrinology* **100**, 1696–1702.

Chen, D. S. and Poisner, A. M. (1976). Direct stimulation of renin release by calcium. *Proc. Soc. Exp. Biol. Med.* **152**, 565–567.

Crass, M. F. and Pang, P. K. T. (1980). Parathyroid hormone: a coronary artery vasodilator. *Science* **207**, 1087–1089.

De Leeuw, P., Blick, W. and Vanes, P. N. (1985). Effet des inhibiteurs calcique sur l'activite sympathique. *Arch. Mal. Coeur.* **78**, 51–54.

DiPette, D. J., Greilich, P. E., Nickols, G. A. Graham, G. A., Green, A., Cooper, C. W. and Holland, O. B. (1990). Effect of dietary calcium supplementation on blood pressure and calciotropic hormones in mineralocorticoid-salt hypertension. *J. Hypertens.* **8**, 515–520.

Ellison, D. H. and McCarron, D. A. (1982). Infusion of bovine parathyroid hormone 1–34 attenuates the pressor response to angiotensin II in spontaneously hypertensive rats. *Clin. Exp. Hypertens.* A**4**, 1637–1647.

Ellison, D. H. and McCarron, D. A. (1984). Structural prerequisites for the hypotensive action of parathyroid hormone. *Am. J. Physiol.* **246**, F551–F556.

Fischer, J. A. and Blum, J. W. (1980). Noncalcium control of parathyroid hormone secretion. *Miner. Electrolyte Metab.* **3**, 158–166.

Fleckenstein, A., Tritthart, H., Doring, H. J. and Byon, K. Y. (1972). Bay a 1040, ein hochaktiver Ca-antagonistischer Inhibitor der elektro-mechanischen Koppelungsprozesse im Warmbluter-Myokard. *Arzneimittelforschung* (*Drug Res.*) **22**, 22–33.

Godfraind, T., Miller, R. C. and Socrates, L. J. (1982). Selective α_1- and α_2-adrenoceptor agonist-induced contractions and ^{45}Ca fluxes in the rat isolated aorta. *Br. J. Pharmacol.* **77**, 597–604.

Goulding, A. and Gold, E. (1986). Effects of dietary sodium chloride loading on parathyroid function, 1,25-dihydroxy, vitamin D, calcium balance, and bone metabolism in female rats during chronic prednisolone administration. *Endocrinology* **119**, 2148–2154.

Grobbee, D. E., Hackeng, W. H. L., Birkenhager, J. C. and Hofman, A. (1986). Intact parathyroid hormone (1-84) in primary hypertension. *Clin. Exp. Hypertens.* A**8**, 299–308.

Hagihara, M., Togari, A., Matsumoto, S. and Nagatsu, T. (1990). Dietary calcium deprivation increased the levels of plasma catecholamines and catecholamine-synthesizing enzymes of adrenal glands in rats. *Biochem. Pharmacol.* **39**, 1229–1231.

Hatton, D. C., Huie, P. E., Muntzel, M. S., Metz, J. A. and McCarron, D. A. (1987). Stress-induced blood pressure responses in SHR: effect of dietary calcium. *Am. J. Physiol.* **252**, R48–R54.

Hidaka, H., Nagatsu, T., Takeya, K., Takeuchi, T., Suda, H., Kojiri, K., Matsuzaki, M. and Umezawa, H. (1969). Fusaric acid, a hypotensive agent produced by fungi. *J. Antibiot.* **22**, 228–230.

Hruska, K. A., Goligorsky, M., Scoble, J., Tsutsumi, M., Westbrook, S. and Moskowitz, D. (1986). Effects of parathyroid hormone on cytosolic calcium in renal proximal tubular primary cultures. *Am. J. Physiol.* **251**, F188–F198.

Huie, P. E., Hatton, D. C., Muntzel, M. S., Metz, J. A. and McCarron, D. A. (1987). Genetic variability in response to dietary calcium. *Life Sci.* **41**, 2185–2193.

Hvarfner, A., Bergstom, R., Morlin, C., Wide, L. and Ljunghall, S. (1987). Relationships between calcium metabolic indices and blood pressure in patients with essential hypertension as compared with a healthy population. *J. Hypertens.* **5**, 451–456.

Itokawa, Y., Tanaka, C. and Fujiwara, M. (1974). Changes in body temperature and blood pressure in rats with calcium and magnesium deficiencies. *J. Appl. Physiol.* **37**, 835–839.

Jones, M. R., Ghaffari, F., Tomerson, B. W. and Clemens, R. A. (1986). Hypotensive effect of a high calcium diet in the Wistar rat. *Miner. Electrolyte Metab.* **12**, 85–91.

Kageyama, Y., Suzuki, H., Hayashi, K. and Saruta, T. (1986). Effects of calcium loading on blood pressure in spontaneously hypertensive rats; attenuation of the vascular reactivity. *Clin. Exp. Hypertens.* **A8**, 355–370.

Kageyama, Y., Suzuki, H., Arima, K. and Saruta, T. (1987). Oral calcium treatment lowers blood pressure in renovascular hypertensive rats by suppressing the renin-angiotensin system. *Hypertension* **10**, 375–382.

Kaneko, T., Ohtani, R., Lewanczuk, R. Z. and Pang, P. K. T. (1989). A novel cell type in the parathyroid glands of spontaneously hypertensive rats. *Am. J. Hypertens.* **2**, 549–552.

Kawashima, H. (1990). Parathyroid hormone causes a transient rise in intracellular ionized calcium in vascular smooth muscle cells. *Biochem. Biophys. Res. Commun.* **166**, 709–714.

Kukreja, S. C., Hargis, G. K., Bowser, E. N., Henderson, W. J., Fisherman, E. W. and Williams, G. A. (1975). Role of adrenergic stimuli in parathyroid hormone secretion in man. *J. Clin. Endocrinol. Metab.* **40**, 478–481.

Kurtz, T. W. and Morris, R. C. (1985). Dietary chloride as a determinant of disordered calcium metabolism in salt dependent hypertension. *Life Sci.* **36**, 921–929.

Kvetňanský, K., Albrecht, I., Torda, T., Saleh, N., Jahnova, E. and Mikulaj, L. (1976). Effect of stress on catecholamine synthesizing and degrading enzymes in control and spontaneously hypertensive rats. In: *Catecholamines and Stress.* E. Usdin, R. Kvetňanský and I. J. Kopin (Eds). Pergamon Press, Oxford, pp. 237–249.

Laffan, R. J., Goldberg, M. E., High, J. P., Schaeffer, T. R., Waugh, M. H. and Rubin, B. (1978). Antihypertensive activity in rats of SQ14,225, an orally active inhibitor of angiotensin 1-converting enzyme. *J. Pharmacol. Exp. Ther.* **204**, 281–288.

Lau, K., Chen, S. and Eby, B. (1984a). Evidence for the role of PO_4 deficiency in antihypertensive action of a high-Ca diet. *Am. J. Physiol.* **246**, H324–H331.

Lau, K., Zikos, D., Spirnak, J. and Eby, B. (1984b). Evidence for an intestinal mechanism in hypercalciuria of spontaneously hypertensive rats. *Am. J. Physiol.* **247**, E625–E633.

Ljunhgall, S., Akerstrom, G., Benson, L., Hetta, J., Rudberg, C. and Wide, L. (1984). Effects of epinephrine and norepinephrine on serum parathyroid hormone and calcium in normal subjects. *Exp. Clin. Endocrinol.* **84**, 313–318.

Luft, F. C., Aronoff, G. R., Sloan, R. S., Fineberg, N. S. and Weinberger, M. H. (1986). Short-term augmented calcium intake has no effect on sodium homeostasis. *Clin. Pharmacol. Ther.* **39**, 414–419.

Mann, J. F. E., Bommer, J., Kreusser, W., Klooker, P., Rambausek, M. and Ritz, E. (1984). Parathormone and blood pressure in the spontaneously hypertensive rat. *Adv. Exp. Med. Biol.* **178**, 291–293.

Massry, S. G., Iseki, K. and Campese, V. M. (1986). Serum calcium, parathyroid hormone, and blood pressure. *Am. J. Nephrol.* **6** (Suppl. 1), 19–28.

McCarron, D. A., Yung, N. N., Ugoretz, B. A. and Krutzik, S. (1981a). Disturbances of calcium metabolism in the spontaneously hypertensive rat. *Hypertension* **3** (Suppl. 1), I162–I167.

McCarron, D. A., Muther, R. S., Plant, S. B. and Krutzik, S. (1981b). Parathyroid hormone: a determinant of post-transplant blood pressure regulation. *Am. J. Kidney Dis.* **1**, 38–44.

McCarron, D. A. (1982a). Low serum concentrations of ionized calcium in patients with hypertension. *N. Engl. J. Med.* **307**, 226–228.

McCarron, D. A. (1982b). Blood pressure and calcium balance in the Wistar-Kyoto rat. *Life Sci.* **30**, 683–689.

McCarron, D. A. and Morris, C. D. (1984). Calcium, parathyroid hormone, and hypertension. *Adv. Nephrol.* **14**, 479–501.

McCarron, D. A., Ellison, D. H. and Anderson, S. (1984). Vasodilation mediated by human PTH 1-34 in the spontaneously hypertensive rat. *Am. J. Physiol.* **246**, F96–F100.

McCarron, D. A. and Morris, C. D. (1985). Blood pressure response to oral calcium in persons with mild to moderate hypertension. *Ann. Intern. Med.* **103**, 825–831.

Nabata, H. (1977). Effects of calcium-antagonistic coronary vasodilators on myocardial contractility and membrane potentials. *Jpn. J. Pharmacol.* **27**, 239–249.

Nagatsu, I., Nagatsu, T., Mizutani, K., Umezawa, H., Matsuzaki, M. and Takeuchi, T. (1971). Adrenal tyrosine hydroxylase and dopamine β-hydroxylase in spontaneously hypertensive rats. *Nature* **230**, 381–382.

Nagatsu, T., Mizutani, K., Nagatsu, I., Umezawa, H., Matsuzaki, M. and Takeuchi, T. (1973). Catecholamine synthetic enzymes of spontaneously hypertensive rats and microbial hypotensive products. *Mol. Cell. Biochem.* **1**, 107–113.

Nakamura, R., Watanabe, T. and Sokabe, H. (1981). Acute hypotensive action of parathyroid hormone-(1-34) fragments in hypertensive rats. *Proc. Soc. Exp. Biol. Med.* **168**, 168–171.

Nicolov, N., Todorova, M., Ilieva, T., Velkov, Z., Lolov, R., Petkova, M., Ancov, V., Sheitanova, S., Tzoncheva, A. and Grigorova, R. (1988). Effect of calcium blocking agent verapamil on blood pressure, ventricular contractility, parathyroid hormone, calcium and phosphorus in plasma, catecholamines, corticosterone and plasma renin activity in spontaneously hypertensive rats. *Clin. Exp. Hypertens* **A10**, 273–288.

Ozaki, M., Suzuki, Y., Yamori, Y. and Okamoto, K. (1968). Adrenal catecholamine content in the spontaneously hypertensive rats. *Jpn. Circ. J.* **32**, 1367–1372.

Pang, P. K. T., Tenner, T. E., Yee, J. A., Yang, M. and Janssen, H. F. (1980). Hypotensive action of parathyroid hormone preparations on rats and dogs. *Proc. Natl. Acad. Sci. USA* **77**, 675–678.

Pang, P. K. T., Yang, M. C. M. and Sham, J. S. K. (1988). Parathyroid hormone and calcium entry blockade in a vascular tissue. *Life Sci.* **42**, 1395–1400.

Pang, P K. T. and Lewanczuk, R. Z. (1989). Parathyroid origin of a new circulating hypertensive factor in spontaneously hypertensive rats. *Am. J. Hypertens.* **2**, 898–902.

Pang, P. K. T., Wang, R., Shan, J., Karpinski, E. and Benishin, C. G. (1990a). Specific inhibition of long-lasting, L-type calcium channels by synthetic parathyroid hormone. *Proc. Natl. Acad. Sci. USA* **87**, 623–627.

Pang, P. K. T., Kaneko, T. and Lewanczuk, R. Z. (1990b). Parathyroid origin of a new hypertensive factor. *Exp. Gerontol.* **25**, 269–277.

Powell, H. R., McCredie, D. A. and Rotenberg, E. (1978). Renin release by parathyroid hormone in the dog. *Endocrinology* **103**, 985–989.

Reid, I. R., Civitelli, R., Halstead, L. R., Avioli, L. V. and Hruska, K. A. (1987). Parathyroid hormone acutely elevates intracellular calcium in osteoblastlike cells. *Am. J. Physiol.* **252**, E45–E51.

Resnick, L. M., Laragh, J. H., Sealey, J. E. and Alderman, M. H. (1983). Divalent cations in essential hypertension; relations between serum ionized calcium, magnesium, and plasma renin activity. *N. Engl. J. Med.* **309**, 888–891.

Resnick, L. M., Nicholson, J. P. and Laragh, J. H. (1986). Calcium metabolism in essential hypertension: relationship to altered renin system activity. *Fed. Proc.* **45**, 2739–2745.

Resnick, L. M. (1987a). Uniformity and diversity of calcium metabolism in hypertension. *Am. J. Med.* **82** (Suppl. 1B), 16–26.

Resnick, L. M. (1987b). Dietary calcium and hypertension. *J. Nutr.* **117**, 1806–1808.

Schleiffer, R., Berthelot, A. and Gairard, A. (1979). Action of parthyroid extract on arterial blood pressure and on contraction and ^{45}Ca exchange in isolated aorta of the rat. *Eur. J. Pharmacol.* **58**, 163–167.

Schleiffer, R., Berthelot, A., Pernot, F. and Gairard, A. (1981). Parathyroids, thyroid and development of hypertension in SHR. *Jpn. Circ. J.* **45**, 1272–1279.

Tenner, T. E. Jr, Buddingh, F., Mei-Ling, Y., Chuen-May, Y. and Pang, P. K. T. (1989). Dietary calcium and development of hypertension in spontaneously hypertensive rats. *Magnesium* **8**, 288–298.

Togari, A., Arai, M., Shamoto, T., Matsumoto, S. and Nagatsu, T. (1989). Elevation of blood pressure in young rats fed a low calcium diet; effects of nifedipine and captopril. *Biochem. Pharmacol.* **38**, 889–893.

Togari, A., Sintani, S., Arai, M., Matsumoto, S. and Nagatsu, T. (1990). Acute effect of parathyroid hormone-(1-34) fragments on blood perssure in rats fed a low calcium diet. *Gen. Pharmacol.* **21**, 547–549.

Umemura, S., Smyth, D. D., Nicar, M., Rapp, J. P. and Pettinger, W. A. (1986). Altered calcium homeostasis in Dahl hypertensive rats: physiological and biochemical studies. *J. Hypertens.* **4**, 19–26.

Umezawa, H., Takeuchi, T., Iinuma, H., Suzuki, K., Ito, M., Matsuzaki, M., Nagatsu, T. and Tanabe, O. (1970). A new microbial product, oudenone, inhibiting tyrosine hydroxylase. *J. Antibiot.* **28**, 514–518.

Wrigh, G. L. and Rankin, G. O. (1982). Concentrations of ionic and total calcium in plasma of four models of hypertension. *Am. J. Physiol.* **243**, H365–H370.

Yamaguchi, D. T., Hahn, T. J., Iida-Klein, A., Kleeman, C. R. and Muallem, S. (1987). Parathyroid hormone-activated calcium channels in an osteoblast-like clonal osteosarcoma cell line. *J. Biol. Chem.* **262**, 7711–7718.

Zawada, E. T., Brickman, A. S., Maxwell, M. H. and Tuck, M. (1980). Hypertension associated with hyperparathyroidism is not responsive to angiotensin blockade. *J. Clin. Endocrinol. Metab.* **50**, 912–915.